Pappus of Alexandria:
Book 4 of the *Collection*

For other titles published in this series, go to
http://www.springer.com/series/4142

Sources and Studies
in the History of Mathematics and
Physical Sciences

Pappus of Alexandria:
Book 4 of the *Collection*

Edited With Translation and Commentary by

Heike Sefrin-Weis

Springer

Heike Sefrin-Weis
Department of Philosophy
University of South Carolina
Columbia SC
USA
sefrinwe@mailbox.sc.edu

Sources Managing Editor:
Jed Z. Buchwald
California Institute of Technology
Division of the Humanities and Social Sciences
MC 101–40
Pasadena, CA 91125
USA

Associate Editors:
J.L. Berggren
Simon Fraser University
Department of Mathematics
University Drive 8888
V5A 1S6 Burnaby, BC
Canada

Jesper Lützen
University of Copenhagen
Institute of Mathematics
Universitetsparken 5
2100 Koebenhaven
Denmark

ISBN 978-1-4471-2564-8   ISBN 978-1-84996-005-2(eBook)
DOI 10.1007/978-1-84996-005-2
Springer London Dordrecht Heidelberg New York

British Library Cataloguing in Publication Data
A catalogue record for this book is available from the British Library

Mathematics Classification Number (2010) 00A05, 00A30, 03A05, 01A05, 01A20, 01A85, 03-03, 51-03
and 97-03

Springer is part of Springer Science+Business Media (www.springer.com)

Dieses Buch
ist für Roger, Hannah und Lisa

Preface

In the seventeenth century, it was a common opinion among those interested in ancient Greek mathematics that there were four great mathematicians in antiquity: Euclid, Archimedes, Apollonius – and Pappus of Alexandria. Especially the fourth and seventh book of the *Collection* were widely read and avidly discussed. They do indeed contain a host of interesting material on ancient Greek geometry that is not attested elsewhere. The fourth book in particular offers quite a few vignettes on "higher" Greek mathematics, treating of the classical problems of squaring the circle, doubling the cube, and trisecting the angle. Also reported is an intriguing little piece on mapping the sequence of natural numbers into a closed configuration of touching circles. Despite his importance as a historical source, Pappus has nowadays become something like the ugly stepchild among the ancient mathematicians. The present edition is intended to provide a basis for giving the writer Pappus another hearing, to rekindle scholarly interest in him. It contains a new edition of the Greek text of *Collectio* IV (based on a fresh transcription from the main manuscript Vat. gr. 218), an annotated translation, and a commentary. The text offers an alternative to Hultsch's standard edition. The commentary provides access to quite a few aspects of the work that have so far been somewhat neglected. Above all, it supports the reconstruction of a coherent plan and vision within *Collectio* IV.

This edition is developed out of a complete revision of my 1997 German dissertation in the Mathematics Department (Section History of Mathematics) at Mainz University (published in microfiche form in 1998). In difference from the earlier version, a Greek text was included, a new translation into English was made, and the commentary was reformulated as well.

Numerous people have supported and helped me over the years with the Pappus project. To all of them, my warmest thanks. In particular, David Rowe, my dissertation advisor, has been extremely generous with his time, support, and advice. Henk Bos, as second reader for the original dissertation, gave encouraging, yet incisive and helpful advice. Bob Berghout discussed Pappus with me, pointed me to Ath Treweek's edition, and let me consult Ath Treweek's material on *Collectio* IV. Siegmund Probst (Leibniz-Archiv, Hannover) introduced me to the art and trade of editing manuscripts. Jeremiah Hackett helped me gain access to the Vat. gr. 218. I should also like to thank the Bibliotheca Apostolica Vaticana for the generous hospitality and the inspiring atmosphere they provide for scholars consulting their collection, and especially for allowing me to study the text and figures of Vat. gr. 218 for the present edition. Ms. Regine Becker proofread the entire Greek text. Ms. Lisa Weis improved my English in numerous places. My deepest gratitude is owed to the members of my family, who have, generously, if not always voluntarily, been living with Pappus for a number of years. That is why this book is dedicated to them.

Mainz/Columbia, August 2008   Heike Sefrin-Weis

Contents

General Introduction

Without a doubt, Pappus of Alexandria's *Collectio* IV is one of the most important source texts for the history of Greek mathematics. Undisputed is specifically the importance of its second half (especially Props. 19–34), because it contains our main, and in most cases even our only direct, access to otherwise lost contributions of the Hellenistic mathematicians to "higher" geometry. Inter alia, Props. 19–22 give us an alternative Archimedean treatment of the plane spiral (one that employs a "mechanical" approach), Props. 26–29 provide the only extant description of the quadratrix, the squaring of the circle with it, and the only two surviving examples for analysis of loci on surfaces, and Props. 31–34 contain the extant ancient sources on angle trisection, among them the only one that gives a complete construction via conics. The Hellenistic solutions for all three classical problems (squaring the circle, doubling the cube, and trisecting the angle) are attested in *Coll.* IV. Likewise undisputed is the fact that *Coll.* IV, together with *Coll.* VII, played a crucial role in the reception and transformation of ancient Greek geometry during the early modern era.

Nevertheless, a complete English translation of this important work is at present still a desideratum. Whereas A. Jones published an edition, translation and commentary of *Coll.* VII in 1986, the only complete translation of the *Coll.* (II–VIII) into a modern language is (Ver Eecke 1933b) into French. A German translation of *Coll.* IV is contained in (Sefrin-Weis 1998)[1]. The main purpose of the present work is to provide a Greek text and an English translation of the complete text of *Coll.* IV, and to make *Coll.* IV accessible as a whole to the modern reader. This purpose is indeed a dual one. A complete translation makes the famous passages accessible, in their original context. It also provides the basis for making sense of those very passages beyond their local mathematical content, as part of a comprehensive picture that emerges out of the book as a whole. Currently, no such comprehensive view of *Coll.* IV exists. The present work will help close this gap, and this is its second purpose.

Perhaps, partly as a result of the fact that *Coll.* IV was not available in an accessible edition, it has so far not been understood, even appreciated, as a thoughtfully planned, coherent composition by a well-informed author. Another reason for this lack of appreciation is the fact that unity and coherence has been sought on the

---

[1] German translation and commentary, listed under primary sources, Pappus.

level of mathematical content. While such a perspective is quite understandable, given that *Coll.* IV contains arguments that are of primary interest because of their mathematical content, it cannot make sense of the work as a whole. It is not, I would contend, the perspective from which the work was composed. *Coll.* IV contains a host of mathematical examples, on various topics and with various degrees of difficulty and complexity. They do not form a thematic unity. Topics reach from constructing the irrational *Minor* to trivial theorems on segments in a circle to squaring the circle to a quadrature of a curved surface in space (see section "Survey of *Coll.* IV"). The sources on which Pappus draws are equally non-uniform and cover a considerable time span. As long as one looks for unity and coherence on the level of mathematical content only, one would have to agree with the negative judgment on *Coll.* IV as it is given by many scholars, and summed up in Jones' verdict that *Coll.* IV lacks an "overall governing plan" (Jones 1986a, p. 6).

The picture changes radically, however, if one takes a different perspective, one that focuses on methodology rather than content. Such a perspective emerges quite naturally when one takes Pappus' own meta-theoretical remarks, which appear about half-way into the book, into account. In a famous passage, he differentiates between three kinds of geometrical research according to the types of curves needed for construction and problem solving. Furthermore, a homogeneity requirement applies, thereby associating to each problem one of these distinct classes, or rather kinds, of geometry: plane (using only circle and straight line), solid (using conics in addition), and linear (using "higher" curves), kept apart on methodological grounds. Within these three distinct kinds, Pappus looks for further subspecifications of methods, strategies, and styles of mathematical argumentation, and illustrates them with attractive, famous, or just methodologically well-designed (easy) examples, so as to be able to tell a story about the character of Classical and Hellenistic geometrical discourse. His examples are thus chosen from the point of view of the methods they are exemplifying, and their content, though it is preferably one that captures the reader's attention, is subordinated to the overall plan of profiling the methods of ancient geometry in a comprehensive way. This is the way I propose *Coll.* IV should be read, and it obviously has consequences also for the evaluation of the well-known passages. My thesis is that *Coll. IV can be read, and was intended to be read, as a unified, coherent and essentially exhaustive survey of the classical geometric tradition from the point of view of methods.* The metatheoretical passage, which was taken seriously in the present work as programmatic, is not to be understood, necessarily, as the expression of a communis opinio among ancient mathematicians. In fact, I rather doubt it was that. It generalizes certain features of Apollonius' analytical work in plane geometry, and probably was not shared by the mathematicians as a description of their views on the methods of mathematical discourse. It was, however, important and useful for Pappus. He brought it into play in *Coll.* III, and repeated it in a shorter version in *Coll.* VII. It served him well as a guiding principle to select, revise, and structure his material in *Coll.* IV, in order to give a coherent profile of the mathematical tradition. It provides a unifying perspective for the text of *Coll.* IV. *Coll.* IV then tells us a reasonably coherent, and astonishingly comprehensive, story about how classical Greek mathematics worked. The present translation and commentary traces and documents this story. It can be useful for a reader coming to Pappus' *Coll.* IV for the first time, trying to appreciate it as a whole, as a coherent account. And it can be useful as a basis for further historical and philosophical investigations that attempt to take Pappus seriously as a writer, not just as a mine. Unlike the great Hellenistic mathematicians Archimedes and Apollonius, the author Pappus did not endeavor to produce new, original, creative work that would develop his field toward the future. He was essentially a commentator, who attempted to preserve, transmit, and explain the tradition. He does not proceed mindlessly or randomly, though. He is a competent mathematician trying to make sense – with respectable success – of the tradition. His goals as a writer simply differ from those of the writers of the "golden age" of Greek mathematics. There is no need to view these goals, or the testimony that results from them, as inferior or of lesser intrinsic value for historical scholarship, as has often been done in the past. Whether or not it makes sense to view the period of late antiquity as an era of intellectual and cultural decline, the current trend in historical scholarship on ancient Greek mathematics, with a renewed emphasis on taking ancient authors (of any period) seriously as writers in their own context and on their own terms could be fruitfully applied to Pappus as well.

1   Life and Works of Pappus of Alexandria

The following remarks draw mostly on Ziegler's article in the RE, with additional information from Jones[1] 1986a, pp. 1–20, to which the reader is referred for further information. My assessment of *Coll.* IV differs from the one given in these sources.

1.1   Pappus' Life and Times

Detailed biographical information on Pappus is extremely scarce. His lifespan falls either into the reign of Theodosius, i.e., the second half of the fourth century AD, or perhaps the reigns of Diocletian and Constantine, i.e., the time around 300 AD. It was a time in which Christianity had become the dominant religion in the Roman Empire. The pagan cultural elite, to which Pappus belonged, was in the defensive, and it was fighting an increasingly hopeless battle for the survival of the cultural tradition that was its heritage. Mathematics was part of that tradition. Before the century was over, Christianity had transformed from a persecuted religion (the last, rather brutal wave of persecutions occurred in 302/303) to the only religion allowed in the State (395). Under its control, the ancient pagan culture was rigorously, sometimes aggressively, suppressed. In some regions, especially in Alexandria, the tides of clashing world views ran particularly high, and had done so already at the end of the third century. In 415, the pagan mathematician Hypatia was murdered

---

[1] see also Jones 1986b, Treweek 1957.

there on account of her commitment to Neoplatonism. In short, it was a period of fundamental transitions, with accompanying widespread social and even existential insecurity. It is perhaps not insignificant that Pappus lived and was probably born at Alexandria, in all likelihood had Neoplatonic leanings (as did most of the educated pagans during that period), and wrote his survey of the classical mathematical tradition under the circumstances just described. For this may help explain why Pappus, when dealing with geometry, looks constantly backward, to the classics in the field, and attempts to make them accessible, just as the culture of which they formed a part was increasingly marginalized, and was soon going to be history. It is almost as if he was trying to make sense of the tradition of his field so that he could leave to posterity a key to it, since there was not going to be a thriving ongoing tradition of instruction. This was perhaps a time to look backward, and save and defend what could be saved rather than a time to build for the future. From this perspective, the much deplored lack of originality, which has been detected in the works of Pappus and others during this time (e.g., Theon of Alexandria, who was the father of Hypatia, and the most influential editor of Euclid's *Elements*), becomes understandable. It should not be equated with a lack of mathematical competence.

1.2 Pappus' Works

Almost all of Pappus' work belongs to the field of mathematics. Four areas can be identified, and they are all "classics": geometry, geography, astronomy, and mechanics. The latter three are "mixed sciences," applied geometry, if you will. In what follows, I will give a brief list of Pappus' works, starting with a few remarks on the *Collectio*. I will be a little more detailed with respect to the geometrical works, and focus again on *Coll.* IV specifically at the end.

The *Collectio* is the only one of Pappus' works that survives in Greek, and therefore our main source for Pappus' mathematics (his commentary on *Elements* X survives in Arabic, see section "Geometry Proper"). Book I of the *Collectio* is lost (unless identical with the commentary on *Elements* X), and of Book II only a part survives. The rest of the collection is more or less preserved intact in the original Greek. There are gaps, e.g., the proem and conclusion of Book IV, or the instrumental constructions in Book VIII, as well as its conclusion. The work was not conceived by its author as a single, closed, coherent opus, as the different addressees mentioned in the proems, the numerous duplicates, the divergent subject matters, and finally some explicit cross references to "a preceding book" attest. In fact, I will list Book VI as an astronomical work, and Book VIII as a work on mechanics. Both are, of course, also works on "applied" geometry.

1.2.1 Geometry Proper

(a) *Commentary on Euclid, Elements X.* This work, perhaps in two books, is preserved in an Arabic translation. An edition with English translation is available (Junge and Thomson 1930). It replaces an earlier German translation by Suter and an even earlier French translation of excerpts by Woepcke. Jones suspects that this commentary could be the otherwise lost Book I of the *Collectio*. It deserves scholarly attention, for it could contribute to our understanding of the ancients' theory of irrationals (cf. *Coll.* IV, Props. 2 and 3).

(b) Geometrical books within the *Collectio*. Book II contains a kind of game in numbers, around a hexameter verse. On the basis of this example, numerical operation with large numbers is illustrated. Perhaps the relevant techniques were those developed by Apollonius. Jones contradicts the common assumption, put forth originally by Heiberg, that the material stems directly from Apollonius' *Ocytokion*.

Book III, addressed to the head of the mathematics division at the Platonic Academy in Athens, a woman called Pandrosion, treats of construction problems that were of special interest to Platonists. Pappus starts off by criticizing three flawed arguments by students of hers, and adds further, more general explanatory remarks on an elementary level. The first problem discussed is cube duplication. Pandrosion's student had attempted a plane solution (circles and straight lines, calculation of ratios). It is refuted by extensive critical analysis, and Pappus then moves on to formulate his meta-theoretical position on the three distinct kinds in geometry and the impossibility to achieve a correct solution from the "wrong" kind. An exact geometrical solution for the cube duplication requires conic sections. The cube duplication thus is an example for a "solid" problem. Pappus is trying to make the case that if one wants to restrict oneself to the argumentative scope provided by circle and straight line, one has to restrict oneself to corresponding problems, i.e., those that are "by nature" plane. The meta-theoretical passage is a duplicate of the meta-theoretical passage in *Coll.* IV. Pappus does not give an exact solution for the cube duplication via conics in *Coll.* III, however. Apparently, he is trying to do justice to the level of mathematical education in his particular audience. Instead, he proposes simple ruler manipulation as a device to establish the crucial *neusis* (not claiming that it is an exact mathematical construction).[1] The next topic is the geometrical representation of the three Pythagorean means.[2] Here, too, Pappus starts from a student argument, which he criticizes, and then proceeds to a discussion and construction of ten means. His approach shows affinities to the methods employed by the Neopythagorean Nicomachus (who is usually seen as the authority in this branch of investigating numbers), but Pappus' ten means are not completely identical with the ones Nicomachus lists. A third part of *Coll.* III is devoted to the construction of a triangle inside a given right-angled triangle

---

[1] A *neusis* is a geometrical construction, in which a straight line of given length has to be placed between two other given lines in such a way that its continuation verges (Greek verb neuein) toward a given point. On *neusis* constructions see the commentary on Props. 23–25 below.

[2] The original Pythagorean means were the arithmetical, geometrical, and the harmonic mean. A special kind of number theory, or number speculation, arose around them, and their number was increased to 10. This kind of consideration was of special importance in the Neopythagorean school. The reader is referred to Nicomachus, *Introductio Arithmetica*, and the literature on it.

under the condition that the sum of the two sides is larger than the sum of hypotenuse and one kathete. Again, Pappus starts from a discussion of a flawed argument presented to him, and moves on to generalizing elementary considerations. As a source for his exposition on the topic, he mentions an otherwise unknown author Erycinus. The book concludes with a constructive inscription of the Platonic solids into a sphere. The presentation is in analytic-synthetic form and differs decisively from the one given in *Elements* XIII. An appendix to *Coll.* III revisits the problem of cube duplication, offering an alternative discussion to the first group of propositions mentioned above.

Book IV, as I hope the present complete translation and commentary will document, is a representative portrait and survey of the methods in the three kinds that make up classical geometry, on the basis of illustrious examples for the most part, exhausting the classical geometrical tradition down to the time of Apollonius. The authors and sources used are Aristaeus, Euclid, Archimedes, Nicomedes, and Apollonius. The book has three parts, in correspondence to the partition of geometry into three kinds: plane, linear, and solid. For a more detailed account of *Coll.* IV see the following sections.

Book V studies isoperimetric figures. Within the collection, it is the book that shows the most signs of careful polishing. It contains two proems, and consists of two parts. Part I discusses isoperimetric figures in the plane. It is based on a work of Zenodorus. Part II discusses isoperimetric figures in space. It is less comprehensive than the first part. Contained are considerations on semiregular polyhedra, as well as the Platonic solids and their relation to each other and the sphere. Perhaps Pappus is using a work by Archimedes as his source.[1] For he ascribes a work on semiregular polyhedra to Archimedes, and numerous arguments in *Coll.* V show a close conceptual connection to arguments in Archimedes, *De Sphaera et Cylindro*.

On Book VI see section "Astronomy/Astrology".

Book VII is a handbook in catalog form. It is addressed to a certain Hermodorus, and was intended for use while going through the works belonging to *the analytical field* in mathematics, as a kind of running commentary.[2] The main, though not the only, author in this field was Apollonius. For each of the works discussed, Pappus gives a list of content, a synopsis, and elementary auxiliary lemmata for intermediate steps, presumably closely following the actual sequence in the texts. His lemmata mostly consist in reductions to, or deductions from, propositions in *Elements* I–VI (elementary geometry). They seem to be intended to facilitate reading the actual texts themselves, and the reader is expected to have them in front of him or her. Unfortunately, most of the texts in the analytical field are lost (notable exceptions are the *Data* and the *Sectio rationis*), so that *Coll.* VII is in fact our main source for a reconstruction of the *treasury of analysis* (not quite its original purpose!). Its value for research in the history of mathe-

---

[1] Jones (1986a, pp. 578–580) argues for a possible connection of a part of these propositions to Aristaeus' lost work on the Platonic solids. Pappus may have used it as an additional source.

[2] See Jones (1986a) for a translation and commentary of *Coll.* VII. The book also contains most valuable information on Pappus, the *Collectio*, and Greek geometrical analysis.

matics therefore lies not so much in its actual content, but in the fact that this content provides an indirect and partial access to the vast field of Greek analytical geometry, which is now lost to us.

On Book VIII see section "Mechanics."

1.2.2 Geography

Two works are attested:

(a) *Chorography of the known world.* It is based on Ptolemy's *Geography* and contains, among other things, curiosities in ethnography. Fragments are preserved in Armenian, in a text dating from the seventh century. Ziegler mentions a French translation, Jones cites a newer one into English by Hewson from 1971.

(b) *Rivers in Libya.* The book is listed in the Suda. No other traces survive.

1.2.3 Astronomy/Astrology

The sheer number of titles, and their projected volume, indicates that this was an area of special interest to Pappus, maybe his main focus. Unfortunately, not much of it is preserved.

(a) *Commentary on Ptolemy's Almagest.* Some remarks on Books V and VI survive in the original Greek. They were edited by Rome within a more comprehensive work which also contained Theon's commentary (Rome (1931–1943), cited from Jones (1986a)). Cross-references and a reference in Eutocius confirm that commentaries on Books I, III, and V existed. According to Jones, it is very likely that Pappus commented on all 13 Books. The commentary must have been a very extensive work.[1]

(b) *Commentary on Ptolemy's Planisphaerium.* This work is not preserved, but Fihrist mentions it in connection with an Arabic translation by Thabit ibn Qurra. Ptolemy's work dealt with stereographical projection, and is preserved, according to Jones, in an Arabic translation.

(c) *Commentary on the Analemma of Diodorus.* This work is mentioned in *Coll. IV*, in connection with the discussion of the conchoid (see Prop. 23). The *analemma* was a method for problem solving in spherical geometry, and derived from astronomical applications in connection with the sun dial. Neither Diodorus' work nor Pappus' commentary survive. Perhaps fragments of the commentary are contained in one of the Bobbio manuscripts (cf. Jones 1986, p. 12).[2]

(d) Book VI of the *Collectio: Astronomical Field.* This work is fully preserved. Like *Coll.* VII (see above), it is a kind of handbook with explanations in catalog form, to accompany the study of the so-called minor astronomical works as a

---

[1] Cf. Neugebauer (1975, p. 966).

[2] On Diodorus and the *Analemma* see also Neugebauer (1975, pp. 840 ff).

running commentary. The text presupposes knowledge of *Elements* I–VI. In a first part, Pappus attempts to correct mistakes commonly made, and gaps unduly left in the usual teaching (!) of Euclid's *Phaenomena*, Theodosius' *Spherics*, and Theodosius' *Days and Nights*. A second part contains excerpts and explanatory remarks on theorems associated with Autolycus and Aristarchus (compared to Hipparchus and Ptolemy), and with Euclid's *Optics*.[1]

(e) *Astrological Almanach*. According to Jones, an excerpt from this work ascribed to Pappus is preserved in a Florentine compilation book of astrological texts. In addition, Jones mentions a reference to an astrological work by Pappus in an astrological manuscript from the thirteenth century.

(f) *Alchemistic Oath*. This is a rather short formulaic text with theological-spiritual content. According to Tannery (1912) and Bulmer-Thomas (in his DSB article), it has to be accepted as essentially authentic. Its main part is strongly Neoplatonic in outlook. The conclusion shows gnostic influence, and therefore Jones argues that this sentence (loosely connected to the main text) should be viewed as an interpolation.

1.2.4 Mechanics

Book VIII of the *Collectio*. This work was circulated in late antiquity independently from the rest of the *Collectio* under a title like *"introduction to mechanics."* It was apparently widely distributed, and was received into the Islamic culture at a relatively early date. Extant Arabic translations could be used to close some gaps in the Greek text as it has come down to us. Because of its special history, a few remarks on its content may be appropriate, even though it is not directly relevant for *Coll.* IV. The introduction characterizes the field of mechanics from a methodological point of view, and differentiates two sub-disciplines, or branches: theoretical and practical. The division is given by reference to Heron of Alexandria, the major authority in the field. The relation of geometry and mechanics is described in terms that are strongly reminiscent of Aristotelian concepts of science in general, and of the "mixed sciences" in particular. Archimedes is named as the founder of the theoretical branch of mechanics; Carpus and Heron are mentioned as important figures as well. Finally, the introduction gives a survey of the book. Most of the material in *Coll.* VIII probably rests, directly or indirectly, on Heron's work in mechanics. A first group of propositions deals with classical problems in ancient (theoretical) mechanics: centers of gravity, motions of a weight on an inclined plane. For the latter, Heron's *Baroulkos* is referred to. A second group of theorems targets instrumental techniques for dealing with (practical?) problems in mechanics. Concrete instruments for construction are discussed. Inter alia, the construction of the *neusis*

---

[1] Heath (1921, II, p. 397) and Neugebauer (1975, p. 767) both give a rather negative judgment of this work of Pappus. Perhaps a more favorable judgment would result, if the "didactic" scope and purpose of the work were taken into account.

used in cube duplication by simple manipulation with a marked ruler (cf. *Coll.* IV, discussion of the conchoid), and the determination of the base of a chipped-off cylinder (broken column?) with the help of an ellipse through five points, are subjects. Constructions with a ruler and a compass with fixed distance can be restored from Arabic translations. The book concludes with extensive excerpts from works by Heron on cogwheels and screws.

2   Survey of Coll. IV

Coll. IV, the subject of the present translation and commentary, belongs to the geometrical works proper (see above). Its beginning, including the proem, is missing. We have no explicit statement of Pappus' intentions and goals in the work and need to reconstruct its subject matter from the text itself. This is not a trivial task. For the text shows no overall thematic coherence on the level of mathematical content. Only weak motivic connections can be detected, and they constitute something like a bare red thread, establishing partial and very loose coherence on the literal level. As a result of this lack, the judgment on *Coll.* IV has so far been that it is just a random collection of diverse, indeed very diverse, mathematical vignettes. In my opinion, this goes too far. An overall governing plan can, after all, be detected in *Coll.* IV. It is to be found not on the level of literal, mathematical content, but on the level of methodology. The famous programmatic statement on the three kinds of geometry serves as a focal point of reference for the material presented by Pappus. In all the vignettes he presents, even where the topic he chooses to exemplify a certain methodological approach is very appealing and interesting in itself (e.g., squaring the circle), it is the methods that are profiled and emphasized, with the content serving as an incentive, to capture the reader's interest. Just as the copyist indicates at the end of *Coll.* IV, the book contains "splendid theorems, plane, solid and linear ones." It divides into three parts. Pappus surveys the methods of all three branches of classical Greek geometry, using examples that are either clearly designed by himself so as to exhibit methodological aspects, or by drawing on famous mathematicians and their results (preferably spectacular ones, like the squaring of the circle or the duplication of the cube or the trisection of the angle): Euclid, Archimedes, Nicomedes, and Apollonius.[1] Despite this restriction to just a few major authors, and the rigorous restriction to relatively short argumentative units as "exemplifiers," Pappus succeeds in presenting a rich and rather differentiated picture of the different styles and traditions within classical Greek geometry. His representation is not exhaustive, and not intended as a complete documentation. Rather, it is consciously and planfully selective. His approach is via exemplary

---

[1] Perhaps Aristaeus' work on solid loci was also used directly. At present, it cannot be decided to what degree Pappus may have drawn directly on pre- Euclidean sources. Compare below, Props. 31–34.

arguments that are didactically sound and representative. He makes an effort to
select arguments that can be made accessible on the basis of a knowledge of
Elements I–VI (elementary geometry), while nevertheless exhibiting the typical
features of a particular style in mathematics.

The three parts of Coll. IV are:

I Plane contributions Props. 1–18
II Linear contributions, Props. 19–30
Meta-theoretical passage on the three kinds of geometry, with homogeneity
    requirement
III Solid contributions, transition from solid to linear, demarcating solid from plane
    problems, Props. 31–44

Props. 1–3 illustrate classical synthetic argumentation in direct connection to
Euclid's *Elements*. All three of them follow the standard pattern of *apodeixis*
(proof) familiar from the *Elements*: a proposition is formulated, set down in the
concrete (*ekthesis*), with ensuing construction (*kataskeue*); a deductive proof, drawing
on the diagram, then leads to the conclusion.[1] Prop. 1, a theorem, gives a generalization of the Pythagorean theorem and is closely modeled on the argument given
in *Elements* I, 47. It illustrates the form of a classical synthetic proof. Props. 2 and
3 are problems[2]; they construct irrational lines in close connection to *Elements* X.[3]
Prop. 2 gives a surprisingly simple construction of the *Minor*; it is closely modeled
on *Elements* XIII, 11. Prop. 3 starts from a configuration that is very similar to that
for Prop. 2 and gives a construction for an irrational that is not contained in
Elements X, but is one level "higher." Perhaps it shows how one can work beyond
Euclid, while remaining firmly within the Euclidean framework.

Props. 4–6 illustrate the structural schema of (plane) analysis-synthesis. The second
part of this two-partite procedure is essentially a synthetic proof, like the ones given
in Props. 1–3. The first part, the analysis, is essentially a heuristic strategy with the
goal of identifying grounds for a deductive proof. One starts from the assumption
that one has already solved the problem at hand (or that the proposition is in fact
true), and then transforms and transposes features in this "target situation," until
one reaches a situation which is indeed already corroborated. This phase of the
analysis is called "*apagoge*" or "*epagoge*." One operates via reductions or deductions, and via suitable extensions of the configuration. In a second phase of the

---

[1] For the format and ingredients of a classical synthetic argument, specifically a synthetic proof,
see the introductory remarks and the proof protocol in the commentary on Prop. 1 with references.
See also Heath (1926, pp. 129–131). A very helpful investigation of classical Greek mathematical
proof is Netz (1999).

[2] In a theorem, one proves a proposition, most often by constructive proof. In a problem, one constructs
a required entity, and then shows that this construction has the required properties. On theorems versus
problems in Greek geometry cf. Heath (1926, pp. 124–129).

[3] For an explanation of the term "irrational lines," and for some information on *Elements* X, including
the *Minor*, see the introduction to Props. 2 and 3 in the commentary.

analysis, one then shows that the end state, as it were, of the *apagoge* is independent from the initial analysis-assumption (the assumption that the problem is solved, or the theorem in fact true). One needs to show that the ingredients of the end state of the *apagoge* are *given* (roughly speaking: determined and constructible)[1] within the original configuration or via suitable input from elsewhere, and determine, if need be, conditions for solvability as well as sub-cases (*diorismos*). This second phase of the analysis is called (in a modern term) *resolutio*. After a successful *resolutio*, the synthesis can pick up and provide a deductive proof, drawing on the material and steps in the analysis. In many cases, the synthesis will be obvious after a successful analysis, echoing the *resolutio*, and then retracing the steps of the analysis backward.[2] Obviously, the analysis carries the burden, contains the creative mathematical work, in an analysis-synthesis. The synthesis nevertheless is the part that carries the proof. The mathematical content of Props. 4–6 is not very spectacular. Instead, the method itself, the interplay of analysis and synthesis, and their respective roles, are made very transparent. Pappus himself is probably the author of these theorems.

Props. 7–10 center on a special case of the so-called Apollonian problem (given three circles, find a fourth one that touches them all). Apollonius' *Tactiones*, though not explicitly quoted, clearly form the background of this group of propositions. The arguments target the strategy of geometrical analysis only (i.e., they are not complete theorems/problems). No complete solution even to the special case discussed is given; the arguments focus specifically on the *resolutio* phase of analysis, and within it, the determination of *given* features. Prop. 7 is unrelated in content to the problem at hand, and illustrates the operation with Euclid's *Data*, to show that for a quadrilateral with all four sides *given* in length, and a right angle at one corner point, the diagonal that does not subtend this angle is *given* as well. Obviously, it is not this content, but rather the methods that are in view. Prop. 8, for which perhaps an argument from Apollonius, *Tactiones* I, 16/17, served as a source (see "Translation and Commentary"), is the most intrinsically interesting proposition in the group. Unfortunately, it is not fully worked out and edited. It appears that Pappus himself has constructed this group of propositions, in the form presented, as well. The group yields a sketch of the methods employed in the analytic field (for plane problems), with an emphasis on illustrating the crucial *resolutio* phase. By integrating Prop. 7, Pappus makes the point that Euclid's *Data* are to be viewed as a basic and central reference work in this area (even against a trend in his source text for Prop. 8). What is sadly missing in this portrait of the *resolutio* strategy is an adequate representation of the *diorismos*. It was central for Apollonius, who was the main authority in this area of plane analytic Greek geometry.

---

[1] This is a technical term: Latin: *data,* Greek: δοθέντα. See the introduction to Prop. 7 in the commentary.

[2] For the schema of analysis-synthesis, its ingredients, its significance, and its relation to classical synthesis, see the introduction to Props. 4–12 in the commentary. Compare also Pappus' general characterization of the method in the proem of *Coll.* VII.

Props. 11 and 12 round off the picture of plane analysis-synthesis, focusing on the *apagoge* phase of the analysis. Again, the content of the propositions is not mathematically relevant. Prop. 11 is purely synthetic. An easily reconstructible analysis would be limited to the non-algorithmic strategy of suitably extending the configuration. In such a situation, the analysis work completely disappears into the *kataskeue* (construction) within the synthesis, and does not leave any traces in the apodeixis. Perhaps this was what Pappus wanted to illustrate with Prop. 11.[1] Prop. 12, the last proposition in Pappus' portrait of plane analysis-synthesis, contains a full analysis and synthesis, just like the first proposition in this group did. In the case of Prop. 12, the *apagoge* consists solely in reduction, and it is purely deductive. The result is that the *resolutio* is minimal, and the synthesis exactly retraces the steps of the analysis, because all the steps used in the *apagoge* are also convertible. Prop. 12 represents the original, and probably also historically original, core of ancient geometrical analysis: the idea of reduction. Unless he drew on otherwise unattested examples from early Greek geometry, these two propositions are constructed by Pappus – presumably with the intention, at least for Prop. 12, to illustrate the analytical strategy in the *apagoge* phase.

Props. 13–18 form by far the largest coherent bit of text in *Coll.* IV. In fact, we get a kind of monograph in miniature format (in an abridged, and therefore somewhat fragmented form). The subject of this charming, clever group of propositions is the arbelos configuration (cf. the figure in Prop. 16). In it, a surprising connection manifests itself between the ratios of diameters and perpendiculars in a finite configuration with an infinite series of inscribed tangent circles on the one hand, and the natural numbers on the other. The mathematical subject matter connects to considerations on points of similarity, and in this sense, it reaches rather deeply and taps into an area that was much later systematically developed in projective geometry. The argumentative means are purely synthetic, and astonishingly simple. The author succeeds in capitalizing ingenuously on means from elementary plane geometry, while presenting his material "locally" in a rather conservative style. The central theorem in the group employs a nuclear form of complete induction. The mathematical content of Props. 13–18 is thus highly attractive, and beautifully, thought-provokingly displayed. The group of theorems has been associated with Archimedes as a potential original author. Despite its fragmented form, one may very well think it (or rather its more extended original) worthy of Archimedes. Even though such an ascription cannot be verified, the arbelos treatise exhibits a well-defined mathematical style, the features of which warrant the label "mathematics, Archimedean style" for this type of plane ancient geometry.

Without explicitly stating so, Props. 19 ff. make the transition from plane to linear geometry. Even without an explicit remark by Pappus, he may have reasonably expected his audience to note that we are dealing now with problems of a different character, and with mathematics of a different kind. The author of Props.

---

[1] But see the commentary on Props. 11 and 12.

19–22 is Archimedes. These propositions deal with the plane spiral in a way that shows some connection to theorems from *Spiral Lines,* but nevertheless radically differs in the investigative and argumentative methods employed. Among other things, Prop. 21 (stating that the area of the spiral is one third of the circle in which it is inscribed) employs Archimedes" mechanical method,[1] and operates with indivisibles (or else an implicit argument via transition to infinity). Prop. 22 gives a theorem on the size of spiral sectors. The propositions are a valuable source for the heuristic background of Archimedes' study of spiral lines. They belong to the context of squaring the circle, and probably illustrate the seminal contribution made by Archimedes toward the mathematical investigation of motion curves (the "linear kind" in Pappus' terminology). Such curves are typically generated by abstract idealized motions, a quasi-mechanical ingredient. When dealing with them, one has to make a transition to mathematically graspable *symptomata,* which then are the basis for the geometrical investigation as such. Obviously, for the development of the mathematics of such curves, one has to mathematize suitably. In antiquity, two paths were pursued for this: either one proceeds by fully exploiting the mechanical metaphor (Archimedes), or by conducting an analysis of loci (Nicomedes, inter alia, see following sections). At least this is the picture that emerges from the developmental story Pappus tells.[2] Both the concern for an adequate mathematically acceptable definition of the curves and the concern for establishing a valid geometry on them are of major interest.

Props. 23–25 target a second motion curve, the conchoid of Nicomedes.[3] The propositions are drawn from Nicomedes' (lost) treatise on the conchoid. The *genesis* of the curve is via motions, but unlike Archimedes' spiral, where the *symptoma* is derived directly from the motions, Nicomedes' conchoid is characterized pointwise, in a quasi-analytical way, as the locus of all points that have a certain *neusis* property: all points on the curve have the property that the straight line from them, verging toward the pole, cut out a line of fixed length between the curve and a given

---

[1] On the ingredients and the significance of Archimedes's "mechanical method" see the introduction to Props. 19–22 in the commentary with bibliographical notes; cf. also Archimedes, *Ephodos* and *Quadratura Parabolae.*

[2] See the introduction to Props. 19–30 in the commentary, on the *symptoma*-mathematics of motion curves, also on the terms "mechanical," "instrumental," *genesis, symptoma,* and on the question how, and to what degree, this field of study is viewed as geometry. Descartes obviously had this part of *Coll.* IV in view when he developed his classification of curves, and excluded what he defined as "mechanical" curves. He seems to have assumed (erroneously) that the ancients dismissed all these curves and all mathematics on them, whereas he dismissed only some. Newton, drawing likewise on *Coll.* IV, came to a different assessment. See the commentary on Props. 19–30 passim.

[3] This curve arises when one moves a ruler that is attached to a fixed point (pole), along a fixed straight line (canon), with the stipulation that any intercept between that straight line and the curve has to have a fixed length. The main branch of the arising curve has the shape of a shell, hence the name for the curve. See the figure in the translation, Prop. 23 with introductory paragraphs.

straight line (the canon).[1] This may very well mark a transition toward a different characterization of the higher curves, via analysis (see Props. 28 and 29). With a *neusis* construction that can be won from the conchoid (construct a line that verges toward a fixed point and creates an intercept of given length inside a given angle, i.e., between two given straight lines, Prop. 23), Props. 24 and 25 establish the cube duplication, indeed the production of a cube that has a given ratio to a cube that is put forth. The text in Props. 23–25 has a partial duplicate in *Coll.* III, and another one in *Coll.* VIII. The most extensive source on ancient cube duplications is, in this case, not Pappus, but Eutocius *In Arch. Sph. et Cyl. II*, pp. 54–106 Heiberg. Eutocius' report also contains Nicomedes' construction (pp. 98–104 Heiberg), and a passage very similar to Prop. 24.

The author Nicomedes forms a bridge to the next group of propositions. Props. 26–29 deal with the quadratrix, its *genesis* (generation), *symptoma* (characterizing mathematical property), discussion of some fundamental problems with the curve (source: Sporus), two *symptoma*-theorems on it (Prop. 26, rectifying the circle, Prop. 27, squaring the circle), and two arguments via analysis of loci on surfaces that seek to show, via analysis, that the quadratrix is uniquely determined, relative to either an Apollonian helix or an Archimedean spiral. The quadratrix itself is a transcendental curve in the plane. It was probably invented in the fifth or fourth century BC (i.e., before Euclid) for the division of an acute angle in a given ratio. It can, however, also be used to square the circle, and it is this property from which the curve takes its name.[2] Nicomedes is explicitly associated with the quadratrix and the quadrature (Props. 26 and 27). He may be the author of Prop. 29 as well. Props. 26–29, along with the minor quadratrix theorems in Props. 35–41, are our only source on the ancient quadratrix, and Props. 26–29 are our only sources on squaring the circle with it. The analytical characterizations in Props. 28 and 29 are, in addition, our only testimonies on analysis of loci on surfaces. Thus, this part of *Coll.* IV is especially interesting in terms of its mathematical content. But even so, it is again the investigative and argumentative methods, this time for linear geometry and its characteristic *symptoma*-mathematics, that are the focus of Pappus' presentation. It remains unclear to what degree the analytical characterization of curves like the quadratrix, as in Props. 28 and 29, is representative, and what the Hellenistic mathematicians who specialized in this area thought and produced on this issue. Among other things, the status of the curves, even after analytical characterization, remains somewhat shady. Yet in Pappus' portrait, results like Props. 28 and 29

---

[1] On the meaning of the term "neusis," and for examples for neusis constructions in ancient geometry, see the commentary on Props. 23–25.

[2] More specifically, the quadratrix directly yields the rectification of the arc of a quadrant. From there, quadrature is immediately available, once one has a theorem like Archimedes, *Circ. mens.* I. Whether the discovery of the quadrature property is pre-Euclidean (Dinostratus, fourth century BC) or post-Archimedean (Nicomedes, third/second century) BC is a matter of dispute. On this issue, and generally on the quadratrix, see the commentary on Props. 26–29.

represent the culmination of a tradition of justifying the foundations of *symptoma*-mathematics of the higher curves via analysis. He explicitly accepts the *symptoma* – theorems on the quadratrix as geometrically valid.

Prop. 30 returns to Archimedes, also the author of the first propositions in Pappus' portrait of the geometry of motion curves. He appears to have been the initiator of this branch of mathematics, while working specifically along a path of investigation that incorporates quasi-mechanical methods. A spiral is generated on the surface of a hemisphere, via two synchronized uniform motions (speeds in the ratio 1: 4). From this *genesis*, the main *symptoma* of the spiral is directly read off. Then an area theorem is proved: The area cut off on the sphere above the spiral is eight times the segment cut off from a quadrant of a maximum circle when one connects the end points of the arc, and the surface cut off on the hemisphere below the spiral is eight times the remaining triangle in the quadrant, i.e., it is equal to the square over the diameter of the sphere. This theorem constitutes the first example for a successful quadrature of a curved surface in space. Prop. 30 is perhaps somewhat harder to read than the other contributions in *Coll.* IV, because of the cumbersome notations it uses. Nevertheless, its result is rather interesting. Connections to *De Sphaera et Cylindro* abound. The argumentative strategy resembles the one used in Prop. 21 (limit process, mechanical method). Pappus' portrait suggests that Archimedes was the main representative of this quasi-mechanical branch of mathematics of higher curves. His successors seem to have favored the analytical path.

At this point, Pappus concludes his portrait of the mathematics of the linear kind. A few further examples on the *symptoma*-mathematics of the quadratrix will come up in part III. Part II focused on the big topics, as it were, especially the foundations, and the methodological horizon, of "higher" mathematics. Pappus moves on to a general remark on the three kinds of geometry, and the methodological consequences to be drawn from this tri-partition. This is the famous meta-theoretical passage. There are to be three non-overlapping kinds of mathematics, determined by the methods used to solve problems and accomplish constructions. First, there is plane geometry. It uses only circles and straight lines. This branch of mathematics and its different sub-branches were the subject of Props. 1–18. Next come problems that cannot be solved with circle and straight line alone, but need one or more conic sections in addition. This kind of mathematics is called solid, because conics "have their *genesis*" in a cone, i.e., a solid figure (circle and straight line are generated in the plane). Pappus' presentation suggests that the geometry of conics developed out of an attempt to solve problems that were unsuccessfully attacked with plane means at first. He specifically points to the angle trisection. In the parallel text in *Coll.* III, the cube duplication plays this role. Schematic though it is, Pappus' account may very well be more or less accurate. His picture concurs with the estimate of most modern scholars on the development of the theory of conic sections. So far, Pappus has not presented any "solid" arguments. They will be the subject of Props. 31–34 and 42–44. A third kind of mathematics is to be called "linear." Its basic curves cannot be characterized as precisely as the circles, straight lines, and conics of the other two kinds. In fact, this kind of mathematics covers all the rest of mathematical curves. It is not as easy to see how they form a single "kind." Instead of a characterization by means of basic curves, Pappus gives a longer description which is not quite uniform. He differentiates two "paths" toward the fundamental curves: they are either generated by motions, or determined via analysis. Pappus mentions a few contributors, but their works, and their curves, are lost, so that we cannot take recourse to any text outside of *Coll.* IV to evaluate Pappus' description of the third kind. Pappus shows some uneasiness with regard to the status of the curves. Nevertheless, he unequivocally counts the mathematics on them, i.e., the *symptoma*-mathematics of the curves, as fully legitimate geometry. Examples for this kind of mathematics were given in Props. 21 and 22, 23, 26 and 27, and 30; further examples on the *symptoma*-mathematics of the quadratrix will be given in Props. 35–41.

Given that geometry is to have these strictly separate kinds (gene), a homogeneity criterion applies. It is required that mathematical problems must be solved with the means that are appropriate to the specific nature of the particular problem, i.e., they must be "akin" to the problem. Obviously, this targets not so much attempts to solve problems with means that are insufficient (a solid problem simply cannot be solved by plane means, since "solid" means "in need of using at least one conic section"). Rather, the requirement targets solutions that use "higher" curves than required, e.g., solid solutions where plane methods would have sufficed, and linear solutions where solid or plane methods would have sufficed. They are rejected, because they fail to capture the object of investigation for what it essentially is. They are not "akin" to it, they come from the wrong genos. This homogeneity requirement is different from modern ways of thinking about appropriate means, even when the same label is used. With Pappus, it is closely connected to an essentialist view on definition and scientific argumentation,[1] not just to the idea of minimalizing the means required. Pappus' homogeneity criterion was noticed, much discussed, and also appropriated in various ways, by the mathematicians in the sixteenth and seventeenth century reading Pappus, e.g., by Vieta, Descartes, and Newton, and developed and transformed from thereon. It is doubtful whether it was operative in this generality in antiquity. However, a similar criterion was developed by Apollonius for the differentiation of plane versus solid *neusis* problems and loci. He did in fact require that plane *neuses* must be constructed with plane means. And he developed a toolbox for differentiating the level of problems, where the question was whether a problem was plane or solid. Apollonius himself may have been more interested in minimizing operational tools and procedures than in an essentialist justification such as the one Pappus employs. Nevertheless, the resulting restriction requirement was his. Apparently, it was used by others after him to scrutinize already existing *neusis* arguments, and other theorems as well. Perhaps the two examples for arguments that fail to meet the homogeneity requirement mentioned by Pappus, a *neusis* from Archimedes's *Spiral Lines*, and Apollonius' construction of a normal to the parabola, come from this very context. An analysis of the Archimedean *neusis*, with the intention of showing that it is solid, is given in Props. 42–44.

---

[1] Cf. Aristotle's theory of science, especially *Posterior Analytics* I, 1–13; see the remarks on the meta-theoretical passage in the commentary.

Props. 31–34 present three different solutions for the problem of angle trisection (divide a given angle into three equal parts). In Pappus' portrait, the angle trisection appears as an exemplary problem of the solid kind. In *Coll.* III, the cube duplication plays this role. In fact, all problems that are solid in Pappus' sense reduce to either of these two problems. Thus, Pappus is quite correct in his assessment of the importance of the trisection. The material in Props. 31–34 consists of four layers. Props. 31–33 give an angle trisection via *neusis*. Within it, an older version in which the *neusis* was not constructed via conics is still present (Props. 31 and 32). The *neusis* in question can be constructed with Nicomedes' conchoid.[1] In the original, possibly pre-Euclidean argument, the *neusis* was probably constructed by simple ruler manipulation. This older layer was then worked over, and the result is an anlytical-synthetical argument. In Prop. 31, the *neusis* is reduced, via analysis, to the construction of a hyperbola through a given point with given asymptotes, and in Prop. 33, this hyperbola is constructed in an analytical-synthetical argument. Prop. 33 may be close to Apollonius' lost analytical-synthetical solution for the angle trisection, though Pappus seems to have played a major role in spelling out the details of the argument as presented in *Coll.* IV. It is noteworthy that a much simpler, purely synthetical solution for Prop. 33 is possible via *Konika* II, 4. Apparently Pappus wants to make the point that the characteristic working strategy in "solid" geometry is analysis-synthesis. All his examples from solid geometry are analytical. Prop. 34 contains two further angle trisections, avoiding the *neusis*. The analysis is emphasized, with the synthesis only sketched for 34a, and left to the reader for 34b. Both arguments employ the same hyperbola, under different representations. Prop. 34b contains an older layer that goes back, in all likelihood, to an argument from Aristaeus' (lost) *Loci on Surfaces*. The hyperbola is determined through its focus-directrix property. Prop. 34a builds on Prop. 34b. It is the simplest of the three solutions via conics, and rests on an analysis reducing the problem to *Konika* I, 21. Pappus was probably the author of Prop. 34a. To what degree he himself revised Aristaeus, or else an intermediate source drawing on Aristaeus, cannot be determined with certainty. Props. 31–34 handle conic sections as loci. A tendency toward algorithmization and reduction to standard configurations can be detected in Pappus' portrait of the typical methods, despite the fact that he chose examples that come from very different time periods.

General angle division is, as Pappus remarks, not a problem that can be solved via conics. In *Coll.* IV, the problem serves to illustrate how a transition from the second to the third kind of geometry takes place when we generalize problems. Props. 35–38 are examples for *symptoma*-mathematics of the quadratrix. They may derive from Nicomedes' work on the curve. Prop. 35 shows how the quadratrix, or, alternatively, the Archimedean spiral, can be used to divide a given angle in a given ratio. Props. 37 and 38 show how it is possible to construct a regular polygon with any given number of sides. These two propositions are visibly analogous to

---

[1] Vice versa, the *neusis* construction via conics, in Props. 31/33, which probably goes back to Apollonius, can be used for the cube duplication.

Elements IV, 10/11. No attempt is made to single out angle divisions that would become plane, or solid.

Props. 39–41 continue the *symptoma*-mathematics of the quadratrix. The focus now is on the rectification property of the curve (Prop. 26). Some perhaps rather unspectacular consequences are drawn from it: As Prop. 26 rectifies the circle, one can also use the quadratrix to find, conversely, a circle the circumference of which is equal to a given straight line (Prop. 39), one can construct a circular arc that has a given ratio to a given line segment, as a chord under it (Prop. 40), and one can define and construct incommensurable and irrational arcs (or angles), drawing on the definition of incommensurable straight lines (Prop. 41). The above-mentioned demarcation problem does not arise for these problems. Circle rectification is, as it were, by nature linear.

In the final group of propositions (Props. 42–44), Pappus comes back to the demarcation question, and specifically to the criticism he has voiced against the *neuses* in Archimedes's *Spiral Lines*: that they are solid, whereas a plane argument would have sufficed. Props. 42–44 pursue two goals, to show via analysis that the *neusis* is indeed solid, and to present an analysis that is useful also for working on numerous other solid problems. The second goal may explain the particular choice of the *neusis*. For Pappus chooses not *SL* 7 or *SL* 8, the ones one might expect, but rather another *neusis*, closely related to *SL* 9 and to a *neusis* employed in a now-lost angle trisection. It is, however, also related to all of *SL* 5–9. Given the connection to the angle trisection, Pappus was correct in claiming that the analysis could be useful for many other solid problems. The analyses in Props. 42–44 show features of typifying and standardization very much like the arguments in Props. 31–34. They do indeed lead to the result that the locus for a solution of the Archimedean *neusis* is, in general, determined by the intersection of a hyperbola and a parabola. Apparently, Pappus believed this shows that the *neusis* is solid. His analytical argument does not show this beyond doubt, though. Specifically, he does not attempt to apply *diorismos* to identify plane cases or impossible cases, and he cannot guarantee that his analysis has exhausted all the information available in the configuration, including information that might make a plane solution possible. Nevertheless, Props. 42–44 are again interesting both for their mathematical content, and for methodological reasons. They contain three of the very few examples for analytical arguments on solid loci that are preserved from antiquity, and they illustrate how one might have used analysis for determining whether a problem is, in general, solid or plane.

3   Summary: *Coll.* IV at a Glance

I Plane Geometry

(a) *1–6 Plane geometry, Euclidean style* (scope: *Elements*)
   1 Generalization of the Pythagorean theorem, synthetic
   2/3 Applications of the theory of irrational magnitudes of the first order, synthetic
   4–6 The basic structure of plane geometric analysis; theorems, analytic-synthetic

(b) *7–12 Plane geometry, Apollonian style* (scope: *Treasury of analysis*, plane problems)
7 Role and use of *Data*; analytic
8–10 *Resolutio* for a special case of the *Apollonian problem*; analytic
11/12 Effects of analysis as mere extension of configuration,[1] and as pure reduction; synthetic/analytic-synthetic
(c) *13–18 Plane geometry, Archimedean style* (scope: *Elements* and beyond)
Arbelos theorem (synthetic; monographic)

II Linear Geometry: Symptoma-Mathematics of Motion curves

(a) 19–22 Plane spiral (Archimedes, quasi-mechanical methods)
(b) 23–25 Conchoid (Multiplying the cube, Nicomedes, quasi-analytical methods)
(c) 26–29 Quadratrix (Squaring the circle, Nicomedes, transition from mechanical to analytical characterization of the *genesis*)
(d) 30 Spherical spiral (Archimedes, quasi-mechanical method)

Meta-Theoretical Passage on the Three Kinds of Geometry: Homogeneity criterion

III Solid Geometry, Transition "Upward," Demarcation "Downward"

(a) 31–34 Angle trisection (solid loci/conic sections, several stages of methodology, pre-Euclidean, Aristaeus, Apollonius, Pappus)
(b) 35–38 Angle division and applications (linear, as a result of generalization; *symptoma* – mathematics of quadratrix and spiral, Nicomedes?)
(c) 39–41 *Symptoma-mathematics of the quadratrix,* rectification property (linear by nature, Nicomedes?)
(d) 42–44 *Analysis of an Archimedean neusis* (analysis of solid loci, determining the level of a proposition or problem; Pappus)

The intermediate status of solid geometry accounts for the position of its examples after the presentation of the other two kinds, and also for the occurrence of issues of transition "upward" and "downward" in this part of *Coll.* IV. Solid geometry is the bridge between the two extreme kinds, and it was the area with regard to which demarcation issues became virulent, propelled the development of techniques, and were investigated systematically (above all, by Apollonius).

---

[1] But see the commentary on Prop. 11. This evaluation of Prop. 11 is highly tentative.

Part I Greek Text and Annotated Translation

Introductory Remarks on Part I

The following remarks draw on (Treweek 1957), (Jones 1986b) and (Jones 1986a), pp. 18–65 for the text and transmission of the *Collectio*. The reader is referred there for further information.

Tradition, Reception, and Editions of the Text of the *Collectio*

The text of the *Collectio*, as it has come down to us, was not conceived, or published, as a single coherent work by Pappus himself. Probably, it was compiled, edited, and published shortly after Pappus' death by an unknown author. An exception is Book VIII, which may have been published by Pappus separately. It has an independent tradition, with clearly documented reception in the Islamic culture. For all the other books of the *Collectio*, no traces of a thorough reception can be documented, neither in late antiquity, nor in Islamic or in Byzantine culture. The extant text goes back to a single Byzantine manuscript (Vat. gr. 218, called **A** here) from the tenth century, in turn at least two, but probably not many more steps removed from Pappus' original autograph. This archetype is found in the Vatican Library today. All later copies, dating from the sixteenth century and later, stem directly or indirectly from this manuscript. There are about 40 such derived complete or partial copies. For a detailed description of the history of **A**, and for a stemma for the manuscript tradition, see Treweek (1957), with additions in Jones (1986a).

As said above, all copies of **A** date from the sixteenth century or later. Thus, the *Collectio* did not have a reception in Western Europe during the Middle Ages that has left significant documented traces. There is one possible exception, the isolation of which rather proves the general point. The manuscript **A** was already in Rome in the thirteenth century. Unguru (1974) made the case that Witelo (thirteenth century) may very well have had indirect access to at least the part of *Collectio* VI that deals with propositions from Euclid's *Optics*. Perhaps William of Moerbeke, who was at Rome during that time, perused **A** and translated such passages from *Collectio* VI for Witelo as seemed useful for his *Perspectiva*.

The first complete printed edition of the *Collectio*, by Commandino (published posthumously), appeared in 1588. It contains a Latin translation, and

H. Sefrin-Weis, *Pappus of Alexandria: Book 4 of the Collection*,
Sources and Studies in the History of Mathematics and Physical Sciences,
DOI 10.1007/978-1-84996-005-2, © Springer-Verlag London Limited 2010

critical as well as explanatory notes. Commandino's work went through several reprints, the last one was Bologna 1660 (with revisions). In the decades after the first publication, an intensive reception and discussion of Pappus' work took place. Above all, the search for the ancient analytical method and also general methodological questions, besides several mathematical vignettes from Pappus, notably the part of *Coll.* IV that deals with "higher curves," inspired the mathematicians of the day to study Pappus and use him toward further mathematical progress. The impact of both *Coll.* VII and *Coll.* IV on seventeenth century mathematics was enormous. This topic would certainly deserve further exploration.[1] During the following centuries, several projects for a complete edition of the Greek original were launched, but none was brought to completion. Jones (1986a) lists the most significant partial editions. Among them, Halley's 1706 edition of the (Arabic translation of the) *Sectio rationis*, with an excerpt from *Coll.* VII (Pappus' commentary on the work), and Torelli's edition of *Coll.* IV, # 30–35 (on the quadratrix), are perhaps worth mentioning here, as illustrations of the general practice. The first complete critical edition was published by Hultsch 1875–1878 (referred to as Hu in part I of the present edition). It contains the Greek text with critical apparatus, and a Latin translation and notes. To this day, it remains the standard text (excepting *Coll.* VII, for which see also Jones (1986a)). Treweek's 1950 new critical edition of *Coll.* II–V (Tr) has unfortunately never been published. The only complete translation of the *Collectio* into a modern language is by Ver Eecke (1933b). For *Coll.* IV, a German translation was given by Sefrin-Weis (1998; see bibliography, primary sources, Pappus). An English translation is provided here.

Remarks on the Greek Text Printed Here

The Greek text of *Collectio* IV that is given here is essentially an edition of Vat. gr. 218 (A), f. 33r–55v.[2] For the text, a transcription was made from photographs of **A**. It was then collated with the original in the Vatican library, and with Hultsch's and Treweek's editions of *Coll.* IV (Hu and Tr). Wherever the text printed here diverges from Hu,[3] I have put the respective readings in italics, noting Hu's and Tr's readings in the apparatus. Additions to the text of **A** are marked by angular brackets (<>), deleted or suspected phrases are put in square brackets ([]). The lines as written in **A** were kept, and the beginning of pages in **A** are indicated with headers in English. In the case of a few very infelicitous hyphenations, I have put the full word in either of the lines and noted the hyphenation in

---

[1] Cf., e.g., Bos (2001).

[2] Manuscript sigla adopted from (Treweek 1957), who incorporated Hultsch's sigla.

[3] For the first part of Prop. 44, the emended version in appendix Hu pp. 1232f., with emendations by Hultsch and Baltzer, was used as a standard of reference.

the apparatus. Thus, the text as given here provides an alternative to both Hu and Tr, while remaining closely connected to these fully critical editions. It can be read in parallel with them, and with **A** as well.

The apparatus in the present edition is a reduced one. It was constituted as follows. Wherever the adopted text diverges from **A**, **A**'s reading is reported in the apparatus, excepting mere orthographic corrections. If a correction or addition can be traced back to the manuscripts **B** or **S**, I have noted this fact as well, in addition to documenting Hu's and Tr's readings (and occasionally, Co's).[1] I have not had the opportunity to collate the manuscripts **B** and **S**, or any of the recentiores for *Coll.* IV, directly. Instead, I have relied on concurrence between Hu and Tr for readings in **B** and **S**. Whenever I report an emendation as attested in Hu, Co and/or Tr, I do not intend to imply that the emendation originates with them. For a more complete documentation of manuscript readings in the recentiores, the reader is referred to Tr. Most of the adopted emendations are clearly required by the mathematical sense, and probably result from scribal errors such as Δ for A, or other erroneous labels for points. One would have resorted to them in any case. Whenever an emendation was not thus clearly justified by the mathematical content, **A**'s reading was adopted, even in the face of grammatical or stylistic irregularities.

Remarks on the Translation

As said above, the Greek text is based on Vat. gr. 218 (A) and closely connected to Hultsch's critical text of *Collectio* IV (Hultsch 1876–1878, Hu). Hultsch's annotated Latin translation in Hu was very helpful for the English translation and commentary presented here as well. Also useful were Commandino's translation and commentary in his 1588 edition and the 1660 revised edition of the same work (Co), Ver Eecke's 1933 French annotated translation, and my own 1998 German annotated translation with commentary. The translation tries to be as close to the original Greek as possible. I have made an effort to render the peculiar formulaic way of expression in the Greek mathematical arguments into English by using standardized phrases corresponding to the Greek formulae. Greek mathematical prose is, however, extremely elliptical. A literal English translation would be incomprehensible. As a compromise, additions were implemented so as to produce tolerably complete English sentences. The additions are put in angular brackets (<>), so that the reader can get an idea of Pappus' own way of expressing mathematical thoughts. For the structuring of the text, I have used Hultsch's separation into propositions (right margin and header in Hu, in the Latin translation), because separation into units of mathematical content facilitates understanding. The

---

[1] Among the recentiores, it is mostly the manuscripts B and S that contain helpful alternative readings throughout *Coll.* IV. Compare the apparatus in Hu and Tr. For a list of the extant manuscripts and their interconnection see Jones (1986a, pp. 56–62), and Treweek (1957).

numbering of propositions is identical with Commandino's in Co, except for Prop. 44, which is missing in Co. In addition, the chapter divisions of the Greek text in **A** (Roman numerals in Hu at the beginning of paragraphs in the Latin translation, Greek numerals in Hu at the beginning of paragraphs in Hultsch's Greek text, labeled as #1ff. here) are included. Wherever possible, the translation is close to Hultsch's text, even where I disagreed with his choice for particular readings. In a few cases, for example in the concluding passage on Sporus' criticism of the quadratrix (Prop. 26), I have translated what I believed to be the correct reading, and have documented my deviation in the notes. The result is, I hope, a readable English version of the Greek text that is rather close to the original, and can be used together with Hultsch's edition and Latin translation, as well as with Ver Eecke's French translation (Ver Eecke follows Hultsch very closely, both in his translation and in his notes). My intention is that Pappus' mathematical argumentation should be accessible from his prose. I have therefore used (very limited) transformations into modern notation and algebraic language only in the notes, or in the commentary.

The notes to the translation do, however, provide references to theorems from Euclid's *Elements* (and some other standard ancient texts, notably Euclid's *Data*, Apollonius' *Konika*, and Archimedes' works, in the standard critical editions), or quasi-algebraic explanations that justify Pappus' intermediate argumentative steps. While I do by no means intend to imply that this is how an ancient mathematician would have proceeded (i.e., justify his steps via explicit reference to Euclid), I do believe that Euclid's *Elements* had the role of a basis and center for geometrical instruction for Pappus. I take it that he wanted his readers to bring the relevant content to bear, and, if unable, to consult Euclid as they were working their way through *Collectio* IV. What I am doing in quoting Euclid is just providing one such possible path of justification. For the references to Euclid's *Elements*, I have used Heath's 1926 English edition instead of the standard critical edition, because (Heath 1926) is widely accessible and very reliable and helpful. References to individual propositions in the *Elements* will be given in Roman numerals, followed by Arabic numerals (e.g., I, 47 refers to *Elements* Book I, Proposition 47 in (Heath 1926)), references to books will be given in Roman numerals alone. The commentary (see Part II) complements the notes to the translation. It will give proof protocols so as to facilitate surveying whole arguments at a glance and identifying crucial ideas and steps in a proof. In addition, it contains historical background information in outline, bibliographical information, and attempts to provide a context for groups of propositions, both with regard to *Collectio* IV, and more generally with regard to the history of Greek mathematics. Furthermore, it locates issues where Pappus' propositions, or groups of propositions, might be useful for further investigations. Explanatory remarks on central keywords, e.g., "analysis-synthesis," "*neusis*" "angle trisection," are provided there as well, again in the form of general outlines with bibliographical information for further in-depth study, (see below, Part II with introductory remarks). In addition to technical mathematical information, the notes to the translation contain some philological remarks. Salient stylistic peculiarities, hapax legomena, and the use of specific Greek words were noted, when I thought their occurrence could be especially significant for further interpretation, though this route is not pursued in detail in the present translation and commentary. I have restricted myself to simple documentation and obvious direct inferences, as a possible basis for further study. For example, in the introduction to Prop. 28, Pappus announces an analysis of the *genesis* of the quadratrix, and consequently uses the word "analuesthai" (subject to geometrical analysis). This is significant, because interpreters starting with Hultsch have read this word, erroneously, as equivalent to "luesthai" (solve), and believe that Pappus is trying to provide a mathematically exact solution for the quadrature of the circle, which, of course, he cannot achieve. Attention to the actual word used can help clarify the meaning of Prop. 28 (see following sections). In the present study, I have restricted myself to just pointing out the use of "analuesthai," and drawing the obvious inference that what we get in Prop. 28 is a geometrical analysis of the *genesis*, not a (failed) solution of the quadrature. My documentation of such "salient Greek terms" is not exhaustive.

Remarks on the Diagrams

In **A**, the figures for individual propositions appear at the end of each argument, as inserts in indented spaces – alongside the beginning of the following proposition. In the present edition, I have relocated the diagrams. They appear as inserts in the translation, wherever possible, directly after the statement of propositions. The lettering is in Latin, in accordance with the translation. For a concordance of Greek diagram letters and their rendering in Latin letters see the final page of this introduction. The diagrams were modeled as closely as possible on the figures in **A**. Gross distortions in comparison to the argument in the mathematical text were ameliorated. For example, points were shifted so as to make congruent angles and lines appear as congruent. For diagrams that contain only circles and straight lines, as for example, in Props. 1–18, minimal intrusion of this sort was sufficient. In **A**, all curves are represented by circular arcs, however. The diagrams for propositions dealing with higher curves were therefore subject to more vigorous revisions (with the exception of propositions on the plane spiral). Thus, the diagrams containing the quadratrix (Props. 26, and 35 ff.) are somewhat close to **A**, but reshaped considerably, and the diagrams containing the conchoid and those referring to three-dimensional objects were completely redrawn. Specifically, the figures for Props. 23–30 are not connected to **A**. The diagram for Prop. 44 was drawn afresh as well, since it is missing in **A**. Descriptions of the changes made in the constitution of the diagrams are given in an appendix. The appendix also contains a drawing of the diagram for the limit case in Prop. 15, which is not dealt with in Pappus' text.

List of Sigla and Abbreviations

Manuscripts

A	Vat. gr. 218	Tenth century

Archetype for all existant manuscripts, sole independent witness to the text of Pappus, *Collectio* IV. In the Vatican library. For a description, see Jones (1986, pp. 30–35), and Treweek (1957).

A2   Corrector's hand in **A**

B   Par. gr. 2440   Sixteenth century, before 1554

Earliest of the known copies of the lost Strasbourg manuscript **R**, which in turn was a copy of **A**.

B2 B3   Corrector's hands in **B**

S   Leiden Scal. 3   after 1562

Copy of the Paris manuscript **C** (1562), which was made for Ramus from a lost manuscript x, in turn a copy of **A**.

Editions and Translations

Co   Commandino (1588, 1660)
                     Latin translation with notes.

Eut.   Eutocius *In Archimedis De sphaera et cylindro* II
                     Contains an alternative version of Prop. 24

Hu   (Hultsch 1876–1878)
                     Critical edition, Latin translation and notes

Hu/Baltzer   Hu appendix p. 1232f.
                     Revised text for first part of Prop. 44

app. Hu   Apparatus in Hu

To   Torelli (1789)
                     Contains # 30–35, cited in app. Hu

Tr   (Treweek 1950)
                     Critical edition. Unpublished dissertation.

Mathematical Symbols and Abbreviations Used in the Notes and Commentary

$=$	Equality of line segments, angles, areas, Equivalence of ratios
$\sim$	Similarity of triangles
$\cong$	Congruent triangles
a:b	The ratio of a to b
(a:b) X (c:d)	The compounded ratio of a:b and c:d
$\triangle$ABC	The triangle ABC
AB²	The square over AB
AB × CD	The rectangle contained by AB and CD
∠ABC	The angle ABC
AB ∥ CD	AB is parallel to CD
AB⊥CD	AB is perpendicular to CD

Concordance of Greek Letters (A) and Latin Equivalents (Translation, Commentary, Diagrams)

A	A
B	B
Γ	C
Δ	D
E	E
Z	Z
H	H
Θ	**T**
I	I
K	K
Λ	L
M	M
N	N
Ξ	X
O	O
Π	P
P	**R**
Σ	S
T	**T'**
Υ	Y
Φ	F
X	**X'**
Ω	**W**

Part Ia
Greek Text

Pappus, Collectio IV. Vat. gr. 218 (Codex A, f. 33r° - 55 v°)

f. 33r (Prop. 1)
Prop. 1
#1 ἐὰν ᾖ τρίγωνον τὸ ΑΒΓ, καὶ ἀπὸ τῶν ΑΒ ΒΓ ἀναγραφῇ τυχόν-
τα παραλληλόγραμμα τὰ [1] *ΑΒ ΔΕ ΒΓ ΖΗ*[2], καὶ αἱ ΔΕ ΖΗ
ἐκβληθῶσιν ἐπὶ τὸ Θ, καὶ ἐπιζευχθῇ ἡ ΘΒ, γίνεται τὰ *ΑΒ*
ΔΕ ΒΓ ΖΗ[3] παραλληλόγραμμα ἴσα τῷ ὑπὸ τῶν ΑΓ ΘΒ
περιεχομένῳ παραλληλογράμμῳ ἐν γωνίᾳ ἥ ἐστιν ἴση συναμφοτέρῳ τῇ ὑπὸ ΒΑΓ ΔΘΒ. ἐκβεβλήσθω[4] γὰρ ἡ ΘΒ ἐπὶ τὸ Κ, καὶ
διὰ τῶν ΑΓ[5] τῇ ΘΚ παράλληλοι ἤχθωσαν αἱ ΑΛ ΓΜ, καὶ ἐπεζεύχθω ἡ ΛΜ. ἐπεὶ[6] παραλληλόγραμμόν ἐστιν τὸ ΑΛΘΒ, αἱ *ΑΛΘΒ*[7]
ἴσαι τέ εἰσιν καὶ παράλληλοι· ὁμοίως καὶ αἱ ΜΓ ΘΒ[8] ἴσαι τέ εἰσιν
καὶ παράλληλοι, ὥστε καὶ αἱ ΛΑ ΜΓ ἴσαι τέ εἰσιν καὶ παράλληλοι. καὶ αἱ ΛΜ ΑΓ ἄρα ἴσαι τε καὶ παράλληλοί εἰσιν· παραλληλόγραμμον ἄρα ἐστὶν τὸ *ΑΛ ΜΓ*[9] ἐν γωνίᾳ τῇ ὑπὸ ΛΑΓ, τουτέστιν[10]

f. 33v (Prop. 1 and 2)
συναμφοτέρῳ τῇ τε ὑπὸ ΒΑΓ[11] καὶ ὑπὸ ΔΘΒ· ἴση γάρ ἐστιν
ἡ ὑπὸ ΔΘΒ τῇ ὑπὸ ΛΑΒ. καὶ ἐπεὶ τὸ [ἀπὸ][12] ΔΑΒΕ παραλληλόγραμμον τῷ *ΛΑ ΘΒ*[13] ἴσον ἐστίν, ἐπί τε γὰρ τῆς αὐτῆς βάσεώς ἐστιν τῆς

---

[1] τυχον τὰ σημεῖα παραλληλόγραμμα Α, corr. Hu,Tr

[2] *ΑΒΔΕ ΒΓΖΗ* Co, Hu, Tr

[3] *ΑΒΔΕ ΒΓΖΗ* Β, Hu, Tr

[4] ἐκβληθῇ γὰρ Α

[5] distinx. BS, Hu, Tr

[6] ἐπι παραλληλόγραμμον Α

[7] *ΑΛ ΘΒ* Hu, Tr

[8] *ΟΒ* Α corr. Co, Hu, Tr

[9] *ΑΛ ΜΓ* AB coniunx. S Hu, Tr

[10] τοῦ - τ ἔστιν Α τουτέστιν Hu

[11] *ΑΒΓ* AB corr. S Hu, Tr

[12] ἀπὸ del. Hu, Tr

[13] *ΛΑΒΘ* Hu, ΛΑΘΒ Tr

H. Sefrin-Weis, *Pappus of Alexandria: Book 4 of the Collection*,
Sources and Studies in the History of Mathematics and Physical Sciences,
DOI 10.1007/978-1-84996-005-2, © Springer-Verlag London Limited 2010

ΑΒ καὶ ἐν ταῖς αὐταῖς παραλλήλοις ταῖς ΑΒ ΔΘ, ἀλλὰ τὸ *ΛΑ ΘΒ*[1]
τῷ *ΛΑ ΚΝ*[2] ἴσον ἐστίν, ἐπί τε γὰρ τῆς αὐτῆς βάσεώς ἐστιν
τῆς ΛΑ[3] καὶ ἐν ταῖς αὐταῖς παραλλήλοις ταῖς ΛΑ ΘΚ, καὶ τὸ
ΑΔ ΕΒ[4] ἄρα τῷ *ΛΑ ΚΝ*[5] ἴσον ἐστίν. διὰ τὰ αὐτὰ καὶ τὸ *ΒΗ ΖΓ*[6] τῷ
ΚΝ ΓΜ[7] ἴσον ἐστίν· τὰ ἄρα *ΔΑ ΒΕ ΒΗ ΖΓ*[8] παραλληλόγραμμα
τῷ *ΛΑ ΓΜ*[9] ἴσα[10] ἐστίν, τουτέστιν τῷ ὑπὸ ΑΓ ΘΒ ἐν γωνίᾳ τῇ
ὑπὸ ΛΑΓ, ἥ ἐστιν ἴση συναμφοτέραις ταῖς ὑπὸ ΒΑΓ ΒΘΔ. καὶ
ἔστι τοῦτο καθολικώτερον[11] πολλῷ τοῦ ἐν τοῖς ὀρθογωνίοις
ἐπὶ τῶν τετραγώνων ἐν τοῖς στοιχείοις δεδειγμένου[12].

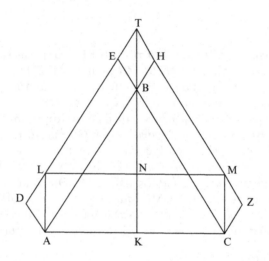

Prop. 2
<u>#2</u> ἡμικύκλιον ἐπὶ τῆς ΑΒ ῥητὴν ἔχον
τὴν διάμετρον, καὶ τῇ ἐκ τοῦ

---

[1] *ΛΑΒΘ* Hu, Tr

[2] coniunx. Hu, Tr

[3] *ΑΒ* Tr

[4] coniunx. Hu, Tr

[5] coniunx. Hu, Tr

[6] coniunx. Hu, Tr

[7] *ΝΚΓΜ* Hu, Tr

[8] *ΔΑΒΕ* ΒΗΖΓ Hu Tr

[9] coniunx. Hu, Tr

[10] *ἴσαι* A corr. Hu, Tr

[11] καὶ ολικώτερον A corr. Hu, Tr

[12] καὶ τῶν ὁμοίων καὶ ὁμοίως ἀναγεγραμμένων *lib. 6 Element.* add. V2 (178, 12 app. Hu)

κέντρου ἴση¹ καὶ < τῇ ΑΒ >² ἐπ' εὐθείας ἡ
ΒΓ, καὶ ἐφαπτομένη ἡ ΓΔ, καὶ
δίχα ἡ ΒΔ περιφέρεια τῷ Ε σημείῳ, καὶ ἐπεζεύχθω ἡ ΓΕ· οὕτως³
ἡ ΓΕ ἄλογός ἐστιν ἡ καλουμένη
ἐλάσσων. εἰλήφθω τὸ Ζ κέντρον
τοῦ ἡμικυκλίου, καὶ ἐπεζεύχθωσαν αἱ ΖΔ ΖΕ. ἐπεὶ ὀρθή ἐστιν
ἡ ὑπὸ ΖΔΓ, ἐν ἡμικυκλίῳ ἐστὶν τῷ ἐπὶ τῆς ΖΓ, οὗ κέντρον ἐστὶν τὸ Β. καὶ τῆς ΒΔ ἐπιζευχθείσης ἰσόπλευρον γίνεται τὸ ΒΖΔ τρίγωνον, ὥστε διμοίρου μέν ἐστι ἡ ὑπὸ ΔΖΒ γωνία, τρίτου δὲ ἡ ὑπὸ ΕΖΒ. ἤχθω κάθετος ἀπὸ τοῦ Ε ἐπὶ τὴν
ΑΒ διάμετρον ἡ ΗΕ· ἰσογώνιον ἄρα τὸ ΓΖΔ τρίγωνον τῷ ΕΖΗ
τριγώνῳ, καὶ ἔστιν ὡς ἡ ΖΓ πρὸς τὴν ΓΔ, ἡ ΕΖ⁴ πρὸς ΖΗ. ἐπίτριτον δὲ τὸ ἀπὸ ΖΓ τοῦ ἀπὸ ΓΔ· ἐπίτριτον ἄρα καὶ τὸ ἀπὸ ΕΖ
τοῦ ἀπὸ ΖΗ· λόγος ἄρα τοῦ ἀπὸ ΕΖ πρὸς τὸ ἀπὸ τῆς ΖΗ ὃν ις¹
πρὸς ιβ¹, τοῦ δὲ ἀπὸ ΖΓ πρὸς < τὸ >⁵ ἀπὸ ΕΖ ὃν ξδ¹ πρὸς ις¹. καὶ τοῦ
ἀπὸ ΖΓ ἄρα πρὸς τὸ ἀπὸ ΖΗ λόγος ἐστὶν ὃν ξδ¹ πρὸς ιβ¹. ἔστω δ'ἡ⁶
ΖΒ τετραπλασία τῆς ΒΘ· καὶ ἔστιν τῆς ΒΖ διπλασίων⁷
ἡ ΖΓ· λόγος ἄρα τῆς ΖΓ πρὸς τὴν ΖΘ ὃν η¹ πρὸς ε¹, καὶ⁸ τῆς

f. 34 (Prop. 2 and 3)

ΖΘ πρὸς ΘΓ ὃν ε¹ πρὸς γ¹· καὶ τοῦ ἀπὸ τῆς ΖΓ ἄρα πρὸς τὸ ἀπὸ
τῆς ΖΘ λόγος ἐστὶν ὃν ξδ¹ πρὸς κε¹. ἐδείχθη δὲ τοῦ ἀπὸ ΓΖ πρὸς τὸ
ἀπὸ ΖΗ λόγος ὃν ξδ¹ πρὸς ιβ¹· καὶ τοῦ ἀπὸ ΘΖ ἄρα πρὸς τὸ ἀπὸ ΖΗ
λόγος ἐστὶν ὡς⁹ κε¹ πρὸς ιβ¹· αἱ ΘΖ ΖΗ ἄρα ῥηταί εἰσιν δυνάμει
μόνον σύμμετροι, καὶ ἡ ΘΖ τῆς ΖΗ μεῖζον δύναται τῷ ἀπὸ ἀ-
συμμέτρου ἑαυτῇ. καὶ ὅλη¹⁰ ἡ ΖΘ σύμμετρός ἐστιν ῥητῇ τῇ ΑΒ·
ἀποτομὴ ἄρα τετάρτη¹¹ ἐστὶν ἡ ΘΗ. ῥητὴ¹² δὲ ἡ ΖΓ καὶ < ἡ >¹³ διπλῆ
αὐτῆς· ἡ ἄρα δυναμένη τὸ < δὶς >¹⁴ ὑπὸ ΖΓ ΗΘ ἄλογός ἐστιν ἡ καλου-

---

¹ ἴση ΑΒ corr. S Hu, Tr
² τῇ ΑΒ add. Co, Hu
³ ὅτι Hu, Tr
⁴ ΕΗ A, corr. Co, Hu, Tr
⁵ τὸ add. S, Hu, Tr
⁶ ἔστω δή Hu ἔσται δὴ A
⁷ διπλασίων S δ////ίων A
⁸ η¹ πρὸς ε¹, καὶ τῆς Hu Η πρὸς //// τῆς A
⁹ ὃν Tr ὡς A Hu
¹⁰ ὅλη A corr. S Hu, Tr
¹¹ τετάρτη A corr. Hu, Tr
¹² ῥητὴ A corr. Hu, Tr
¹³ ἡ add. Hu, Tr
¹⁴ δὶς add. Co, Hu, Tr

μένη ἐλάσσων [ἐστιν][1]. καὶ δύναται τὸ δὶς ὑπὸ ΓΖ ΗΘ ἡ ΓΕ· ἐλάσσων ἄρα ἐστὶν ἡ ΓΕ. ὅτι δὲ ἡ ΓΕ δύναται τὸ δὶς ὑπὸ ΓΖ ΗΘ,
οὕτως ἔσται δῆλον· ἐπεζεύχθω ἡ ΕΘ. ἐπεὶ[2] τὸ ἀπὸ ΕΓ ἴσον ἐστὶν
τοῖς ἀπὸ τῶν ΕΘ ΘΓ καὶ τῷ δὶς ὑπὸ ΓΘ ΘΗ, ἔστιν δὲ καὶ τὰ[2]
ἀπὸ ΕΘ ΘΖ ἴσα τῷ ἀπὸ ΕΖ καὶ τῷ δὶς ὑπὸ ΖΘ ΘΗ. [ἀνάλο-
γον[3] ἄρα ἐστὶν ὡς τὸ ἀπὸ ΓΕ πρὸς τὰ[4] ἀπὸ ΕΘ ΘΓ μετὰ τοῦ δὶς
ὑπὸ ΓΘΗ, οὕτως τὰ[5] ἀπὸ ΕΘ ΘΖ πρὸς τὸ ἀπὸ ΕΖ μετὰ τοῦ δὶς
ὑπὸ ΖΗΘ[6]. καὶ ὡς ἓν πρὸς ἕν, πάντα < πρὸς πάντα >[7]. καὶ ἴσον ἐστὶν τὸ ἀπὸ
ΓΕ τοῖς ἀπὸ ΕΘΓ καὶ τῷ δὶς ὑπὸ ΓΘΗ[8]], ἴσα ἄρα καὶ τὰ
ἀπὸ ΓΕ ΕΘ ΘΖ τοῖς ἀπὸ ΕΘ ΘΓ ΕΖ καὶ τῷ δὶς ὑπὸ ΓΗΘ[9]
μετὰ τοῦ δὶς ὑπὸ ΖΘΗ, τουτέστιν τῷ δὶς ὑπὸ ΓΖ ΗΘ. κοινὸν
ἀφῃρήσθω[10] τὸ ἀπὸ ΕΘ· λοιπὰ ἄρα τὰ ἀπὸ ΕΓ ΖΘ ἴσα ἐστὶν
τοῖς ἀπὸ ΕΖ ΘΓ καὶ τῷ δὶς ὑπὸ ΓΖ ΗΘ. ὧν τὸ ἀπὸ ΖΘ ἴσον
τοῖς ἀπὸ τῶν ΕΖ ΘΓ, τὸ μὲν γὰρ ἀπὸ τῆς ΖΘ ἐστὶν κε[ι], τὸ δὲ
ἀπὸ τῆς ΘΓ θ[11], καὶ τὸ ἀπὸ ΕΖ ιϛ· λοιπὸν ἄρα τὸ ἀπὸ ΓΕ
ἴσον ἐστὶν τῷ δὶς ὑπὸ ΖΓ ΗΘ[12].

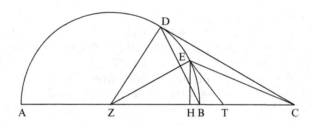

Prop. 3

[1] ἐστιν del. Hu, Tr
[2] τὸ A corr BS Hu, Tr
[3] ἀνάλογον... ἀπὸ ΕΘΓ καὶ τὸ δὶς ὑπὸ ΓΘΗ del. Hu
[4] τὸ A corr. Co, Hu, Tr
[5] τὸ A corr. S Hu, Tr
[6] ΖΘΗ Hu, Tr
[7] πρὸς πάντα add. Tr
[8] ΓΕΘΗ A corr. Hu, Tr
[9] ΓΘΗ Hu, Tr
[10] ἀ-φῃρήσθω A ἀφῃρήσθαι Tr
[11] ΘΓΕ A corr. Co, Hu, Tr
[12] ΖΓΝΘ A, corr. Hu

 # 3 ἡμικύκλιον τὸ ἀπὸ¹ τῆς ΑΓ
ῥήτην ἔχον τὴν διάμετρον,
καὶ τῇ ἐκ τοῦ κέντρου ἴση
ἔστω ἡ ΓΔ, καὶ ἐφαπτομένη
ἡ ΔΒ, καὶ δίχα τετμήσθω ἡ
ὑπὸ ΓΔΒ γωνία [η]² ὑπὸ τῆς ΔΖ· οὕτως³ ἡ ΔΖ ὑπεροχή ἐστιν ἣ
ὑπερέχει ἡ ἐκ δύο ὀνομάτων τῆς μετὰ ῥητοῦ μέσον τὸ ὅλον
ποιούσης. εἰλήφθω γὰρ τὸ Η κέντρον τοῦ ἡμικυκλίου, καὶ
ἐπεζεύχθω ἡ ΒΗ, καὶ ἐπὶ τῆς ΗΔ γεγράφθω ἡμικύκλιον τὸ

f. 34v (Prop. 3)
ΗΒ < Δ >⁴, καὶ ἐκβεβλήσθω ἡ ΖΔ < ἐπὶ τὸ > Κ⁵· ἴση ἄρα ἐστὶν ἡ ΒΚ περιφέρεια < τῇ ΚΗ περιφερείᾳ >⁶. ἤχθω
κάθετος ἐπὶ τὴν ΑΓ ἡ ΚΛ. καὶ ἐπεὶ ἐξαγώνου ἐστὶν πλευρὰ ἡ ΒΗ,
ἡμίσεια δὲ τῆς τοῦ⁷ ἐξαγώνου ἡ ΚΛ, ἐκβαλλομένη γὰρ τὴν διπλῆν
τῆς ΚΗ περιφερείας ὑποτείνει, διπλασία ἄρα ἡ ΒΗ τῆς ΚΛ,
τουτέστιν ἡ ΓΚ τῆς ΚΛ. καὶ ἔστιν ὀρθὴ ἡ ὑπὸ ΚΛΓ· ἐπίτριτον ἄρα
ἐστὶν⁸ τὸ ἀπὸ ΚΓ τοῦ ἀπὸ ΓΛ, τουτέστιν < τὸ ἀπὸ >⁹ ΔΓ τοῦ ἀπὸ ΓΛ· αἱ ΔΓ ΓΛ
ἄρα ῥηταί εἰσιν δυνάμει μόνον σύμμετροι, καὶ ἡ ΔΓ τῆς ΓΛ
μεῖζον δύναται τῷ ἀπὸ συμμέτρου¹⁰ ἑαυτῇ, καὶ ἡ μείζων ἡ
ΔΓ σύμμετρός ἐστι ῥητῇ¹¹ τῇ ΑΓ· ἐκ δύο ὀνομάτων ἄρα πρώτη ἐστὶν ἡ ΛΔ, ῥητὴ δὲ ἡ ΗΔ· ἡ ἄρα τὸ ὑπὸ τῶν ΗΔΛ χωρίον δυναμένη ἄλογός ἐστιν ἡ καλουμένη ἐκ δύο ὀνομάτων. δύναται δὲ
αὐτὸ ἡ ΔΚ¹², διὰ γὰρ τὸ ἰσογώνιον εἶναι τὸ ΗΔΚ τρίγωνον τῷ ΔΛΚ¹³
τριγώνῳ ἐστὶν ὡς ἡ ΗΔ πρὸς ΔΚ, ἡ ΚΔ πρὸς ΔΛ· ἡ [δὲ]¹⁴ ΔΚ ἄρα ἐκ
δύο ὀνομάτων ἐστίν. καὶ ἐπεὶ¹⁵ διμοίρου ἐστὶν ἡ ὑπὸ ΒΗΓ γωνία καὶ
ἴση < ἡ >¹⁶ ΗΒ τῇ ΗΓ, ἰσόπλευρον ἄρα ἐστὶν τὸ ΒΗΓ τρίγωνον. ἤχθω

---

¹ἐπὶ Co, Hu, Tr
²η del. Hu
³ὅτι Co, Hu, Tr
⁴τὸ *ΗΒ** A τὸ *ηβ* S Co τὸ *ΗΒΔ* B Hu, Tr
⁵ἡ *ΖΔΚ* ABS ἡ *ΔΖΚ* Co ἡ *ΔΖ* ἐπὶ τὸ *Κ* Hu, Tr
⁶τῇ *ΚΗ περιφερείᾳ* add. Tr τῇ *ΚΗ* add. Hu
⁷τοῦ om. Hu
⁸ἐστὶν om. Tr
⁹τὸ ἀπὸ add. Hu, Tr
¹⁰ἀπὸ ἀσυμμέτρου AS ἀπὸ τῆς ἀσυμμέτρου B corr. Co, Hu, Tr
¹¹ῥητῇ AB corr. S Hu, Tr
¹²*Η ΔΚ* A corr. Hu, Tr
¹³*ΗΛΚ* A corr. Co, Hu, Tr
¹⁴δὲ del. Hu, Tr
¹⁵ἐπι AB corr. S Hu, Tr
¹⁶ἴση *** τῇ A1 ἴση * ΗΒ τῇ A2 ἴση ἡ ΗΒ τῇ B Hu, Tr

δὴ κάθετος ἡ ΒΘ[1]· διπλῆ ἄρα ἐστὶν ἡ ΗΓ, τουτέστιν ἡ ΔΓ, τῆς
ΓΘ. καὶ ἐδείχθη τὸ ἀπὸ ΔΓ τοῦ ἀπὸ ΓΛ ἐπίτριτον· τὸ ἄρα
ἀπὸ ΛΓ τριπλάσιόν ἐστιν τοῦ ἀπὸ ΓΘ· αἱ ΛΓ ΓΘ ἄρα ῥηταί
εἰσιν δυνάμει μόνον σύμμετροι, καὶ ἡ ΛΓ τῆς ΓΘ μεῖζον δύναται
τῷ ἀπὸ ἀσυμμέτρου ἑαυτῇ, καὶ τὸ ἔλασσον ὄνομα τὸ ΓΘ σύμ-
μετρόν ἐστιν ῥητῇ τῇ ΑΓ· ἡ ΛΘ ἄρα ἀποτομή ἐστιν πέμπτη.
καὶ ἐπεὶ τὸ μὲν ὑπὸ ΔΗΘ[2] ἴσον ἐστὶν τῷ ἀπὸ ΒΗ διὰ τὸ ἰσογώνια[3]
εἶναι τὰ ΒΗΘ ΒΗΔ τρίγωνα, τὸ δὲ ὑπὸ ΔΗΛ[4] ἴσον ἐστὶν τῷ
ἀπὸ ΚΗ[5] διὰ τὸ ἰσογώνια εἶναι τὰ ΚΗΛ ΚΗΔ τρίγωνα, ἔστιν ἄρα[6]
ὡς τὸ ὑπὸ *ΔΗΘ* πρὸς τὸ ἀπὸ ΒΗ, οὕτως τὸ < ὑπὸ >[7] *ΔΗΛ* πρὸς τὸ ἀπὸ ΚΗ. ἐναλλὰξ δὲ ὡς τὸ[8] ὑπὸ *ΔΗΘ* πρὸς τὸ ὑπὸ *ΔΗΛ*, < οὕτως τὸ ἀπὸ *ΒΗ* πρὸς τὸ ἀπὸ
ΚΗ. ὡς δὲ τὸ ὑπὸ *ΔΗΘ* πρὸς τὸ ὑπὸ *ΔΗΛ*, >[9] οὕτως ἡ ΘΗ πρὸς τὴν
ΗΛ, κοινὸν[10] γὰρ ὕψος τὸ *ΔΗ*· καὶ ὡς ἄρα ἡ ΘΗ πρὸς τὴν ΗΛ, οὕτως τὸ ἀπὸ ΒΗ, τουτέστιν τὸ ἀπὸ ΖΗ, πρὸς τὸ < ἀπὸ >[11] ΗΚ· διελόντι ἄρα
ἔσται[12] ὡς ἡ *ΘΛ* πρὸς *ΛΗ*, οὕτως[13] τὸ ὑπὸ τῶν *ΔΗ ΛΘ* πρὸς τὸ ὑπὸ *ΔΗΛ*.
καὶ ἐδείχθη ἴσον τὸ ὑπὸ τῶν *ΔΗΛ* τῷ ἀπὸ ΗΚ· ἴσον ἄρα καὶ
τὸ ὑπὸ τῶν *ΔΗ ΛΘ* τῷ ἀπὸ ΚΖ. καὶ ἔστιν ἡ μὲν ΛΘ ἀποτομὴ
πέμπτη, ἡ δὲ *ΔΗ* ῥητή· ἡ ἄρα ΚΖ ἡ μετὰ ῥητοῦ μέσον τὸ ὅλον
ποιοῦσά ἐστιν. ἐδείχθη δὲ καὶ ἡ ΔΚ ἐκ δύο ὀνομάτων· ἡ ἄρα[14]

f. 35 (Prop. 3 and 4)
ΔΖ[15] ὑπεροχή ἐστιν ἡ ὑπερέχει < ἡ > ἐκ δύο ὀνομάτων[16] τῆς μετὰ ῥητοῦ
μέσον τὸ
ὅλον ποιούσης.

---

[1] *κάθετος ΗΒΔΘ* AB, corr. Hu, Tr

[2] *ὑπὸ ΛΗΘ* AB1S corr. B3 Hu, Tr

[3] *ἰσογώνι-α* A

[4] *ὑπὸ ΒΗΛ* A corr. Hu, Tr

[5] *ἀπὸ ΚΛ* A corr. Hu, Tr

[6] *ἄρα ἐστὶν* A corr. Hu, Tr

[7] *ὑπὸ* add. Hu, Tr

[8] ΚΗ. καὶ ἐναλλάξ. ὡς δὲ τὸ Hu

[9] *οὕτως; ΔΗΛ* add. Tr

[10] *κοινὸν γὰρ ὕψος τὸ ΔΗ* del. Hu

[11] *τὸ ΑΗΚ* A corr. Hu, Tr

[12] *ἔστιν* Hu, Tr

[13] *οὕτως τὸ ἀπὸ ΚΖ πρὸς τὸ ἀπὸ ΗΚ. καὶ* Hu

[14] *λοιπὴ ἄρα ἡ* Hu

[15] *ΛΖ* A corr. Hu, Tr

[16] *ἡ ὑπερέχει ἐκ δύο ὀνομάτων* A corr. Hu, Tr

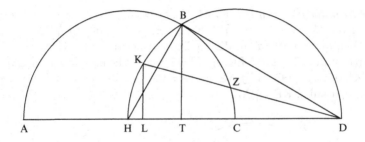

Prop. 4

<u>#4</u> ἔστω κύκλος ὁ ΑΒΓ, οὗ κέντρον
ἔστω τὸ Ε, διάμετρος δὲ ἡ ΒΓ, καὶ
ἐφαπτομένη ἡ ΑΔ συμπί-
πτουσα τῇ ΒΓ κατὰ τὸ Δ, καὶ
διήχθω ἡ ΔΖ, καὶ ἐπιζευ-
χθεῖσα ἡ ΑΕ ἐκβεβλήσθω ἐπὶ
τὸ Η, καὶ ἐπεζεύχθωσαν αἱ ΖΚΗ ΗΛΘ· οὕτως[1] ἴση ἐστὶν ἡ ΕΚ τῇ ΕΛ.
γεγονέτω, καὶ ἤχθω τῇ ΚΛ παράλληλος ἡ ΘΞΜ· ἴση ἄρα καὶ ἡ ΜΞ
τῇ ΞΘ. ἤχθω ἀπὸ τοῦ Ε ἐπὶ τὴν ΖΘ κάθετος ἡ ΕΝ· ἴση ἄρα ἐστὶν
ἡ ΖΝ τῇ ΝΘ. ἦν δὲ καὶ ἡ ΜΞ τῇ ΞΘ· παράλληλος[2] ἄρα ἐστὶν ἡ ΝΞ τῇ ΜΖ·
οὕτως ἄρα ἴση ἐστὶν ἡ ὑπὸ τῶν ΘΝΞ τῇ ὑπὸ τῶν ΝΖΜ, τουτέστι[3]
τῇ ὑπὸ τῶν ΘΑΞ· οὕτως ἄρα ἐν κύκλῳ ἐστὶ τὰ ΑΝΞΘ[4] σημεῖα·
οὕτως ἄρα ἴση ἐστὶν ἡ ὑπὸ τῶν ΑΝΘ γωνία τῇ ὑπὸ τῶν ΑΞΘ,
τουτέστιν τῇ ὑπὸ τῶν ΑΕΛ· οὕτως ἄρα ἐν κύκλῳ ἐστὶ τὰ Α Δ Ε Ν[5] σημεῖα. ἔστιν δέ· ὀρθὴ γάρ ἐστιν ἑκατέρα τῶν ὑπὸ τῶν ΕΑΔ ΕΝΔ. συντεθήσεται δὴ οὕτως. ἐπεὶ ὀρθή ἐστιν ἑκατέρα τῶν ὑπὸ τῶν ΕΑΔ ΕΝΔ,
ἐν κύκλῳ ἐστὶ τὰ Α Δ ΕΝ[6] σημεῖα· ἴση ἄρα ἐστὶν ἡ ὑπὸ ΑΝΔ τῇ ὑπὸ
ΑΕΔ. ἀλλ᾿ ἡ ὑπὸ ΑΕΔ ἴση ἐστὶν τῇ ὑπὸ ΑΞΘ διὰ[7] τὰς παραλλήλους τὰς ΕΔ ΞΘ[8]· ἐν κύκλῳ ἄρα τὰ ΑΝΞΘ[9] σημεῖα· ἴση ἄρα ἐστὶν ἡ
ὑπὸ ΘΑΞ γωνία τῇ ὑπὸ ΘΝΞ. ἀλλ᾿ ἡ[10] ὑπὸ ΘΑΞ ἴση ἐστὶν τῇ ὑπὸ

---

[1] ὅτι Hu, Tr

[2] ἴση ΑΒ corr. S Hu, Tr

[3] τοῦτ᾿ ἔστ Α τουτέστι BS τουτέστιν Hu, Tr

[4] Α Ν Ξ Θ distinx. Β Hu

[5] ΑΕΝ Α distinx. Β Α Ν Ε Δ Hu Α Δ Ε Ν Tr

[6] ΑΔ ΕΝ AS Α Δ Ε Ν Β Hu, Tr

[7] διὸ τὰς παραλλήλους ΑΒ corr. S Hu, Tr

[8] τὰ ΕΞΘ Α corr. Co Hu, Tr

[9] Α Ν Ε Θ Hu, Tr

[10] αλλη Α corr. Hu, Tr

< ΘΖΜ· ἴση ἄρα ἐστὶν ἡ ὑπὸ ΘΖΜ τῇ ὑπὸ > [1]
ΘΝΞ· παράλληλος ἄρα ἐστὶν ἡ ΖΜ[2] τῇ ΝΞ. καὶ ἔστιν ἴση ἡ ΖΝ
τῇ ΝΘ· ἴση ἄρα ἐστὶν καὶ ἡ ΜΞ τῇ ΞΘ. καὶ ἔστιν ὡς ἡ ΞΗ πρὸς
ΗΕ, οὕτως ἡ μὲν ΞΜ πρὸς ΕΚ, ἡ δὲ[3] ΘΞ πρὸς ΛΕ· καὶ ὡς ἄρα ἡ ΞΜ
πρὸς ΕΚ[4], οὕτως ἡ ΘΞ πρὸς ΛΕ. καὶ ἐναλλάξ. καὶ ἴση ἐστὶν ἡ ΜΞ τῇ
ΞΘ· ἴση ἄρα καὶ ἡ ΚΕ τῇ ΛΕ.

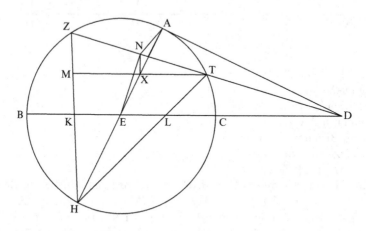

Prop. 5
#5 ἔστω κύκλος ὁ ΑΒΓ, καὶ ἐφαπτό-
μεναι αἱ ΑΔ ΔΓ, καὶ ἐπεζεύχθω ἡ ΑΓ, καὶ διήχθω ἡ ΕΖ, ἔστω
ἴση ἡ ΕΗ τῇ ΗΖ[5]· οὕτως[6] καὶ
ἡ ΘΗ < τῇ Η > Κ[7] ἐστὶν ἴση. ἤχθω

f. 35v (Prop. 5 and 6)
τῇ ΑΓ παράλληλος ἡ ΕΜ, καὶ εἰλήφθω τὸ [Κ][8] κέντρον τοῦ κύκλου
τὸ Λ, καὶ ἐπεζεύχθωσαν αἱ ΛΑ ΛΖ ΛΓ ΛΜ < ΛΕ >[9] ΛΗ. ἐπεὶ ἴση ἐστὶν ἡ ΕΗ
τῇ ΗΖ, ἴση ἐστὶν καὶ ἡ ΜΓ τῇ ΓΖ. καὶ ἔστιν πρὸς ὀρθὰς[10] τῇ ΓΛ·
ἴση ἄρα ἐστὶν ἡ ΛΖ τῇ ΛΜ. καὶ ἐπεὶ ἴση ἐστὶν ἡ ΑΔ τῇ ΔΓ, ἴση

---

[1] ΘΖΜ· ἴση; τῇ ὑπὸ add. Tr. τῇ ὑπὸ ΘΖΜ· παράλληλος pro τῇ ὑπὸ ΘΝΞ· παράλληλος Hu
[2] ΖΗΜ A corr. Co, Hu, Tr
[3] τῇ δὲ A corr. Hu, Tr
[4] ΘΚ AB corr. S Co, Hu, Tr
[5] ἔστω δὲ ἡ ΕΗ ἴση τῇ ΗΖ Hu ἔστω δὲ ἡ εη ἴση τῇ ηζ S
[6] ὅτι Co, Hu, Tr
[7] ///η ἤχθω A suppl. Hu, Tr
[8] del. Hu, Tr
[9] ΛΕ add. S Hu, Tr.
[10] καὶ ἔστιν ἡ ΜΓ πρὸς ὀρθὰς S Hu

ἐστὶν ἡ ΑΕ τῇ ΜΓ. ἔστιν δὲ καὶ ἡ ΑΛ τῇ ΛΓ ἴση, καὶ ὀρθὴ ἡ ὑπὸ
τῶν ΕΑΛ ὀρθῇ τῇ ὑπὸ τῶν ΜΓΛ ἐστιν ἴση· ἴση ἄρα ἐστὶν καὶ ἡ
ΕΛ τῇ ΛΜ, τουτέστιν τῇ ΛΖ. ἀλλὰ καὶ ἡ ΕΗ τῇ ΗΖ ἐστιν ἴση· ἡ ΗΛ
ἄρα κάθετός ἐστιν ἐπὶ τήν ΕΖ· ἴση ἄρα ἐστὶν ἡ ΘΗ τῇ ΗΚ.

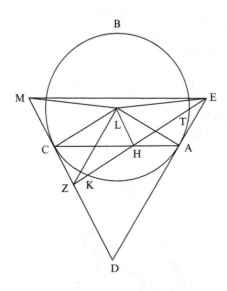

Prop. 6

<u>#6</u> ἔστω κύκλος ὁ ΑΒΓ, καὶ ἐφα-
πτόμεναι < αἱ >[1] ΑΔ ΔΓ, καὶ ἐπεζεύχθω ἡ ΑΓ, καὶ διήχθω
ἡ ΕΖ, καὶ ἔστω ἴση ἡ ΗΘ τῇ
ΗΚ· οὕτως[2] καὶ ἡ ΕΗ τῇ ΗΖ
ἐστιν ἴση. εἰλήφθω τὸ κέντρον τοῦ κύκλου τὸ Λ, καὶ
ἐπεζεύχθωσαν αἱ ΕΛ ΛΑ ΛΓ
ΛΗ < ΛΖ >[3]. ἐπεὶ ὀρθή ἐστιν ἑκατέρα τῶν ὑπὸ τῶν[4] ΕΑΛ ΕΗΛ,
< ἐν κύκλῳ ἐστὶν τὰ Ε Α Η Λ σημεῖα· >[5] ἴση < ἄρα >[6] ἐστὶν ἡ ὑπὸ
τῶν ΗΑΛ γωνία τῇ ὑπὸ τῶν ΗΕΛ γωνία. πάλιν ἐπεὶ ὀρθή

---

[1] αἱ add. Hu, Tr

[2] ὅτι Co, Hu, Tr

[3] ΛΖ add. Tr ΕΛ ΛΑ ΛΗ ΛΖ ΛΓ Hu

[4] ἑκατέρα τῶν αὐτῶν AB corr. S Hu, Tr

[5] ἐν κύκλῳ...σημεῖα add. Co, Hu, Tr

[6] ἄρα add. Co, Hu, Tr

ἐστιν ἑκατέρα τῶν ὑπὸ τῶν *ΛΗΚ*[1] ΛΓΖ, ἐν κύκλῳ[2] ἐστὶν τὰ *ΛΗΖΓ*[3]
σημεῖα· ἴση ἄρα ἐστὶν ἡ ὑπὸ τῶν ΗΓΛ[4] γωνία, τουτέστιν ἡ ὑπὸ
τῶν ΗΑΛ[5], τουτέστιν ἡ ὑπὸ τῶν ΗΕΛ, τῇ ὑπὸ τῶν ΗΖΛ· ἴση
ἄρα ἐστὶν καὶ ἡ ΕΛ τῇ ΛΖ. καὶ ἔστιν κάθετος ἡ ΛΗ· ἴση ἄρα ἐστὶν
ἡ ΕΗ τῇ ΗΖ.

Apollonian problem

ἐὰν ὦσιν τρεῖς κύκλοι, τῇ
θέσει καὶ τῷ μεγέθει δεδό-
μενοι καὶ ἐφαπτόμενοι ἀλ-
λήλων, καὶ ὁ περιλαμβά-
νων αὐτοὺς κύκλος δοθεὶς
ἔσται τῷ μεγέθει. προγρά-
φεται δὲ τάδε.

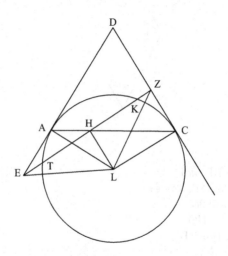

Prop. 7

#_7_ τετράπλευ-
ρον τὸ ΑΒΓΔ ὀρθὴν ἔχον τὴν ὑπὸ ΑΒΓ γωνίαν καὶ δοθεῖσαν
ἑκάστην τῶν ΑΒ ΒΓ ΓΔ ΔΑ εὐθειῶν· δεῖξαι δοθεῖσαν τὴν ἐπιζευ-

f. 36 (Prop. 7)

---

[1] *ΛΗΖ* Hu, Tr
[2] *κύκλων* A *ἐν κύκλῳ* Co, Hu, Tr
[3] distinx. BS Hu, Tr
[4] *ΗΛΓ* A corr. Hu, Tr
[5] *ΗΑΚ* ABS corr. Hu, Tr

γνύουσαν τὰ Δ Β σημεῖα τὴν ΒΔ¹. ἐζεύχθω² ἡ ΑΓ καὶ κάθετοι ἤχθωσαν ἐπὶ μὲν τὴν ΓΔ ἡ ΑΗ, ἐπὶ δὲ τὴν < ΑΓ ἡ ΒΕ. ἐπεὶ οὖν
ἑκατέρα >³ τῶν ΑΒ ΒΓ
δοθεῖσά ἐστιν [ἢ ἐν ἀριθμοῖς]⁴, καὶ ὀρθή ἐστιν ἡ ὑπὸ ΑΒΓ, καὶ
κάθετός ἐστιν ἡ ΒΕ, δοθεῖσα ἄρα ἔσται καὶ ἑκάστη τῶν ΑΕ ΕΓ
ΑΓ ΒΕ, καὶ γὰρ τὸ ὑπὸ ΑΓΕ ἴσον ὂν τῷ ἀπὸ ΒΓ γίνεται δοθέν·
καὶ δοθεῖσά ἐστιν ἡ ΑΓ, ὥστε ἑκάστη τῶν ΑΕ ΕΓ ΒΕ⁵ ἔσται δοθεῖσα.
πάλιν ἐπεὶ δοθεῖσά ἐστιν ἑκάστη⁶ τῶν ΑΓ ΓΔ ΔΑ εὐθειῶν, καὶ
κάθετός ἐστιν ἡ ΑΗ⁷, δοθεῖσά ἐστι καὶ ἑκάστη τῶν ΔΗ ΗΓ < ΑΗ >⁸, καὶ γὰρ ἡ
ὑπεροχὴ τοῦ ἀπὸ ΑΓ πρὸς τὸ ἀπὸ ΔΑ παρὰ τὴν ΓΔ παραβλη-
θεῖσα ποιεῖ δοθεῖσαν τὴν τῆς ΓΔ⁹ πρὸς ΗΔ ὑπεροχήν, ὡς ἔστι
λῆμμα· ὥστε καὶ ἑκάστην τῶν ΔΗ ΗΓΑΗ δεδόσθαι. καὶ ἐπεὶ ἰσογώνιόν ἐστι τὸ ΑΗΓ τρίγωνον τῷ ΓΕΖ τριγώνῳ, ἔστιν ὡς ἡ ΗΓ
πρὸς ΓΕ οὕτως ἥ τε ΑΓ πρὸς ΓΖ καὶ ἡ ΑΗ πρὸς τὴν ΕΖ. καὶ ἔστι
δοθεὶς ὁ τῆς ΗΓ πρὸς ΓΕ λόγος· δοθεῖσα < ἄρα >¹⁰ ἔσται καὶ ἑκάστη¹¹ τῶν ΓΖ
ΖΕ. ἀλλὰ καὶ ἑκάστη¹² τῶν ΕΒ ΒΓ· καὶ ἑκάστη ἄρα τῶν ΖΒ ΒΓ ΓΖ¹³
δοθεῖσα. ἤχθω δὴ κάθετος ἐπὶ τὴν ΓΖ ἡ ΒΘ· δοθεῖσα ἄρα ἐστὶν
ἑκάστη τῶν ΖΘ ΘΓ ΒΘ· ὥστε καὶ ἑκατέρα τῶν ΔΘ ΘΒ δοθεῖσά
ἐστι. καὶ ὀρθή ἐστιν ἡ ὑπὸ ΒΘΔ· δοθεῖσα ἄρα ἐστὶν ἡ ΒΔ.

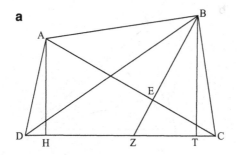

¹ τὴν ΒΔ del. Co, Hu
² ἐζεύχθω ABS ἐπεζεύχθω Hu, Tr
³ ΑΓ ἡ ΒΕ· ἑκατέρα add. Hu, Tr
⁴ ἢ ἐν ἀριθμοῖς del. Hu, Tr
⁵ τῶν ΔΕ ΕΓ ἔσται A τῶν ΑΕ ΕΓ ἔσται Co τῶν ΑΕ ΕΓ ΒΕ Hu, Tr
⁶ ἑκάστη forsan ἑκάστη A corr. Hu, Tr
⁷ ἡ ΛΗ A corr. Co, Hu, Tr
⁸ ΑΗ add. Hu, Tr
⁹ ΓΗ A corr. Hu, Tr
¹⁰ ἄρα add. Hu, Tr
¹¹ ἑκάστη A Tr ἑκατέρα Hu
¹² ἑκάστη A Tr ἑκατέρα Hu
¹³ ΓΔ A corr. Hu, Tr

ἄλλως.

#8 ἤχθω κάθετος ἐπὶ τὴν ΑΓ
ἡ ΔΕ, καὶ ἐκβεβλήσθω ἐπὶ
τὸ Ζ. ἐπεὶ δοθεῖσά ἐστιν ἑκάστη
τῶν ΑΔ ΔΓ ΓΑ, καὶ κάθετος
ἡ ΔΕ, δοθεῖσα ἔσται ἑκατέρα[1]
τῶν ΑΕ ΕΓ. καὶ ἐπεὶ ἰσογώνιόν
ἐστιν τὸ ΑΒΓ τρίγωνον τῷ ΓΕΖ τριγώνῳ, ἔστιν ὡς ἡ ΓΕ πρὸς ΕΖ,
ἡ ΓΒ πρὸς ΒΑ. δοθεὶς δὲ ὁ τῆς ΓΒ πρὸς ΒΑ λόγος· δοθεὶς ἄρα καὶ ὁ
τῆς ΓΕ πρὸς ΕΖ λόγος. καὶ δοθεῖσά ἐστιν ἡ ΓΕ· δοθεῖσα ἄρα καὶ
ἡ ΕΖ. ἦν δὲ καὶ ἡ ΔΕ δοθεῖσα· καὶ ὅλη ἄρα ἡ ΔΖ ἔσται δοθεῖσα.
κατὰ ταὐτὰ δοθήσεται καὶ ἑκατέρα τῶν ΒΖ ΖΓ, ὡς γὰρ ἡ ΑΓ
πρὸς ΒΓ, οὕτως ἡ ΖΓ πρὸς ΓΕ. καὶ δοθεὶς ὁ τῆς ΑΓ πρὸς ΓΒ λόγος.
ἤχθω δὴ πάλιν ἀπὸ τοῦ Δ κάθετος ἡ ΔΗ· δοθεῖσα ἄρα ἑκατέρα τῶν *ΓΗ ΗΖ*[2], ὥστε καὶ ἑκατέρα τῶν ΒΗ ΗΔ δοθεῖσά ἐστιν[3]. καὶ

f. 36v (Prop. 7 and 8)
ὀρθή ἐστιν ἡ Η γωνία· δοθεῖσα ἄρα ἐστὶν καὶ ἡ ΒΔ.

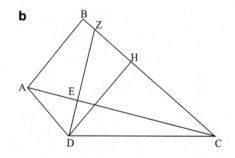

Prop. 8
#9 ἴσοι κύκλοι τῇ θέσει καὶ τῷ μεγέθει
δοθέντες, ὧν κέντρα τὰ *ΑΒ*[4], καὶ
δοθὲν σημεῖον τὸ Γ, καὶ διὰ τοῦ Γ
ἐφαπτόμενος τῶν κύκλων, ὧν

---

[1] ἔσται καὶ ἑκατέρα Co, Hu

[2] *ΖΗ ΗΓ* Hu

[3] ἔστι Hu

[4] AB AB distinx. S Hu, Tr

κέν- τρα τὰ *AB*[1], γεγράφθω ὁ **ΓΕΖ**· ὅτι
δοθεῖσά ἐστιν αὐτοῦ ἡ διάμετρος.
ἐπεζεύχθωσαν αἱ **ΕΖΗ ΓΖΘ ΓΜΠ ΑΒ ΓΕ ΠΖΚ ΘΚ ΘΗ**[2]· γίνεται δὴ
παράλληλος ἡ **ΗΘ** τῇ **ΓΕ** διὰ τὸ τὰς κατὰ κορυφὴν γωνίας τὰς
ὑπὸ **ΕΖΓ ΗΖΘ** ἴσας εἶναι, καὶ ὁμοίας τὰς **ΕΠΖ ΗΚΖ**[3] περιφερείας καὶ τὸ **ΕΓΖ** τρίγωνον ἰσογώνιον τῷ **ΖΗΘ** τριγώνῳ. καὶ
διὰ τὰ αὐτὰ καὶ ἡ **ΘΚ** τῇ **ΠΓ**[4] ἐστὶν παράλληλος. καὶ ἴσοι εἰσὶν
οἱ κύκλοι, ὧν τὰ κέντρα τὰ *AB*[5]· ἴση ἄρα ἡ **ΖΗ** τῇ **ΔΕ**. ἤχθωσαν
κάθετοι αἱ **ΑΣ ΒΛ**· ἴση ἄρα ἡ **ΑΣ** τῇ **ΒΛ**· ὥστε καὶ ἡ μὲν **ΒΜ** τῇ
ΜΑ ἐστὶν ἴση, ἡ δὲ **ΛΜ** τῇ **ΜΣ**, δύο γὰρ τρίγωνά ἐστι τὰ **ΒΛΜ**
ΑΣΜ τὰς δύο γωνίας τὰς κατὰ κορυφὴν ἴσας ἔχοντα καὶ
τὰς πρὸς τοῖς **Λ Σ** σημείοις ὀρθάς, ἔχει δὲ καὶ μίαν πλευρὰν μία
πλευρᾷ[6] ἴσην τὴν **ΒΛ** [καὶ κάθετον][7] τῇ **ΑΣ**. καὶ δοθεῖσά ἐστιν
ἑκάστη τῶν **ΜΛ ΛΒ ΜΣ ΣΑ**[8] [οὕτως καὶ ἡ **ΖΗ ΔΕ** καὶ **ΒΛ ΛΣ**][9]· δοθεῖσα ἄρα καὶ ἑκατέρα τῶν **ΒΜ ΜΑ** εὐθειῶν. ἀλλὰ καὶ ἑκατέρα
τῶν **ΑΓ ΓΒ** δοθεῖσά ἐστι [*ΑΓ ΒΓ δοθεῖσά ἐστιν*][10], θέσει [*εὐθεῖα*][11] γὰρ τὰ **Α**
Β Γ σημεῖα· δοθέν ἄρα τὸ **ΑΒ < Γ >**[12] τρίγωνον τῷ εἴδει· καὶ ἡ **ΓΜ**
ἄρα δοθεῖσα ἔσται καθέτου ἀχθείσης ἀπὸ τοῦ **Γ** ἐπὶ τὴν **ΑΒ**.
καὶ ἐπεὶ δοθεῖσά ἐστιν ἡ **ΝΡ** διάμετρος τοῦ **ΗΘΚ** κύκλου, ἀλλὰ καὶ
ἡ **ΜΑ** δοθεῖσα, καὶ λοιπὴ ἄρα ἡ **ΜΡ** δοθεῖσά ἐστιν. καὶ ἐπεὶ[13] δοθέν ἐστιν τὸ ὑπὸ **ΝΜΡ**, δοθὲν ἄρα καὶ τὸ ὑπὸ **ΗΜΖ**, *τοῦτ'* ἐστιν[14]
τὸ ὑπὸ **ΕΜΖ**, < τουτέστιν τὸ >[15] ὑπὸ τῶν **ΓΜΠ**. καὶ δοθεῖσά ἐστιν ἡ **ΓΜ**· δοθεῖσα
ἄρα καὶ ἡ **ΓΠ**. ἐπεὶ οὖν θέσει καὶ μεγέθει ἐστὶν κύκλος, οὗ κέντρον
τὸ **Α**, καὶ δοθεῖσα τῇ θέσει καὶ τῷ μεγέθει ἡ **ΓΠ**, καὶ διηγμέναι
αἱ **ΠΖΚ ΓΖΘ**, ὥστε παράλληλον εἶναι τῇ **ΓΠ** τὴν[16] **ΚΘ**, δοθεῖσά ἐστιν ἡ διάμετρος τοῦ περὶ τὸ **ΓΖΠ** τρίγωνον κύκλου,
τουτέστιν τοῦ **ΓΕΖ**.

---

[1] *AB* A distinx.BS Hu, Tr

[2] *αἱ ΕΖΗ ΓΖΘ ΓΜΠ ΑΒΓ Ε Π Ζ Κ Θ Κ γίνεται* A corr. Hu, Tr

[3] *ΗΝΘ* A corr. Co, Hu, Tr

[4] *ΠΤ* A corr. Co, Hu, Tr

[5] distinx. Hu, Tr

[6] *μία πλευρὰ* A corr. Hu, Tr

[7] *καὶ κάθετον* del. Hu, Tr

[8] *ΣΛ* A corr. Hu, Tr

[9] *οὕτως· ΒΛ ΛΣ* del. Hu

[10] bis scripta del. B Hu, Tr

[11] *θέσει* A2 in marg. *εὐθεῖα* A1 *θέσει εὐθεῖα* S corr. Hu, Tr

[12] *Γ* add. Co, Hu, Tr

[13] *ἐπὶ* A corr. Hu, Tr

[14] *τουτέστιν* Hu, Tr

[15] *τουτέστιν τὸ* add. Hu, Tr

[16] *τῇ ΚΘ* A corr. Hu, Tr

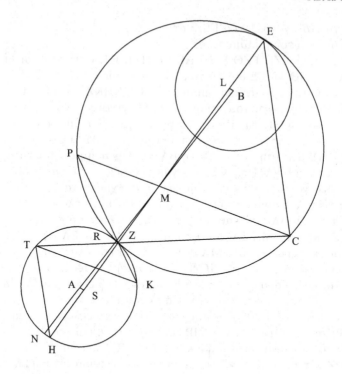

f. 37 (Prop. 9, 10, and 11)

Prop. 9

#10 τρίγωνον τὸ ΑΒΓ ἔχον ἑκάστην τῶν
πλευρῶν δοθεῖσαν, καὶ σημεῖον ἐντὸς[1]
τὸ Δ, καὶ ᾧ ὑπερέχει ἡ ΑΔ τῆς
ΓΔ, τούτῳ ὑπερεχέτω καὶ ἡ ΓΔ
τῆς ΔΒ, καὶ ἔστω ὑπεροχὴ δο-
θεῖσα· ὅτι ἑκάστη τῶν ΑΔ ΔΓ
ΔΒ δοθεῖσά ἐστιν. ἐπεὶ ἡ τῶν ΑΔ
ΔΓ ὑπεροχὴ δοθεῖσά ἐστιν, ἔστω
τῇ ὑπεροχῇ ἴση ἑκατέρα
τῶν ΑΕ ΒΖ· αἱ τρεῖς ἄρα αἱ ΕΔ ΔΓ ΔΖ ἴσαι ἀλλήλαις εἰσίν. γεγράφθω περὶ κέντρον τὸ Δ κύκλος ὁ ΓΕΖ· διὰ δὴ τὸ προγεγραμμένον
δοθεῖσά ἐστιν ἡ ΔΖ. ἧς ἡ ΒΖ[2] ἐστὶν δοθεῖσα [ἐστιν][3]· ἡ λοιπὴ ἄρα ἡ

---

[1] ἐν τοῖς Α corr. Β Hu, Tr
[2] ἧς ἡ BZ Hu ὧν ἡ BZ Α Tr
[3] bis scriptum del. Hu, Tr

ΒΔ ἐστὶν δοθεῖσα. ἀλλὰ καὶ ἡ τῶν ΑΔ ΔΓ < ὑπεροχή ἐστιν δοθεῖσα· ὥστε
καὶ ἑκατέρα τῶν ΑΔ ΔΓ δοθεῖσά ἐστιν. ἑκάστη ἄρα τῶν ΑΔ > ΔΓ ΔΒ ἐστὶν
δοθεῖσα.[1]

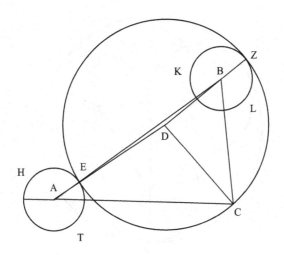

Prop. 10

#11 τὰ μὲν οὖν λήμματα ταῦτα,
τὸ δὲ *ἀρχικόν*[2]· τρεῖς κύκλοι
ἄνισοι ἐφαπτόμενοι ἀλλήλων
δοθείσας ἔχοντες τὰς διαμέ-
τρους, ὧν κέντρα τὰ *ΑΒΓ*[3], καὶ
περὶ αὐτοὺς κύκλος ἐφα-
πτόμενος αὐτῶν ὁ ΔΕΖ, οὗ δέον
ἔστω εὑρεῖν τὴν διάμετρον. ἔστω δὲ αὐτοῦ τὸ κέντρον τὸ Ν[4], καὶ
ἐπὶ τὰ κέντρα τὰ ΑΒΓ[5] ἐπεζεύχθωσαν αἱ ΑΒ ΑΓ ΓΒ καὶ ἔτι
αἱ *ΝΑΔ ΝΒΖ ΝΓ < Ε >*[6]. ἐπεὶ οὖν αἱ διάμετροι τῶν κύκλων, ὧν κέντρα
τὰ *ΑΒΓ*[7], δοθεῖσαί εἰσιν[8], γενήσεται καὶ ἑκάστη τῶν ΑΒ ΒΓ ΓΑ
δοθεῖσα. καὶ αἱ τῶν *ΑΝ ΝΓ ΝΒ*[9] διαφοραὶ δοθεῖσαι· διὰ ἄρα τὸ

---

[1] ἀλλὰ καὶ ἡ τῶν ΑΔ ΔΓ ΔΓ ΔΒ ἔστιν δοθεῖσα Α ὑπεροχή...τῶν ΑΔ. add. Tr ἀλλὰ καὶ ἑκατέρα
τῶν ΑΔ ΔΓ δοθεῖσά ἐστιν· ἑκάστη ἄρα τῶν ΑΔ ΔΓ ΔΒ ἐστὶν δοθεῖσα Hu

[2] *ἀρχαϊκόν* Α, Hu *ἀρχικόν* Hu appendix, Tr

[3] distinx. BS Hu, Tr

[4] *Ν* Α Tr *Η* Hu

[5] distinx. Hu, Tr

[6] *ΝΑΔ ΝΒΖ ΝΓ* Α *ΝΑΔ ΝΒΖ ΝΓΕ* Tr *ΗΑΔ ΗΒΖ ΗΓΕ* Hu

[7] distinx. Β Hu, Tr

[8] *ἐστιν* Α

[9] *ΑΝ ΝΓ ΝΒ* ABS Tr *ΑΗ ΗΓ ΗΒ* Hu

προγεγραμμένον δοθεῖσά ἐστιν ἡ ΑΝ[1]. ἀλλὰ καὶ ἡ ΑΔ δοθεῖσά
ἐστιν, ὥστε δοθεῖσά ἐστιν ἡ διάμετρος τοῦ ΔΕΖ κύκλου. καὶ τοῦτο μὲν
ἐνθάδε μοι πέρας ἔχει, τὰ δὲ λοιπὰ ὑπογράψω.

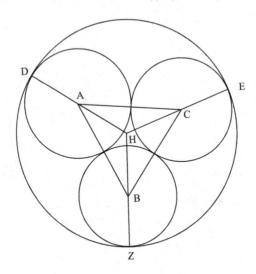

Prop. 11
#12 ἔστω ἡμικύκλιον τὸ ΑΒΓ, < καὶ > [2] κεκλά-
σθω ἡ ΓΒΑ, καὶ διήχθω ἡ ΓΔ, καὶ
ἔστω ἡ ΒΓ ἴση[3] συναμφοτέρῳ
τῇ ΑΒ ΓΔ, καὶ κάθετοι ἤχθωσαν
< ἐπὶ τὴν ΑΓ > [4] αἱ ΒΕ ΔΖ· ὅτι ἡ ΑΖ διπλασίων

f. 37v (Prop. 11 and 12)
ἐστὶν τῆς ΒΕ[5]. [καὶ][6] κείσθω γὰρ τῇ μὲν ΑΕ ἴση ἡ ΕΗ, τῇ δὲ ΑΒ ἴση
ἡ ΒΘ[7], καὶ ἐπεζεύχθωσαν[8] εὐθεῖαι αἱ ΑΘ ΘΗ ΘΖ, καὶ κάθετος ἤχθω
ἡ ΘΚ, καὶ ἐπεζεύχθω[9] ἡ ΒΚ. ἐπεὶ ἡ ΓΒ[10] ἴση ἐστὶν συναμφοτέρῳ
τῇ ΑΒ ΔΓ, ὧν ἡ ΒΘ τῇ ΒΑ ἐστὶν ἴση, λοιπὴ ἄρα ἡ ΘΓ λοιπῇ
τῇ ΓΔ ἐστὶν ἴση· καὶ τὸ ἀπὸ τῆς ΓΔ ἄρα ἴσον ἐστὶν τῷ ἀπὸ
τῆς ΓΘ. τῷ δὲ ἀπὸ τῆς ΔΓ ἴσον ἐστὶν τὸ ὑπὸ τῶν ΑΓΖ· καὶ

---

[1] ΑΝ Tr ΑΗ Α Hu
[2] καὶ add. Hu, Tr
[3] καὶ ἔστω ἡ ΒΓ ἴση Tr καὶ ἴση ἔστω ἡ ΓΒ Hu lacunae in A
[4] ἐπὶ τὴν ΑΓ add. Hu
[5] ΒΗ Α corr. Co, Hu, Tr
[6] καὶ del. Hu
[7] ΒΟ Α corr. S Hu, Tr
[8] ἐπιζεύχθωσαν Α corr. BS Hu, Tr
[9] ἐπιζεύχθω Α corr. BS Hu, Tr
[10] ΓΖ Α corr. Co, Hu, Tr

τὸ ὑπὸ τῶν ΑΓΖ ἄρα ἴσον ἐστὶν τῷ ἀπὸ τῆς ΓΘ· ἴση ἄρα ἐστὶν
ἡ ὑπὸ τῶν ΖΘΓ γωνία τῇ ὑπὸ τῶν ΘΑΗ γωνίᾳ. πάλιν ἐπεὶ
τὸ ὑπὸ τῶν ΓΑΕ ἴσον ἐστὶν τῷ ἀπὸ τῆς ΑΒ, καὶ τὸ δὶς ἄρα
ὑπὸ τῶν ΓΑΕ, τουτέστιν τὸ ὑπὸ τῶν ΓΑΗ, ἴσον ἐστὶν τῷ[1] δὶς ἀπὸ
τῆς ΑΒ, τουτέστιν τῷ ἀπὸ τῆς ΑΘ· ἴση ἄρα ἐστὶν ἡ ὑπὸ τῶν
ΑΘΗ γωνία τῇ ὑπὸ τῶν ΘΓΖ γωνίᾳ[2]. ἔστιν δὲ καὶ ἡ ὑπὸ τῶν ΘΑΗ
ἴση τῇ ὑπὸ τῶν ΖΘΓ· λοιπὴ ἄρα ἡ ὑπὸ τῶν ΑΗΘ λοιπῇ
τῇ ὑπὸ τῶν ΘΖΓ ἐστὶν ἴση· καὶ ἡ ΗΘ ἄρα τῇ ΘΖ ἐστὶν ἴση[3].
καὶ κάθετος ἦκται ἡ ΘΚ· ἴση ἄρα ἐστὶν ἡ ΖΚ τῇ ΚΗ. καὶ ἐπεὶ
ὀρθή ἐστιν ἑκατέρα τῶν ὑπὸ τῶν ΑΒΘ ΑΚΘ, καὶ < ἐν > κύκλῳ[4] ἐστὶν
τὸ *ΑΒ ΘΚ*[5] τετράπλευρον· ἴση < ἄρα >[6] ἐστὶν ἡ ὑπὸ τῶν ΒΘΑ γωνία τῇ
ὑπὸ τῶν ΒΚΑ. ἡμίσους δέ ἐστιν ἡ ὑπὸ τῶν ΒΘΑ· ἡμίσους ἄρα
ἐστὶν καὶ ἡ ὑπὸ τῶν ΒΚΑ. ὀρθὴ δέ ἐστιν ἡ ὑπὸ τῶν ΒΕΚ· ἴση ἄρα
ἐστὶν ἡ ΒΕ τῇ ΕΚ. τῆς δὲ ΕΚ διπλῆ ἐστιν ἡ ΑΖ, ἐπείπερ ἡ μὲν ΑΕ
τῇ ΕΗ ἐστὶν ἴση, ἡ δὲ ΖΚ τῇ ΚΗ· καὶ τῆς ΕΒ ἄρα διπλῆ ἐστιν
ἡ ΑΖ, ὅπερ ※

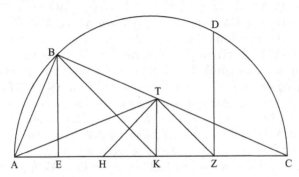

Prop. 12
#13 ἔστω ἡμικύκλιον τὸ ΑΒΓ, καὶ
κεκλάσθω ἡ ΑΒΔ, καὶ ἔστω
ἴση ἡ ΑΒ τῇ ΒΔ, καὶ τῇ ΒΔ
πρὸς ὀρθὰς ἤχθω ἡ ΔΕ, καὶ
ἐπεζεύχθω ἡ ΒΕ, καὶ αὐτῇ πρὸς
ὀρθὰς ἤχθω ἡ ΕΖ, καὶ τὸ κέντρον τὸ Η, καὶ ἔστω ὡς ἡ ΑΗ πρὸς
ΗΔ οὕτως ἡ ΔΘ πρὸς ΘΖ, καὶ ἐπεζεύχθω ἡ ΘΕ· ὅτι ἡ ὑπὸ τῶν
ΒΕΔ γωνία ἴση ἐστὶν τῇ ὑπὸ τῶν ΔΕΘ γωνίᾳ. ἤχθω ἀπὸ τοῦ Η
ἐπὶ τὴν ΒΕ κάθετος ἡ ΗΚ· ἴση ἄρα ἐστὶν ἡ ΒΚ τῇ ΚΕ. καὶ ἔστιν

---

[1] ἴσον ἐστὶν τῶν A corr. Hu, Tr

[2] τῇ...γωνία A corr. Hu, Tr

[3] καὶ ἡ ΗΘ...ΘΖ ἐστὶν A, Tr καὶ ἡ ὑπὸ ΘΗΖ ἄρα τῇ ὑπὸ ΘΖΗ ἐστὶν Hu

[4] καὶ κύκλῳ A ἐν add. Tr ἐν κύκλῳ Hu

[5] coniunx. Hu, Tr

[6] ἄρα add. Hu

ὀρθὴ ἡ ὑπὸ τῶν ΒΔΕ· αἱ τρεῖς ἄρα αἱ ΒΚ ΚΔ ΚΕ ἴσαι ἀλλήλαις
εἰσίν. καὶ παράλληλός ἐστιν ἡ ΗΚ τῇ ΕΖ, καὶ ἐπεὶ ζητῶ τὴν[1]

f. 38 (Prop.12)
ὑπὸ τῶν ΚΕΔ γωνίαν τῇ ὑπὸ τῶν ΔΕΘ γωνίᾳ[2] ἴσην. καὶ ἔστιν ἴση
ἡ ΔΚ τῇ ΚΕ, ὅτι ἄρα ἴση ἐστὶν ἡ ὑπὸ ΚΕΔ γωνία τῇ ὑπὸ ΚΔΕ,
ὅτι ἄρα καὶ ἡ ὑπὸ ΚΔΕ τῇ ὑπὸ ΔΕΘ[3] ἴση ἐστίν, ὅτι ἄρα παράλληλός ἐστιν ἡ ΔΚ τῇ ΕΘ. ἤχθω καὶ τῇ ΔΕ παράλληλος ἡ ΚΛ καὶ ἐκβεβλήσθω ἡ ΓΔ[4] ἐπὶ τὸ Λ, καὶ ἐπεζεύχθω ἡ ΒΛ. ἐπεὶ οὖν ἡ μὲν ΚΛ
τῇ ΔΕ ἐστὶν παράλληλος, ἡ δὲ ΚΗ τῇ ΕΖ, ζητεῖται δὲ καὶ ἡ ΚΔ
τῇ ΕΘ παράλληλος, ὅτι ἄρα διὰ τὸ ἰσογώνιον εἶναι τὸ μὲν ΚΛΗ
τρίγωνον τῷ ΕΔΖ τριγώνῳ, τὸ δὲ ΔΚΗ τῷ ΕΘΖ, ἔστιν ὡς μὲν
ἡ ΛΗ πρὸς ΗΚ, ἡ ΔΖ πρὸς ΖΕ, ὡς [τε][5] δὲ ἡ ΚΗ πρὸς ΗΔ, ἡ ΕΖ πρὸς
ΖΘ· ὅτι ἄρα καὶ ὡς ἡ ΛΗ πρὸς ΗΔ, οὕτως ἡ ΔΖ πρὸς ΖΘ, δι' ἴσου
γάρ· ὅτι ἄρα καὶ ὡς ἡ ΛΔ πρὸς τὴν ΔΗ, οὕτως ἡ ΔΘ πρὸς τὴν ΘΖ,
διελόντι γάρ. ὑπέκειτο δὲ καὶ ὡς ἡ ΔΘ πρὸς ΘΖ, οὕτως ἡ ΑΗ
πρὸς ΗΔ· ὅτι ἄρα ἐστὶν ὡς ἡ ΛΔ πρὸς ΔΗ, οὕτως ἡ ΔΘ πρὸς ΘΖ,
τουτέστιν ἡ ΑΗ πρὸς ΗΔ· ὅτι ἄρα ἴση ἐστὶν ἡ ΛΔ τῇ ΑΗ[6]· ὅτι ἄρα
καὶ ἡ ΛΑ τῇ ΔΗ ἐστὶν ἴση. ἀλλὰ καὶ ἡ ΑΒ τῇ ΒΔ ἐστὶν ἴση· ὅτι
ἄρα καὶ ἡ ΛΒ τῇ ΒΗ ἐστὶν ἴση. ἀλλὰ ἡ ΒΗ ἑκατέρα τῶν ΛΔ
ΑΗ ἐστὶν ἴση· ὅτι ἄρα καὶ ἡ ΒΛ τῇ ΛΔ ἐστὶν ἴση. ἔστιν δέ· ἐπεὶ
γὰρ παράλληλός ἐστιν ἡ ΚΛ τῇ ΔΕ, καὶ ἔστιν ἴση ἡ ΔΚ τῇ ΚΕ, ἴ-
ση ἐστὶν καὶ ἡ ὑπὸ τῶν ΒΚΛ γωνία τῇ ὑπὸ τῶν ΛΚΔ. ἐπεὶ
οὖν ἴση ἐστὶν ἡ ΒΚ τῇ ΚΔ καὶ γωνία ἡ ὑπὸ τῶν ΒΚΛ γωνίᾳ
τῇ ὑπὸ τῶν ΔΚΛ ἐστὶν ἴση, καὶ ἡ ΒΛ ἄρα τῇ ΛΔ ἐστὶν ἴση.
καὶ ἡ σύνθεσις ἀκολούθως τῇ ἀναλύσει. ἐπεὶ γὰρ ἴση ἐστὶν
ἡ ΔΚ τῇ ΚΕ, ἴση καὶ γωνία ἡ ὑπὸ ΚΔΕ τῇ ὑπὸ ΚΕΔ. ἀλλ' ἡ μὲν
ὑπὸ ΚΔΕ τῇ ὑπὸ ΔΚΛ[7] ἐστὶν ἴση, ἡ δὲ ὑπὸ ΚΕΔ[8] τῇ ὑπὸ ΒΚΛ
ἐστὶν ἴση διὰ τὰς ΚΛ ΕΔ παραλλήλους· καὶ ἡ ὑπὸ ΒΚΛ ἄρα τῇ
ὑπὸ ΔΚΛ[9] ἐστὶν ἴση. ἔστιν δὲ καὶ ἡ ΒΚ εὐθεῖα τῇ ΚΔ ἴση· καὶ
βάσις ἄρα ἡ ΒΛ βάσει τῇ ΛΔ ἐστὶν ἴση, ὥστε καὶ ἡ[10] γωνία
ἡ ὑπὸ τῶν ΛΒΔ τῇ ὑπὸ ΒΔΑ[11], τουτέστιν τῇ ὑπὸ ΔΑΒ,

---

[1] scriptura non satis perspicua in A ζητῶ τὴν Tr ἐζήτουν τὴν Hu
[2] γωνίαν A corr. Hu, Tr
[3] ΔΕΣ AB corr. S Hu, Tr
[4] ΓΔ A Hu ΓΑ Tr
[5] τε del. A2, Hu, Tr
[6] ΛΗ AB corr. S Hu, Tr
[7] ΒΚΛ A corr. Hu, Tr
[8] ΚΔΕ AB corr. S Hu, Tr
[9] ΚΛΔ AB corr. S Hu, Tr
[10] ἡ ABS del. Hu
[11] ΒΔ ΔΑ AB corr. S Hu

τουτέστιν τῇ ὑπὸ ΑΒΗ. κοινὴ ἀφῃρήσθω ἡ ὑπὸ ΑΒΔ· λοιπὴ
ἄρα ἡ ὑπὸ ΛΒΑ λοιπῇ τῇ ὑπὸ ΔΒΗ ἐστὶν ἴση. ἀλλὰ καὶ ἡ
ὑπὸ ΒΔΗ τῇ ὑπὸ ΒΑΛ ἐστὶν ἴση· δύο δὴ τρίγωνά ἐστιν τὰ ΒΔΗ
ΒΑΛ τὰς δύο γωνίας ταῖς δύο γωνίαις ἴσας ἔχοντα καὶ μίαν πλευρὰν
τὴν ΑΒ τῇ ΒΔ[1]· ἴση ἄρα ἡ μὲν ΒΗ τῇ ΒΛ, ἡ δὲ ΔΗ τῇ ΛΑ, ὥστε καὶ

f. 38v (Prop. 12, arbelos theorem, and Prop. 13)

ἡ ΛΔ τῇ ΑΗ ἐστὶν[2] ἴση. ἐπεὶ οὖν ὑπόκειται ὡς ἡ ΑΗ πρὸς ΗΔ, ἡ ΔΘ
πρὸς ΘΖ, ἴση δὲ ἡ ΑΗ τῇ ΛΔ, ἔστιν ἄρα ὡς ἡ ΛΔ πρὸς ΔΗ, ἡ ΔΘ
πρὸς ΘΖ· συνθέντι < ἄρα >[3] ὡς ἡ ΛΗ πρὸς ΗΔ, ἡ ΔΖ πρὸς ΖΘ. ἔστιν δὲ καὶ
ὡς ἡ ΛΗ πρὸς ΗΚ, ἡ ΔΖ πρὸς ΖΕ· < δι' ἴσου ἄρα >[4] καὶ ὡς ἡ ΚΗ πρὸς ΗΔ,
ἡ ΕΖ πρὸς ΖΘ.
καὶ ἔστιν ἴση ἡ ὑπὸ ΕΖΘ τῇ ὑπὸ ΚΗΔ[5] διὰ τὸ παραλλήλους εἶναι
τὰς ΕΖ ΚΗ· ἴση ἄρα καὶ ἡ ὑπὸ ΕΘΖ τῇ ὑπὸ ΚΔΗ[6]· παράλληλος ἄρα
ἐστὶν καὶ ἡ ΚΔ τῇ ΕΘ· ἴση ἄρα ἐστὶν ἡ < ὑπὸ >[7] ΚΔΕ, τουτέστιν ἡ ὑπὸ ΚΕΔ,
γωνία τῇ ὑπὸ ΔΕΘ.

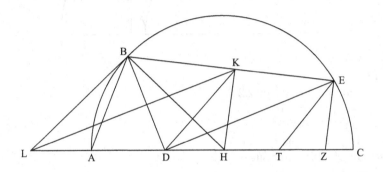

Arbelos Theorem

<u>#14</u> φέρεται ἔν τισιν ἀρχαία
πρότασις τοιαύτη[8]· ὑπο-
κείσθω τρία ἡμικύκλια[9]
ἐφαπτόμενα ἀλλή-
λων τὰ ΑΒΓ ΑΔΕ ΕΖΓ, καὶ
εἰς τὸ μεταξὺ τῶν περιφερειῶν αὐτῶν χωρίον, ὃ δὴ καλοῦσιν ἄρβηλον,

---

[1] τὴν *ΑΒ* τῆς ΒΔ A corr. B Hu, Tr

[2] τῇ *ΑΒ* ἐστὶν ΑΒ corr. S Hu, Tr

[3] ἄρα add. Hu

[4] ἐξ ἴσου ἄρα add. Hu corr. p. 1227 appendix Hu

[5] *ΚΗΑ* ABS corr. Co, Hu, Tr

[6] *ΚΔ* η A corr. Hu, Tr

[7] ὑπὸ add. Hu, Tr

[8] τοιαύτη A corr. Hu, Tr

[9] ἡμικύκλι - α A

ἐγγεγράφθωσαν κύκλοι ἐφαπτόμενοι τῶν τε ἡμικυκλίων καὶ ἀλλήλων ὁσοιδηποτοῦν, ὡς οἱ περὶ κέντρα τὰ [Z] *Η ΘΚΛ*[1]· δεῖξαι τὴν
μὲν ἀπὸ τοῦ Η κέντρου κάθετον ἐπὶ τὴν ΑΓ ἴσην τῇ διαμέτρῳ
τοῦ περὶ τὸ Η κύκλου, τὴν δ' ἀπὸ τοῦ Θ κάθετον διπλασίαν τῆς
διαμέτρου τοῦ περὶ τὸ Θ κύκλου, τὴν δ' ἀπὸ τοῦ Κ κάθετον τριπλα-
σίαν, καὶ τὰς ἑξῆς καθέτους τῶν οἰκείων διαμέτρων πολλαπλα-
σίας κατὰ τοὺς ἑξῆς μονάδι ἀλλήλων ὑπερέχοντας ἀριθμοὺς
ἐπ' ἄπειρον γενομένης[2] τῆς τῶν κύκλων ἐγγραφῆς. δειχθήσεται δὲ
[τὰ][3] πρότερον τὰ λαμβανόμενα.

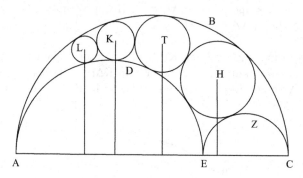

Prop. 13

#15 ἔστωσαν δύο κύκλοι οἱ ΖΒ
ΒΜ περὶ κέντρα τὰ ΑΓ[4]
ἐφαπτόμενοι ἀλλήλων
κατὰ τὸ Β, καὶ μείζων ἔστω ὁ
ΒΜ, ἄλλος δέ τις ἐφαπτόμενος αὐτῶν κατὰ τὰ ΚΛ[5]
περὶ κέντρον τὸ Η ὁ ΚΛ[6],
καὶ ἐπεζεύχθωσαν αἱ ΓΗ ΑΗ. πεσοῦνται δὴ διὰ τῶν ΚΛ[7], καὶ ἡ ἐπὶ τὰ
ΚΛ[8] ἐπιζευγνυμένη εὐθεῖα ἐκβαλλομένη τεμεῖ μὲν τὸν ΖΒ κύκλον[9],
συμπίπτει δὲ τῇ διὰ τῶν ΑΓ[10] κέντρων ἐκβαλλομένη εὐθείᾳ διὰ[11]

---

[1] *ΖΗ ΘΚΛ* A distinx. BS Η Θ Κ Λ Co, Hu, Tr

[2] *γινομένης* Hu

[3] *τὰ* del. Hu, Tr

[4] distinx. BS Hu, Tr

[5] distinx. BS Hu, Tr

[6] *ΘΚΛ* A corr. Hu, Tr

[7] distinx. BS Hu, Tr

[8] distinx. BS Hu, Tr

[9] *τὸν /Β κύκλον* A *τὸν ΖΒ κύκλον* Hu *τὸν ΓΒ κύκλον* Tr

[10] distinx. BS Hu, Tr

[11] *εὐθεῖα ///* A *διὰ* add. BS Hu, Tr

f. 39 (Prop. 13)

τὸ μείζονα εἶναι τὴν ΑΚ πλευρὰν τῆς ΓΔ τοῦ *ΑΚ ΔΓ τραπεζείου*[1].
συμπιπτέτω οὖν κατὰ τὸ Ε τέμνουσα τὸν[2] κύκλον κατὰ τὸ Δ·
δεῖξαι ὅτι ἐστὶν ὡς ἡ ΑΒ πρὸς ΒΓ, οὕτως ἡ ΑΕ πρὸς ΕΓ. ἔστιν δὲ
φανερὸν ἐπιζευχθείσης τῆς ΓΔ[3]· γίνεται γὰρ ἰσογώνια τὰ
ΓΔΛ ΛΚΗ τρίγωνα τὰς κατὰ κορυφὴν γωνίας *πρὸς τῷ Λ*[4] ἴσας ἔχοντα
καὶ περὶ τὰς ΓΗ[5] γωνίας τὰς πλευρὰς ἀνάλογον *ἔχοντα*[6], ὥστε
ἴσας εἶναι τὰς ὑπὸ ΔΓΗ[7] ΓΗΑ γωνίας ἐναλλάξ, καὶ παράλληλον τὴν ΓΔ [καὶ][8] τῇ ΑΗ[9], καὶ ὡς τὴν ΑΕ πρὸς τὴν ΕΓ, τὴν ΑΚ
πρὸς ΓΔ, τουτέστιν τὴν ΑΒ πρὸς ΒΓ. καὶ τὸ ἀναστρόφιον δὲ
φανερόν ἐστιν. ἐὰν γὰρ ᾖ ὡς ἡ ΑΒ πρὸς ΒΓ, οὕτως ἡ ΑΕ πρὸς ΕΓ, ἡ
ΚΔ ἐπ᾽ εὐθείας γίνεται τῇ ΔΕ. παράλληλός τε γάρ ἐστιν ἡ ΑΚ τῇ
ΓΔ καὶ ἔστιν ὡς ἡ ΑΒ πρὸς ΒΓ, τουτέστιν ὡς ἡ ΑΕ πρὸς ΕΓ, ἡ ΑΚ
πρὸς ΓΔ· ἐπ᾽ εὐθείας ἄρα ἐστὶν ἡ ΚΔ[10] τῇ ΔΕ. εἰ γὰρ ἡ διὰ τῶν ΚΕ[11]
οὐχ ἥξει καὶ διὰ τοῦ Δ, ἀλλὰ διὰ τοῦ Θ, γίνεται ὡς ἡ ΑΕ πρὸς ΕΓ, ἡ
ΑΚ πρὸς ΓΘ, ὅπερ ἀδύνατον. ὁμοίως οὐδὲ τοῦ Δ ἐκτὸς ἥξει τέμνουσα τὴν ΓΔ ἐκβληθεῖσαν, οἷον κατὰ τὸ Ν· ἔσται γὰρ πάλιν
ὡς ἡ ΑΕ πρὸς ΕΓ, ἡ ΑΚ πρὸς ΓΝ, ὅπερ ἀδύνατον· ἔστιν γὰρ πρὸς
τὴν ΓΔ. ἢ οὕτως. διὰ τοῦ Κ τῇ ΑΕ παράλληλος ἡ ΚΝ ἤχθω, καὶ
γίνεται παραλληλόγραμμον τὸ *ΑΓΚΝ*[12], καὶ ἴση ἡ ΑΚ τῇ ΓΝ.
καὶ ἐπεί ἐστιν ὡς ἡ ΑΕ πρὸς ΕΓ, οὕτως ἡ ΑΚ, τουτέστιν ἡ ΓΝ, πρὸς
ΓΔ, διελόντι ὡς ἡ ΑΓ πρὸς ΓΕ, ἡ ΝΔ πρὸς ΔΓ. ἐναλλὰξ ὡς ἡ ΑΓ,
τουτέστιν ὡς ἡ ΚΝ, πρὸς ΝΔ, οὕτως ἡ ΕΓ πρὸς ΓΔ. καὶ περὶ τὰς ἴσας
γωνίας τὰς πρὸς τοῖς *ΝΓ*[13] αἱ πλευραὶ ἀνάλογόν εἰσιν· ὅμοιον ἄρα
ἐστὶν τὸ ΕΔΓ τρίγωνον τῷ ΔΝΚ τριγώνῳ· ἴση ἄρα ἐστὶν ἡ ὑπὸ
ΕΔΓ γωνία τῇ ὑπὸ ΝΔΚ. καὶ ἔστιν εὐθεῖα ἡ ΓΝ· εὐθεῖα ἄρα καὶ
ἡ ΚΔΕ. λέγω δὴ ὅτι καὶ τὸ ὑπὸ ΚΕΛ ἴσον ἐστὶ τῷ ἀπὸ ΕΒ. ἐπεὶ γὰρ
ὡς ἡ ΑΕ πρὸς ΕΓ, οὕτως ἡ ΑΒ πρὸς ΒΓ, τουτέστιν πρὸς ΓΖ, ἔσται καὶ
ἡ λοιπὴ ἡ ΒΕ πρὸς λοιπὴν τὴν ΕΖ ὡς ἡ ΑΕ πρὸς ΕΓ, τουτέστιν
ὡς ἡ ΚΕ πρὸς ΕΔ. ἀλλ᾽ ὡς μὲν ἡ ΚΕ πρὸς ΕΔ, οὕτως τὸ ὑπὸ ΚΕΛ

---

[1] *τοῦ ΑΚΔΓ τραπεζίου* Hu, Tr

[2] *τὴν* A corr. BS Hu, Tr

[3] *ἐπιζευχθείσης τῆς ΓΔ* del. Hu

[4] *πρὸς τῷ Λ* del. Hu

[5] distinx. BS Hu, Tr

[6] *ἔχοντα* del. Hu

[7] *ΔΗΓ* A corr. Hu, Tr

[8] *καὶ* del. B Hu, Tr

[9] *ΑΚ* A corr. Hu, Tr

[10] *ΓΔ* A corr. Co, Hu, Tr

[11] distinx. Hu, Tr

[12] *ΑΓΚΝ* ABS *ΑΓΝΚ* Hu, Tr

[13] *ΝΓ* AS distinx. B Hu, Tr

πρὸς τὸ ὑπὸ ΛΕ ΕΔ, ὡς δὲ ἡ ΒΕ πρὸς ΕΖ, οὕτως τὸ ἀπὸ τῆς ΒΕ
πρὸς τὸ ὑπὸ ΒΕΖ, καὶ ἔστιν ἴσον τὸ ὑπὸ ΛΕ ΕΔ τῷ ὑπὸ ΒΕ ΕΖ· ἴσον ἄρα
καὶ τὸ ὑπὸ ΚΕΛ τῷ ἀπὸ ΕΒ.

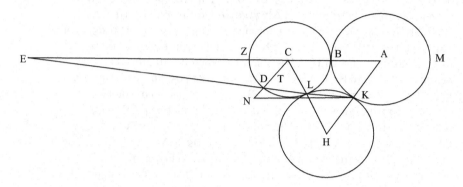

f. 39v (Prop. 14)
Prop. 14
#16 δύο ἡμικύκλια τὰ ΒΗΓ[1] ΒΕΔ, καὶ ἐφαπτόμενος αὐτῶν κύκλος ὁ *EZ*
HΘ[2], ἀπὸ δὲ τοῦ κέντρου αὐτοῦ τοῦ Α κάθετος ἤχθω ἐπὶ τὴν ΒΓ βάσιν
τῶν ἡμικυκλίων ἡ ΑΜ· ὅτι ἐστὶν < ὡς >[3] ἡ ΜΒ πρὸς τὴν ἐκ[4] τοῦ κέντρου τοῦ
ΕΖΗΘ κύκλου, οὕτως ἐπὶ μὲν τῆς πρώτης καταγραφῆς < συν – > ἀμφότερος[5]
ἡ ΓΒ ΒΔ πρὸς τὴν ὑπεροχὴν αὐτῶν τὴν ΓΔ, ἐπὶ δὲ τῆς δευτέρας καὶ
τρίτης οὕτως ἡ τῶν ΓΒ ΒΔ ὑπεροχὴ πρὸς συναμφότερον τὴν
ΓΒ ΒΔ, < τουτέστιν τὴν ΓΔ >[6]. ἤχθω διὰ τοῦ Α τῇ ΒΓ παράλληλος ἡ ΘΖ.
ἐπεὶ οὖν δύο
κύκλοι οἱ ΒΗΓ *EZ HΘ*[7] ἐφάπτονται ἀλλήλων κατὰ τὸ Η, καὶ διάμετροι
ἐν αὐτοῖς παράλληλοί εἰσιν αἱ ΒΓ ΖΘ[8], εὐθεῖα ἔσται ἥτε διὰ τῶν *ΗΘΒ*[9]
καὶ < ἡ >[10] διὰ τῶν *ΗΖΓ*[11]. πάλιν ἐπεὶ δύο κύκλοι οἱ ΒΕΔ *EZ HΘ*[12] ἐφάπτονται
ἀλλήλων κατὰ τὸ Ε, καὶ ἐν αὐτοῖς παράλληλοι διάμετροί εἰσιν αἱ ΘΖ

[1] *HBΓ* AB corr. Co, Hu, Tr

[2] coniunx. B Hu, Tr

[3] ὡς add. Hu, Tr

[4] *EK* A corr. Hu, Tr

[5] συν- add. Tr commendavit Hu appendix p. 1227

[6] τουτέστιν τὴν ΓΔ add. Hu

[7] coniunx. BS Hu, Tr

[8] *ZE* A corr. Hu, Tr

[9] distinx. Hu, Tr

[10] ἡ add. S Hu, Tr

[11] distinx. B Hu, Tr

[12] coniunx. BS Hu, Tr

ΒΔ, εὐθεῖα ἔσται ἥτε διὰ τῶν **ΖΕΒ**[1] καὶ ἡ διὰ τῶν **ΘΕΔ**[2]. ἤχθωσαν καὶ
ἀπὸ τῶν **ΘΖ**[3] σημείων κάθετοι αἱ ΘΚ ΖΛ· ἔσται δὴ διὰ μὲν τὴν ὁ-
μοιότητα τῶν ΒΗΓ ΒΘΚ τριγώνων ὡς ἡ ΒΓ πρὸς ΒΗ, οὕτως ἡ ΒΘ
πρὸς τὴν ΒΚ[4], καὶ τὸ ὑπὸ ΓΒ ΒΚ περιεχόμενον χωρίον ἴσον τῷ[5] ὑπὸ
ΗΒ ΒΘ, διὰ δὲ τὴν ὁμοιότητα τῶν ΒΖΛ ΒΕΔ τριγώνων ὡς ἡ ΔΒ πρὸς
τὴν ΒΕ, οὕτως ἡ ΒΖ πρὸς ΒΛ, καὶ τὸ ὑπὸ ΔΒ ΒΛ ἴσον τῷ ὑπὸ ΖΒ ΒΕ,
καὶ ἔστιν ἴσον τὸ ὑπὸ ΗΒ ΒΘ τῷ ὑπὸ ΖΒ ΒΕ· ἴσον ἄρα καὶ τὸ ὑπὸ ΓΒ
ΒΚ τῷ ὑπὸ ΔΒ ΒΛ, ἂν δὲ ἡ ἀπὸ τοῦ Ζ κάθετος ἐπὶ τὸ Δ πίπτῃ, τῷ ἀ-
πὸ τῆς ΒΔ. ἐπὶ μὲν ἄρα τῆς πρώτης καταγραφῆς ὡς ἡ ΓΒ πρὸς ΒΔ,
οὕτως ἡ ΛΒ πρὸς τὴν ΒΚ, ὥστε καὶ < ὡς >[6] συναμφότερος ἡ ΓΒ ΒΔ πρὸς τὴν
ὑπεροχὴν αὐτῶν τὴν < ΓΔ, οὕτως καὶ συναμφότερος ἡ ΛΒ ΒΚ πρὸς τὴν
ὑπεροχὴν αὐτῶν τὴν > ΚΛ[7]. καὶ ἔστι συναμφοτέρου μὲν τῆς ΛΒ ΒΚ ἡμί-

f. 40 (Prop. 14)

σεια ἡ ΒΜ, διὰ τὸ ἴσην εἶναι τὴν ΚΜ τῇ ΜΛ, τῆς δὲ ΛΚ ἡμίσεια ἡ
ΜΚ[8]· καὶ ὡς ἄρα συναμφότερος ἡ ΓΒ ΒΔ πρὸς τὴν ΓΔ, οὕτως ἡ ΒΜ πρὸς ΜΚ,
τουτέστι[9] πρὸς τὴν ἐκ τοῦ κέντρου τοῦ *ΕΖ ΗΘ*[10] κύκλου. ἐπὶ δὲ τῆς δευτέρας καὶ
τρίτης καταγραφῆς, ἐπεὶ τὸ ὑπὸ ΓΒΚ ἴσον ἐδείχθη [καὶ κοινῶς][11] τῷ ὑπὸ
ΔΒΛ, ὡς ἄρα ἡ ΓΒ πρὸς ΒΔ, οὕτως ἡ ΛΒ πρὸς τὴν ΒΚ. συνθέντι ὡς ἡ ΓΔ πρὸς
ΔΒ, ἡ ΚΛ πρὸς ΚΒ· ὥστε καὶ ὡς ἡ ΓΔ πρὸς τὴν τῶν ΓΒ ΒΔ ὑπεροχήν, οὕτως
ἡ ΚΛ πρὸς τὴν τῶν ΛΒ ΒΚ ὑπεροχήν. καὶ ἔστι τῆς μὲν ΚΛ ἡμίσεια < ἡ >[12] ἐκ
τοῦ κέντρου < τοῦ >[13] *ΕΖ ΗΘ*[14] κύκλου [ἀντὶ τῆς ΛΜ][15], ἡ δὲ ΒΜ ἡμίσεια τῆς τῶν ΛΒ
ΒΚ ὑπεροχῆς διὰ τὸ ἴσην εἶναι τὴν ΛΜ τῇ ΜΚ[16], ὥστε καὶ ὡς ἡ ΜΒ πρὸς
τὴν ἐκ τοῦ κέντρου τοῦ ΕΖΗΘ κύκλου, οὕτως ἐπὶ[17] μὲν τῆς πρώτης καταγραφῆς συναμφότερος ἡ ΓΒ ΒΔ πρὸς τὴν ὑπεροχὴν αὐτῶν τὴν ΓΔ,

---

[1] distinx. B Hu, Tr

[2] distinx. B Hu, Tr

[3] distinx. S Hu, Tr

[4] *ΘΚ* AB corr. S Hu, Tr

[5] *ἴσον τὸ* AB corr. S Hu, Tr

[6] *ὡς* add. Tr

[7] *ΓΔ...τὴν* add. Hu, Tr

[8] *ἡμίσειαν τὴν* ΜΚ ABS corr. Hu, Tr

[9] *τουτέστιν* B Hu

[10] coniunx. BS Hu, Tr

[11] *καὶ κοινῶς* del. Hu

[12] *ἡ* add. Hu, Tr

[13] *τοῦ* add. Hu, Tr

[14] coniunx. BS Hu, Tr

[15] *ἀντὶ τῆς* ΛΜ del. Hu

[16] *ΑΚ* ABS corr. Hu, Tr *ΑΖ* Co

[17] *ὅπως ἡ* A corr. Co, Hu, Tr

ἐπὶ δὲ τῆς δευτέρας καὶ τῆς τρίτης ἡ < τῶν > [1] ΓΒ ΒΔ ὑπεροχὴ πρὸς συναμφότερον τὴν ΓΒΔ, τουτέστιν < τὴν > [2] ΓΔ [ἀνάπαλιν γάρ][3]. συνθεωρεῖται τάδ[1], ὅτι[4]
καὶ τὸ ὑπὸ τῶν ΒΚ ΛΓ ἴσον ἐστὶν τῷ ἀπὸ τῆς ΑΜ[5]. διὰ γὰρ τὴν ὁμοιότητα
τῶν ΒΘΚ ΖΛΓ τριγώνων ἐστὶν ὡς ἡ ΒΚ πρὸς ΚΘ, οὕτως ἡ ΖΛ πρὸς τὴν ΛΓ, καὶ
τὸ ὑπὸ ΒΚ ΛΓ[6] ἴσον τῷ ὑπὸ ΘΚ ΖΛ, τουτέστιν τῷ ἀπὸ τῆς ΑΜ. γίνεται δὲ
καὶ[7] διὰ μὲν τὸ εἶναι ὡς τὴν ΒΓ πρὸς τὴν ΓΔ, οὕτως τὴν ΒΛ πρὸς ΚΛ, τὸ
ὑπὸ ΒΓ καὶ τῆς ΚΛ, τουτέστιν τῆς τοῦ κύκλου διαμέτρου, ἴσον τῷ ὑπὸ
ΒΛ ΔΓ[8], διὰ δὲ τὸ εἶναι ὡς τὴν ΒΔ πρὸς τὴν ΓΔ, οὕτως τὴν ΒΚ πρὸς ΚΛ, τὸ ὑπὸ
τῆς ΒΔ καὶ τῆς ΚΛ, τουτέστιν τῆς τοῦ κύκλου διαμέτρου, ἴσον τῷ ὑπὸ ΒΚ ΔΓ.

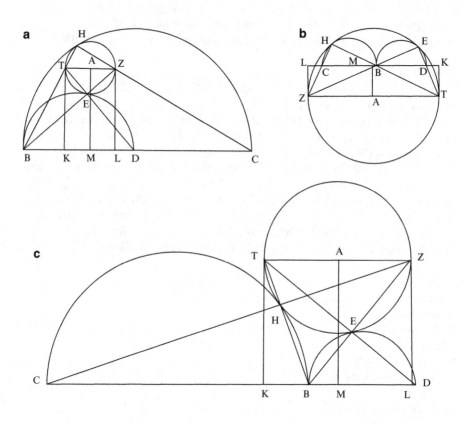

[1] τῶν add. Hu

[2] τὴν add. Hu, Tr

[3] ἀνάπαλιν γάρ del. Hu

[4] συνθεωρεῖται δ", ὅτι Hu

[5] τὸ ἀπὸ τῆς ΑΜ A corr. BS Hu, Tr

[6] τὸ ὑπὸ ΒΚ ΔΓ A corr. Co, Hu, Tr

[7] καὶ om. Tr

[8] ΒΓ ΛΓ ABS corr. Co, Hu, Tr

f. 40v (Prop. 15)
Prop. 15

<u>#17</u> τῶν αὐτῶν ὑποκειμένων γεγράφθω κύκλος ὁ ΘΡΤ ἐφαπτό-
μενος τῶν τε ἐξ ἀρχῆς ἡμικυκλίων καὶ τοῦ ΕΗΘ κύκλου κατὰ
τὰ *ΘΡΤ*[1] σημεῖα, καὶ ἀπὸ τῶν *ΑΠ*[2] κέντρων κάθετοι ἤχθωσαν
ἐπὶ τὴν ΒΓ βάσιν αἱ ΑΜ ΠΝ· λέγω ὅτι ἐστὶν ὡς ἡ ΑΜ μετὰ τῆς διαμέτρου τοῦ ΕΗ κύκλου πρὸς τὴν διάμετρον αὐτοῦ, οὕτως ἡ ΠΝ πρὸς τὴν
τοῦ ΘΡΤ κύκλου διάμετρον. ἤχθω τῇ ΒΔ πρὸς ὀρθὰς ἡ ΒΖ· ἐφάπτεται ἄρα τοῦ ΒΗΓ ἡμικυκλίου. καὶ ἐπιζευχθεῖσα ἡ ΑΠ ἐκβεβλήσθω ἐπὶ τὸ Ζ. ἐπεὶ διὰ τὸ προδειχθὲν ὡς συναμφότερος ἡ
ΓΒΔ πρὸς τὴν ὑπεροχὴν αὐτῶν τὴν ΓΔ, οὕτως καὶ ἡ ΒΜ ἐπὶ < μὲν >[3] τῆς
πρώτης καταγραφῆς < πρὸς τὴν ἐκ τοῦ κέντρου τοῦ ΕΗΘ κύκλου >[4], ἐπὶ
δὲ τῆς δευτέρας < καὶ τρίτης >[5] ὡς ἡ ὑπεροχὴ αὐτῶν
πρὸς συναμφότερον, τουτέστιν ὡς ἡ τῶν ΓΒ ΒΔ ὑπεροχὴ πρὸς
τὴν ΓΔ, οὕτως ἡ ΜΒ πρὸς τὴν ἐκ τοῦ κέντρου τοῦ ΕΗΘ κύκλου, καὶ
ἡ ΒΝ πρὸς τὴν ἐκ τοῦ κέντρου τοῦ ΘΡΤ κύκλου, ἔσται ἄρα καὶ ἐναλλὰξ ὡς ἡ ΜΒ πρὸς τὴν ΒΝ, ἡ ΑΘ ἐκ τοῦ κέντρου τοῦ ΕΗΘ κύκλου
πρὸς τὴν ΘΠ ἐκ τοῦ κέντρου < τοῦ >[6] ΘΡΤ κύκλου. ἀλλ' ὡς ἡ ΜΒ[7] πρὸς
ΒΝ, ἡ ΑΖ πρὸς ΖΠ. ἐπιζευχθείσης γὰρ τῆς ΖΜ ἔσται ὡς ἡ ΜΒ
πρὸς τὴν ΒΝ, οὕτως ἡ ΜΖ πρὸς τὴν ΖΞ. καὶ < ὡς ἄρα >[8] ἡ ΑΖ πρὸς τὴν
ΖΠ, οὕτως
ἡ ΑΘ ἐκ τοῦ κέντρου τοῦ ΕΗΘ κύκλου πρὸς < τὴν >[9] ΘΠ ἐκ τοῦ κέντρου τοῦ ΘΡΤ κύκλου[10]. καὶ τῶν ΕΗΘ ΡΟΤ κύκλων ἐφάπτεταί
τις κύκλος ὁ ΒΡΕΔ κατὰ τὰ *ΡΕ*[11] σημεῖα· διὰ[12] ἄρα τὸ προδειχθὲν ιε'
θεώρημα < ἡ >[13] τὰ *ΡΕ*[14] σημεῖα ἐπιζευγνύουσα εὐθεῖα

---

[1] distinx. BS Hu, Tr
[2] distinx. BS Hu, Tr
[3] *μὲν* add. Hu
[4] *πρὸς...τοῦ ΕΗΘ κύκλου* add. Co, Hu
[5] *καὶ τρίτης* add. Co, Hu, Tr
[6] *τοῦ* add. BS Hu, Tr
[7] *ΜΕ* A corr. Co, Hu, Tr
[8] *ὡς ἄρα* add. Hu, Tr
[9] *τὴν* add. Hu, Tr
[10] *κύκλω* A corr. Hu, Tr
[11] distinx. BS Hu, Tr
[12] *διὰ ἄρα τὸ προδειχθὲν ιε' θεώρημα τὰ ΡΕ σημεῖα* add. A1 in margine ιε' θεώρημα inter-
polatori attribuit Hu
[13] *ἡ* add. Hu, Tr
[14] distinx. Hu, Tr

ἐκβαλλομένη ἐπὶ τὸ Ζ¹ σημεῖον πεσεῖται, καὶ ἴσον ἔσται τὸ
ὑπὸ ΕΖΡ περιεχόμενον ὀρθογώνιον τῷ ἀπὸ τῆς ΘΖ τετραγώνῳ. ἔστιν δὲ καὶ τῷ ἀπὸ τῆς ΖΒ τετραγώνῳ ἴσον τὸ ὑπὸ
ΕΖΡ· ἴσον ἄρα καὶ τὸ ἀπὸ ΖΒ τῷ ἀπὸ ΖΘ· ἴση ἄρα ἡ ΒΖ τῇ
ΖΘ. ἐπεὶ² δὲ καὶ ἡ μὲν ΜΑ ἐκβληθεῖσα τέμνει τὴν τοῦ ΕΗΘ
κύκλου περιφέρειαν κατὰ τὸ Σ, ἡ δὲ ΠΝ τέμνει τὴν τοῦ ΘΡΤ
κύκλου περιφέρειαν κατὰ τὸ Ο σημεῖον, ἴση < ἡ > μὲν³ ΑΘ τῇ ΑΣ, ἡ δὲ
ΠΟ τῇ ΠΘ, καὶ < ἡ >⁴ τὰ ΟΣ⁵ σημεῖα ἐπιζευγνύουσα ἥξει διὰ τοῦ Θ·
ἴση γάρ ἐστιν ἡ ὑπὸ ΘΑΣ γωνία τῇ ὑπὸ ΘΠΟ γωνίᾳ ἐναλλάξ, καὶ ἰσογώνιόν ἐστιν τὸ ΑΘΣ τρίγωνον τῷ ΠΘΟ τριγώνῳ, καὶ ἔστιν εὐθεῖα ἡ ΑΠ· εὐθεῖα ἄρα ἐστὶν καὶ ἡ διὰ τῶν Σ Θ Ο⁶
σημείων ἀπαγομένη. ἥξει δὲ καὶ διὰ τοῦ Β⁷· εὐθεῖα γὰρ ἡ ΘΟΒ διὰ
τὸ εἶναι ὡς τὴν ΒΖ πρὸς ΖΘ, οὕτως τὴν ΟΠ πρὸς τὴν ΠΘ, ἴσων

f. 41 (Prop. 15)
οὐσῶν τῶν ὑπὸ ΒΖΘ ΟΠΘ γωνιῶν ἐν παραλλήλοις ταῖς ΒΖ
ΟΠ· καὶ τοῦτο γὰρ προδέδεικται ιε΄. ἐπιζευχθεῖσα δὲ καὶ ἡ
ΒΠ ἐκβεβλήσθω καὶ συμπιπτέτω τῇ ΜΑ ἐκβληθείσῃ⁸
κατὰ τὸ Κ. ἐπεὶ οὖν ἦν ὡς ἡ ΜΒ πρὸς ΒΝ, τουτέστιν ὡς ἡ ΚΒ πρὸς
τὴν ΒΠ, οὕτως ἡ ΑΖ πρὸς ΖΠ καὶ ἡ ΑΘ πρὸς ΘΠ, [οὕτως ἡ ΑΖ πρὸς ΖΠ
καὶ ἡ ΑΘ πρὸς ΘΠ,]⁹ ἔσται καὶ ὡς ἡ ΚΒ πρὸς ΒΠ, ἡ ΑΣ πρὸς
ΠΟ, καὶ ἡ ΣΚ < πρὸς ΠΟ· ἴση ἄρα ἡ ΑΣ τῇ >¹⁰
ΣΚ. ἐπεὶ οὖν ὅλη ἡ ΑΚ ὅλῃ τῇ διαμέτρῳ τοῦ ΕΗΘ κύ-
κλου ἐστὶν ἴση, καὶ ἔστιν ὡς ἡ ΚΜ πρὸς ΚΣ, οὕτως < ἡ >¹¹ ΝΠ πρὸς ΟΠ,
ἔσται καὶ
ὡς ἡ ΜΚ πρὸς τὴν ΚΑ, τουτέστιν ὡς ἡ ΜΑ μετὰ τῆς διαμέτρου τοῦ
ΕΗΘ κύκλου πρὸς τὴν διάμετρον, οὕτως ἡ ΝΠ πρὸς τὴν τοῦ ΘΡΤ κύκλου διάμετρον, ὅπερ ※

(f. 41v: diagrams Prop. 15)

---

¹τὸ Η σημεῖον A corr. Co, Hu, Tr
²ἔστιν δὲ καὶ A ἔστι δὲ καὶ BS ἐπεὶ δὲ καὶ Hu, Tr ἔτι δὲ καὶ coni. Tr
³ἴση μὲν add. A2 in marg. ἡ add. Tr ἴση ἄρα ἐστὶν ἡ μὲν Hu
⁴ἡ add. Hu, Tr
⁵distinx. BS Hu, Tr
⁶σΘΟ super evanidam scripturam A2 Σ Θ Ο BS Hu, Tr
⁷ΒΕ A corr. Co, Hu Β[Ε] Tr
⁸ἐκβληθείσης A corr. Hu, Tr
⁹bis scripta del. Co, Hu, Tr
¹⁰πρὸς...ἡ ΑΣ τῇ add. Hu, Tr
¹¹ἡ add. Hu, Tr

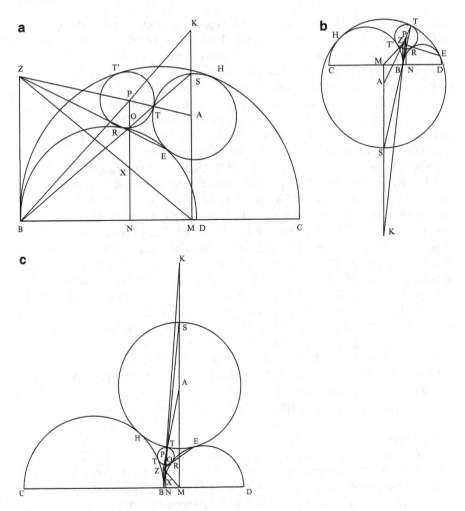

f. 42 (Prop. 16)

Prop. 16

<u>#18</u> τούτων προτεθεωρημένων ὑποκείσθω ἡμικύκλιον τὸ ΒΗΓ, καὶ ἐπὶ
τῆς βάσεως αὐτοῦ τυχὸν σημεῖον εἰλήφθω τὸ Δ, καὶ ἐπὶ τῶν ΒΔ
ΔΓ ἡμικύκλια γεγράφθω τὰ ΒΕΔ ΔΥΓ, καὶ ἐγγεγράφθωσαν εἰς τὸν
μεταξὺ τόπον τῶν τρίων περιφερειῶν τὸν καλούμενον ἄρβηλον κύκλοι ἐφαπτόμενοι τῶν ἡμικυκλίων καὶ ἀλλήλων ὅσοι –
δηποτοῦν, ὡς οἱ[1] περὶ τὰ κέντρα τὰ ΑΠΟ[2], καὶ ἀπὸ τῶν κέντρων αὐτῶν κάθετοι ἐπὶ τὴν ΒΓ ἤχθωσαν αἱ ΑΜ ΠΝ ΟΣ· λέγω ὅτι

---

[1] ὡς ὁ ABS corr. Hu, Tr
[2] distinx. BS Hu, Tr

ἡ μὲν ΑΜ ἴση ἐστὶν τῇ διαμέτρῳ τοῦ περὶ τὸ Α κύκλου, ἡ δὲ ΠΝ διπλῆ ἐστι¹ τῆς διαμέτρου τοῦ περὶ τὸ Π κύκλου, ἡ δὲ ΟΣ τριπλῆ
τῆς διαμέτρου τοῦ περὶ τὸ Ο² κύκλου, καὶ αἱ ἑξῆς [αἱ]³ κάθετοι
τῶν οἰκείων διαμέτρων πολλαπλάσιαι κατὰ τοὺς ἑξῆς μονάδι
ἀλλήλων ὑπερέχοντας ἀριθμούς. ἤχθω διάμετρος ἡ ΘΖ παράλληλος τῇ ΒΓ, καὶ κάθετοι αἱ ΘΚ ΖΛ· ἔσται δὴ κατὰ τὰ προγεγραμ-
μένα τὸ μὲν ὑπὸ ΓΒ ΒΚ περιεχόμενον ὀρθογώνιον ἴσον τῷ ὑπὸ
ΛΒ ΒΔ, τὸ δὲ ὑπὸ ΒΓ ΓΛ⁴ τῷ ὑπὸ ΚΓΔ. καὶ διὰ τοῦτο ὡς ἡ ΒΚ πρὸς
ΚΛ, οὕτως ἡ ΚΛ [πρὸς] πρὸς⁵ ΛΓ· ἑκάτερος γὰρ λόγος ὁ αὐτός ἐστιν
τῷ τῆς ΒΔ πρὸς ΔΓ, ἐπεὶ γὰρ τὸ ὑπὸ ΓΒ ΒΚ ἴσον ἐστὶν τῷ ὑπὸ
ΛΒ ΒΔ, ἔστιν ἄρα ὡς ἡ ΓΒ πρὸς ΒΛ, οὕτως ἡ ΔΒ πρὸς ΒΚ· ἐναλλὰξ
ὡς ἡ ΓΒ πρὸς ΒΔ, οὕτως ἡ ΛΒ πρὸς ΒΚ· διελόντι ὡς ἡ ΓΔ πρὸς
ΔΒ, ἡ ΛΚ πρὸς ΚΒ· ἀνάπαλιν ὡς ἡ ΒΔ πρὸς ΔΓ, ἡ ΒΚ πρὸς ΚΛ.
πάλιν ἐπεὶ τὸ ὑπὸ ΒΓ ΓΛ ἴσον ἐστὶν τῷ ὑπὸ ΚΓ ΓΔ, ἔστιν ἄρα ὡς
ἡ ΒΓ πρὸς ΓΚ, οὕτως ἡ ΔΓ⁶ πρὸς ΓΛ· ἐναλλὰξ ὡς ἡ ΒΓ πρὸς τὴν
ΓΔ, ἡ ΚΓ πρὸς τὴν ΓΛ· διελόντι [ὡς]⁷ ἄρα ἐστὶν ὡς ἡ ΒΔ πρὸς ΔΓ, οὕτως
ἡ ΚΛ πρὸς τὴν ΛΓ· ἦν δὲ καὶ ὡς ἡ ΒΔ πρὸς τὴν ΓΔ, ἡ ΒΚ πρὸς τὴν
ΚΛ· καὶ ὡς ἄρα ἡ ΒΚ πρὸς τὴν ΚΛ, < ἡ ΚΛ >⁸ πρὸς τὴν ΛΓ. ἴσον ἄρα τὸ ὑπὸ
τῶν ΒΚ ΓΛ τῷ ἀπὸ τῆς ΚΛ. προδέδεικται⁹ δὲ τὸ ὑπὸ ΒΚ ΛΓ ἴσον
καὶ τῷ ἀπὸ ΑΜ· ἴση ἐστὶν ἄρα¹⁰ ἡ ΑΜ τῇ ΚΛ, τουτέστιν τῇ ΖΘ
διαμέτρῳ τοῦ περὶ τὸ Α κύκλου. ἐπεὶ δὲ καὶ τοῦτο προδέδεικται, ὅτι ἐστὶν ὡς ἡ ΑΜ μετὰ τῆς ΖΘ πρὸς τὴν ΖΘ, οὕτως ἡ
ΠΝ πρὸς τὴν τοῦ περὶ τὸ Π κύκλου διάμετρον, καὶ ἔστιν < ἡ >¹¹ ΑΜ μετὰ
τῆς ΖΘ διπλῆ τῆς ΖΘ, ἔσται καὶ ἡ ΠΝ τῆς διαμέτρου τοῦ
περὶ τὸ Π κύκλου διπλῆ. ἡ ΠΝ ἄρα μετὰ τῆς διαμέτρου

f. 42v (Prop. 16)
τοῦ περὶ τὸ Π κύκλου τριπλασία τῆς διαμέτρου, καὶ ἔστιν ἐν τῷ
αὐτῷ λόγῳ ἡ ΟΣ πρὸς τὴν διάμετρον τοῦ περὶ τὸ Ο κύκλου· καὶ ἡ ΟΣ
ἄρα τριπλασία τῆς διαμέτρου τοῦ περὶ τὸ Ο κύκλου. καὶ ὁμοίως καὶ ἡ τοῦ ἑξῆς κύκλου κάθετος τῆς διαμέτρου τετραπλασία,
καὶ < αἱ >¹² ἑξῆς κάθετοι τῶν καθ' αὐτὰς διαμέτρων εὑρεθήσονται πολλα-

---

¹ἐστι del. Hu

²περὶ τὸ Θ Α corr. Hu, Tr

³αἱ del. Hu, Tr

⁴τὸ ὑπὸ ΛΒ ΒΔ, τὸ δὲ ὑπὸ ΒΓ ΓΑ Α1 τῷ corr. Α2 ὑπὸ ΒΓ ΓΛ Hu, Tr

⁵bis scriptum (sed alterum πρὸς expunctum) del. Hu, Tr

⁶ἡ ΓΔ Hu

⁷ὡς del. Hu, Tr

⁸ἡ ΚΛ add. Tr οὕτως ἡ ΚΛ Hu

⁹προσδέδεικται Α corr. Hu, Tr

¹⁰ἴση ἄρα ἐστὶν Hu

¹¹ἡ add. Hu, Tr

¹²αἱ add. Hu, Tr

πλάσιαι κατὰ τοὺς ἑξῆς μονάδι ἀλλήλων ὑπερέχοντας ἀριθμούς,
καὶ τοῦτο συμβαῖνον ἐπὶ τὸ ἄπειρον ἀποδειχθήσεται[1]. ἂν < δ' >[2] ἀντὶ τῶν ΒΗΓ ΔΥΓ περιφερειῶν εὐθεῖαι ὦσιν ὀρθαὶ πρὸς τὴν ΒΔ[3],
ὡς ἐπὶ τῆς τρίτης[4] ἔχει καταγραφῆς, τὰ αὐτὰ συμβήσεται περὶ
τοὺς ἐγγραφομένους κύκλους· αὐτόθεν[5] γὰρ ἡ ἀπὸ τοῦ Α κέντρου κάθετος ἐπὶ τὴν ΒΔ[6] ἴση γίνεται τῇ τοῦ περὶ τὸ Α κύκλου
διαμέτρῳ[7]. ἂν δὲ αἱ μὲν[8] ΒΗΓ ΒΕΔ μένωσιν περιφέρειαι, ἀντὶ δὲ τῆς
ΔΥΓ περιφερείας εὐθεῖα ὑποτεθῇ, ὡς ἐπὶ τῆς τετάρτης
ἔχει καταγραφῆς, ἡ ΔΖ[9] ὀρθὴ πρὸς τὴν ΒΓ, τῆς μὲν ΒΓ πρὸς
τὴν ΓΔ τετραγωνικὸν ἐν ἀριθμοῖς λόγον ἐχούσης, σύμμετρος[10]
ἔσται ἡ ἀπὸ τοῦ Α κάθετος τῇ διαμέτρῳ τοῦ περὶ τὸ Α κύκλου, εἰ δὲ μή, ἀσύμμετρος. καθόλου γὰρ ὃν ἔχει λόγον ἡ ΒΓ πρὸς τὴν
ΓΔ, τοῦτον ἔχει τὸν λόγον δυνάμει ἡ ΔΖ[11] πρὸς τὴν διάμετρον τοῦ
περὶ τὸ Α κύκλου, ὡς ἑξῆς δείκνυται. οἷον ἐὰν ᾖ τετραπλασία μήκει ἡ ΒΓ τῆς ΓΔ, γίνεται διπλῆ μήκει ἡ ΔΖ[12], τουτέστιν
ἡ ἀπὸ τοῦ Α κάθετος, τῆς διαμέτρου τοῦ περὶ τὸ Α κύκλου, καὶ
ἡ μὲν ἀπὸ τοῦ Π τριπλῆ, ἡ δ' ἀπὸ τοῦ Ο τετραπλῆ, καὶ ἑξῆς
κατὰ τοὺς ἑξῆς ἀριθμούς.

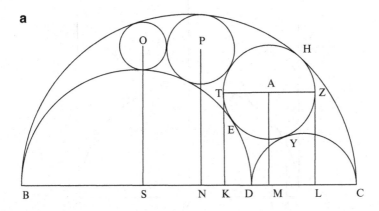

[1] ἀποδειχθήσονται ABS corr. Hu, Tr
[2] δ' add. Hu, Tr
[3] ΒΓ A corr. Hu, Tr
[4] Γ A corr. Hu, Tr
[5] αὐτό* θεν A
[6] ΒΓ A corr. Hu, Tr
[7] διαμέτρου ABS corr. Hu, Tr
[8] μὲν superscriptum A2
[9] ΔΞ A corr. Co, Hu, Tr
[10] σύμμετρον A corr. Hu, Tr
[11] ΔΞ A corr. Hu, Tr
[12] ΔΞ A corr. Hu, Tr

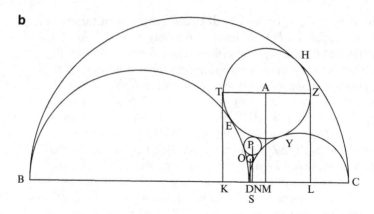

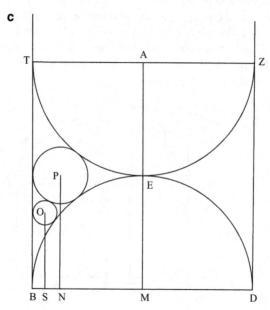

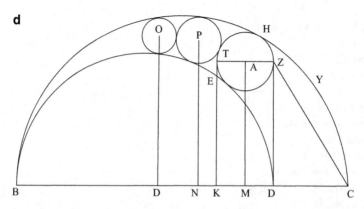

f. 43 (Prop. 17)
Prop. 17

#19 τὸ ὑπερτεθὲν λῆμμα. ἡμικύκλια τὰ ΒΗΓ ΒΑΔ, καὶ ὀρθὴ ἡ ΔΕ, καὶ κύκλος ἐφαπτόμενος ὁ *ΘΗ ΖΑ*[1]· ὅτι ἐστὶν < ὡς >[2] ἡ ΒΓ πρὸς τὴν ΓΔ μήκει, οὕτως ἡ ΔΖ πρὸς τὴν διάμετρον τοῦ *ΘΗ ΖΑ*[3] < κύκλου >[4] δυνάμει. ἤχθω διάμετρος ἡ ΘΖ· εὐθεῖαι ἄρα αἱ ΖΑΒ ΘΑΔ. κάθετος ἤχθω ἡ ΘΚ· ἔσται ἄρα διὰ < τὰ >[5] προδεδειγμένα τὸ ὑπὸ τῶν ΓΒ ΒΚ περιεχόμενον χωρίον ἴσον τῷ ἀπὸ τῆς ΒΔ τετραγώνῳ· ὡς ἄρα ἡ ΒΓ πρὸς ΓΔ, οὕτως ἡ ΒΔ πρὸς ΔΚ, τουτέστιν πρὸς ΘΖ. ὡς δὲ ἡ ΒΔ πρὸς ΘΖ, ἡ ΔΑ πρὸς ΘΑ, ὡς δὲ ἡ ΔΑ πρὸς ΑΘ, οὕτως τὸ ἀπὸ τῆς ΖΔ πρὸς τὸ ἀπὸ τῆς ΘΖ. ὀρθο-

f. 43v (Prop. 17, 18, and 19)
γώνιον γάρ ἐστιν τὸ ΘΖΔ, καὶ κάθετος ἐπὶ τὴν ὑποτείνουσαν ἡ ΖΑ. καὶ ὡς ἄρα ἡ ΒΓ πρὸς ΓΔ, οὕτως τὸ ἀπὸ τῆς ΖΔ πρὸς τὸ ἀπὸ τῆς διαμέτρου τοῦ *ΘΗ ΖΑ*[6] κύκλου.

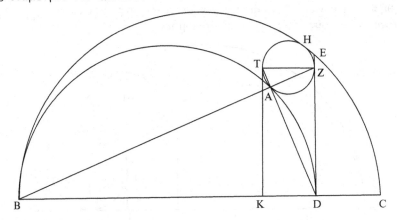

Prop. 18
#20 ἔτι καὶ τοῦτο[7] διὰ τῶν προ-
γεγραμμένων λημμάτων
τεθεώρηται. ἔστω ἡμικύ-
κλια τὰ ΑΒΓ ΑΔΕ, καὶ
γεγράφθωσαν ἐφαπτό-
μενοι τῶν περιφερειῶν
αὐτῶν κύκλοι οἱ περὶ τὰ κέντρα τὰ *ΖΗΘ*[8], καὶ οἱ συνεχεῖς αὐτοῖς

---

[1] coniunx. BS Hu, Tr
[2] ὡς add. Hu, Tr
[3] coniunx. Hu, Tr
[4] κύκλου add. Hu
[5] τὰ add. S Hu, Tr
[6] coniunx. Hu, Tr
[7] τούτω A corr. Hu, Tr
[8] distinx. BS Hu, Tr

ὡς ἐπὶ τὸ Α. ὅτι μὲν οὖν ἡ ἀπὸ τοῦ Ζ κάθετος ἐπὶ τὴν ΑΓ ἴση
ἐστὶ τῇ ἐκ τοῦ κέντρου τοῦ περὶ τὸ Ζ κύκλου δῆλον· λέγω δ' ὅτι καὶ
ἡ μὲν ἀπὸ τοῦ Η κάθετος τριπλασία τῆς ἐκ τοῦ κέντρου τοῦ
περὶ τὸ Η κύκλου, ἡ δὲ ἀπὸ τοῦ Θ πενταπλασία, καὶ < αἱ > [1] ἑξῆς κάθετοι τῶν ἐκ τῶν κέντρων πολλαπλάσιαι[2] κατὰ τοὺς ἑξῆς
περισσοὺς ἀριθμούς. ἐπεὶ γὰρ προδέδεικται ὡς ἡ ἀπὸ τοῦ Ζ κάθετος μετὰ τῆς διαμέτρου πρὸς τὴν διάμετρον, οὕτως ἡ ἀπὸ τοῦ
Η κάθετος πρὸς τὴν ἰδίαν διάμετρον, καὶ ἔστιν ἡ ἀπὸ τοῦ Ζ κάθετος μετὰ τῆς διαμέτρου ἡμιολία τῆς διαμέτρου, τῆς ἄρα ἐκ
τοῦ κέντρου ἔσται τριπλασία. πάλιν ἐπεί ἐστιν ὡς ἡ ἀπὸ τοῦ Η
κάθετος μετὰ τῆς διαμέτρου πρὸς τὴν διάμετρον, οὕτως ἡ ἀπὸ τοῦ
Θ κάθετος πρὸς τὴν διάμετρον, ἡ δ' ἀπὸ τοῦ Η κάθετος μετὰ τῆς
διαμέτρου πρὸς τὴν διάμετρον λόγον ἔχει ὃν ἔχει τὰ πέντε πρὸς τὰ
δύο, ἕξει καὶ ἡ ἀπὸ τοῦ Θ κάθετος πρὸς τὴν διάμετρον τὸν αὐτὸν
λόγον· τῆς ἄρα ἐκ τοῦ κέντρου ἔσται πενταπλασία. ὁμοίως
δειχθήσονται καὶ αἱ ἑξῆς κάθετοι τῶν ἐκ τῶν κέντρων πολλαπλάσιαι κατὰ τοὺς ἑξῆς περισσοὺς ἀριθμούς.

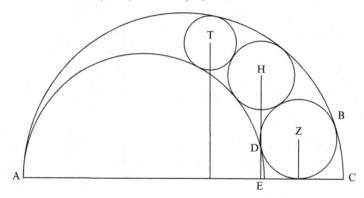

Prop. 19
<u>#21</u> τὸ ἐπὶ τῆς ἕλικος τῆς
ἐν ἐπιπέδῳ γραφομέ-
νης θεώρημα προὔτεινε
μὲν Κόνων ὁ Σάμιος[3] γεω-
μέτρης, ἀπέδειξεν δὲ
᾽Αρχιμήδης[4] θαυμαστῇ

f. 44 (Prop. 19, 20, and 21)
τινι χρησάμενος ἐπιβολῇ. ἔχει δὲ γένεσιν ἡ γραμμὴ τοιαύτην. ἔστω
κύκλος οὗ κέντρον μὲν τὸ Β, ἡ δὲ ἐκ τοῦ κέντρου ἡ ΒΑ. κεκινή-

---

[1] αἱ add. Hu. Tr

[2] πολλαπλασια A corr. Hu, Tr

[3] κώνων ὁ σάμιος A corr. Hu, Tr

[4] ἀρχιμήδης A corr. Hu, Tr

σθω[1] ἡ ΒΑ εὐθεῖα οὕτως ὥστε τὸ μὲν Β μένειν, τὸ δὲ Α ὁμαλῶς φέρεσθαι
κατὰ τῆς τοῦ κύκλου περιφερείας, ἅμα δὲ αὐτῇ ἀρξάμενόν τι
σημεῖον ἀπὸ τοῦ Β φερέσθω κατ᾽ αὐτῆς ὁμαλῶς ὡς ἐπὶ τὸ Α,
καὶ ἐν ἴσῳ χρόνῳ τό τε *Β σημεῖον*[2] τὴν ΒΑ διερχέσθω καὶ τὸ
Α τὴν τοῦ κύκλου περιφέρειαν· γράψει δὴ τὸ κατὰ τὴν[3] ΒΑ κινού-
μενον σημεῖον ἐν τῇ περιφορᾷ γραμμὴν οἵα ἐστὶν ἡ ΒΕΖΑ[4], καὶ
ἀρχὴ μὲν αὐτῆς ἔσται τὸ Β σημεῖον, ἀρχὴ δὲ τῆς *περιφερείας*[5]
ἡ ΒΑ. *αυτη*[6] δὲ ἡ γραμμὴ ἕλιξ καλεῖται. καὶ τὸ ἀρχικὸν αὐτῆς
ἐστι σύμπτωμα τοιοῦτον. ἥτις γὰρ ἂν διαχθῇ πρὸς αὐτὴν ὡς ἡ ΒΖ
καὶ ἐκβληθῇ, ἔστιν ὡς ἡ ὅλη τοῦ κύκλου περιφέρεια πρὸς τὴν
ΑΔΓ περιφέρειαν, οὕτως ἡ ΑΒ εὐθεῖα πρὸς τὴν ΒΖ. τοῦτο δὲ συνιδεῖν
ῥᾴδιον ἐκ τῆς γενέσεως· ἐν ᾧ μὲν γὰρ τὸ Α σημεῖον τὴν ὅλην
κύκλου περιφέρειαν διέρχεται, ἐν τούτῳ καὶ τὸ *Β*[7] τὴν ΒΑ, ἐν ᾧ δὲ
τὸ Α τὴν ΑΔΓ περιφέρειαν, ἐν τούτῳ καὶ τὸ Β [*τὴν Β*][8] τὴν ΒΖ εὐθεῖαν. καὶ εἰσὶν αἱ κινήσεις *αὐταὶ*[9] ἑαυταῖς ἰσοταχεῖς, ὥστε καὶ
ἀνάλογον εἶναι.

Prop. 20

φανερὸν δὲ καὶ τοῦτο, ὅτι αἵτινες ἂν διαχθῶσιν[10]
ἀπὸ τοῦ Β πρὸς τὴν γραμμὴν εὐθεῖαι ἴσας περιέχουσαι γωνίας, τῷ ἴσῳ ἀλλήλων < ὑπερέχουσιν >[11].

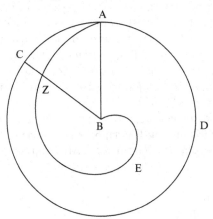

---

[1] *κεκινήσθω* A Hu *κινείσθω* Tr

[2] *τό τε Β σημεῖον* A Tr *τὸ ἀπὸ τοῦ Β σημεῖον* Hu

[3] *τὴν* A Hu *τῆς* appendix Hu p. 1229, Tr

[4] *ΒΖ ΕΛ* A corr. Co, Hu, Tr

[5] *τῆς περιφερείας* A Tr *τῆς περιφορᾶς* Hu

[6] *αυτη* sine acc. A *αὐτὴ* Hu *αὕτη* Tr

[7] *τὸ Β* A Tr *τὸ ἀπὸ τοῦ Β* Hu

[8] *καὶ τὸ Β τὴν Β* A *τὴν Β* del. Hu, Tr *καὶ τὸ ἀπὸ τοῦ Β* Hu

[9] *αυται* sine acc. A *αὐται* Hu *αὐταὶ* Tr

[10] *δειχθῶσιν* A corr. Hu, Tr

[11] *ὑπερέχουσι* add. S *ὑπερέχουσιν* Hu, Tr

Prop. 21

#22 δείκνυται δὲ τὸ περιεχόμενον σχῆμα
ὑπό τε τῆς ἕλικος καὶ τῆς εὐθείας
[τῆς ἕλικος καὶ τῆς εὐθείας]¹
τῆς ἐν ἀρχῇ τῆς περιφορᾶς
τρίτον μέρος τοῦ περιλαμβάνον-
τος αὐτὴν κύκλου. ἔστω γὰρ ὅτε
κύκλος καὶ ἡ προειρημένη²
γραμμή, καὶ ἐκκείσθω παραλληλόγραμμον ὀρθογώνιον τὸ *KN ΛΠ*³,
καὶ ἀπειλήφθω ἡ μὲν AΓ⁴ περιφέρεια μέρος [ἐστὶν]⁵ τι τῆς τοῦ
κύκλου περιφερείας, ἡ δὲ KP εὐθεῖα τῆς KΠ τὸ αὐτὸ μέρος, καὶ
ἐπεζεύχθωσαν ἥτε ΓΒ καὶ ἡ *KΛ*⁶, καὶ τῇ μὲν KN παράλληλος
ἡ PT, τῇ δὲ *KΠ*⁷ ἡ ΩΜ, καὶ περὶ τὸ Β κέντρον περιφέρεια ἡ ZH.
ἐπεὶ οὖν ἐστιν ὡς ἡ ΑΒ εὐθεῖα πρὸς ΑΗ, τουτέστιν ἡ ΒΓ πρὸς ΓΖ, < οὕτως ἡ >⁸ ὅλη

f. 44v (Prop. 21)

τοῦ κύκλου περιφέρεια πρὸς τὴν ΓΑ, τοῦτο γάρ ἐστιν τὸ ἀρχικὸν
τῆς ἕλικος σύμπτωμα, ὡς δὲ ἡ τοῦ κύκλου περιφέρεια πρὸς τὴν ΓΑ,
ἡ ΠΚ πρὸς ΚΡ, ὡς δὲ ἡ ΠΚ πρὸς τὴν ΚΡ, ἡ ΛΚ πρὸς τὴν ΚΩ, τουτέστιν
ἡ PT πρὸς τὴν ΡΩ, καὶ ὡς ἄρα ἡ ΒΓ πρὸς τὴν ΓΖ, ἡ ΤΡ πρὸς ΡΩ. καὶ
ἀναστρέψαντι καὶ ὡς ἄρα τὸ ἀπὸ τῆς ΒΓ πρὸς τὸ ἀπὸ τῆς ΒΖ, οὕτως
τὸ ἀπὸ τῆς PT πρὸς τὸ ἀπὸ τῆς ΤΩ. ἀλλ' ὡς μὲν τὸ ἀπὸ τῆς ΒΓ
πρὸς τὸ ἀπὸ τῆς ΒΖ, οὕτως ὁ ΑΒΓ τομεὺς πρὸς τὸν ΖΒΗ⁹ τομέα. ὡς δὲ
τὸ ἀπὸ PT πρὸς τὸ ἀπὸ ΤΩ, οὕτως ὁ ἀπὸ τοῦ ΚΤ παραλληλογράμμου κύλινδρος περὶ ἄξονα τὸν ΝΤ πρὸς τὸν ἀπὸ τοῦ ΜΤ παραλληλογράμμου κύλινδρον περὶ τὸν αὐτὸν ἄξονα· καὶ ὡς ἄρα ὁ
ΓΒΑ τομεὺς πρὸς τὸν ΖΒΗ¹⁰ τομέα, οὕτως < ὁ >¹¹ ἀπὸ τοῦ ΚΤ παραλληλογράμμου κύλινδρος περὶ ἄξονα τὸν ΝΤ πρὸς τὸν ἀπὸ τοῦ ΜΤ
παραλληλογράμμου κύλινδρον περὶ τὸν αὐτὸν ἄξονα. ὁμοίως δὲ
ἐὰν τῇ μὲν ΑΓ ἴσην θῶμεν τὴν ΓΔ, τῇ δὲ KP ἴσην τὴν PX,
καὶ τὰ αὐτὰ κατασκευάσωμεν, ἔσται ὡς ὁ ΔΒΓ τομεὺς πρὸς τὸν *ΕΘΒ*¹²,

---

¹bis scripta ABS del. Hu, Tr

²*προειρημμένη* A corr. Hu, Tr

³*τὸ KN ΛΠ* A coniunx. S Hu, Tr

⁴*ΑΒΓ* A corr. Hu, Tr

⁵*ἐστὶν* del. Hu, Tr

⁶*KA* A *BA* Co, Hu *KΛ* Tr

⁷*KM* A corr. Hu, Tr

⁸*οὕτως ἡ* add. S man. rec. Hu, Tr

⁹**ZB.η* A corr. Hu, Tr

¹⁰*τὸν ZBH* A2 ex *τὸν *BH*

¹¹*ὁ* add. S Hu, Tr

¹²*τὸν EΘB* A Tr *τὸν EBΘ* Co, Hu

οὕτως ὁ ἀπὸ τοῦ ΡΦ παραλληλογράμμου κύλινδρος περὶ ἄξονα
τὸν ΤΦ πρὸς τὸν ἀπὸ τοῦ ΞΦ παραλληλογράμμου κύλινδρον περὶ τὸν αὐτὸν ἄξονα. τῷ δ᾽ αὐτῷ τρόπῳ ἐφοδεύσαντες δείξομεν
ὡς ὅλον τὸν κύκλον πρὸς πάντα τὰ ἐγγεγραμμένα τῇ ἕλικι
ἐκ τομέων σχήματα, οὕτως τὸν ἀπὸ τοῦ ΝΠ παραλληλογράμμου
κύλινδρον περὶ ἄξονα τὸν ΝΛ πρὸς πάντα τὰ τῷ ἀπὸ τοῦ
ΚΝΛ τριγώνου[1] περὶ τὸν ΛΝ ἄξονα κώνῳ ἐγγραφόμενα ἐκ κυλίνδρων
σχήματα, καὶ πάλιν ὡς τὸν κύκλον πρὸς πάντα τὰ περιγραφόμενα τῇ
ἕλικι ἐκ τομέων σχήματα, οὕτως τὸν κύλινδρον πρὸς πάντα τὰ τῷ αὐτῷ
κώνῳ ἐκ κυλίνδρων περιγραφόμενα σχήματα· ἐξ οὗ φανερὸν ὅτι ὡς ὁ
κύκλος πρὸς τὸ μεταξὺ τῆς ἕλικος καὶ τῆς ΑΒ εὐθείας σχήμα[2], οὕτως ὁ
κύλινδρος πρὸς τὸν κῶνον. τριπλάσιος δὲ ὁ κύλινδρος τοῦ κώνου· τριπλάσιος ἄρα καὶ ὁ κύκλος τοῦ εἰρημένου σχήματος.

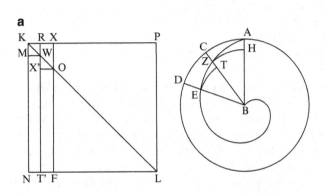

f. 45 (Props. 21 and 22)

#23 τῷ δ᾽ αὐτῷ τρόπῳ δείξομεν ὅτι, κἂν διαχθῇ τις εἰς τὴν ἕλικα ὡς ἡ
ΒΖ καὶ διὰ τοῦ Ζ περὶ < τὸ >[3] κέντρον τὸ Β γραφῇ κύκλος, τὸ περιεχόμενον σχῆμα [γραφη][4] ὑπό τε τῆς ΖΕΒ ἕλικος καὶ τῆς ΖΒ εὐθείας τρίτον μέρος
ἐστὶν τοῦ περιεχομένου σχήματος ὑπό τε τῆς ΖΗΘ περιφερείας τοῦ
κύκλου καὶ τῶν ΖΒΘ[5] εὐθειῶν. ἡ μὲν οὖν ἀπόδειξις τοιαύτη τίς
ἐστιν, ἑξῆς δὲ γράφομεν θεώρημα περὶ τὴν αὐτὴν γραμμὴν ὑπάρχον ἱστορίας ἄξιον.

---

[1] τριγώνῳ A corr. Hu, Tr

[2] σχήματα A corr. Hu, Tr forsan τὰ σχήματα

[3] τὸ add. Hu

[4] γραφη del. Co, Hu, Tr

[5] ΖΒΗ ABS corr. Hu, Tr

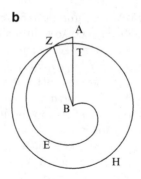

b

Prop. 22

<u>#24</u> ἔστω γὰρ ὅτε κύκλος ὁ προειρη-
μένος ἐν τῇ γενέσει, καὶ ἡ ἕλιξ
αὐτὴ ἡ *AZ EB*[1]· λέγω ὅτι[2], ἥτις ἂν
διαχθῇ ὡς ἡ BZ, ἔστιν ὡς τὸ ὑπὸ
τῆς ὅλης ἕλικος καὶ τῆς
ΑΒ εὐθείας περιεχόμενον σχῆμα πρὸς τὸ ὑπὸ τῆς ZEB ἕλι-
κος καὶ τῆς BZ εὐθείας περιεχόμενον, οὕτως ὁ ἀπὸ τῆς ΑΒ κύβος
πρὸς τὸν ἀπὸ τῆς ZB[3] κύβον. γεγράφθω γὰρ διὰ τοῦ Z κύκλος
περὶ κέντρον τὸ Β ὁ ΖΗΘ. ἐπεὶ οὖν ἐστιν ὡς τὸ ὑπὸ τῆς ΑΖΕΒ
γραμμῆς καὶ τῆς ΑΒ εὐθείας περιεχόμενον σχῆμα πρὸς τὸ ὑπὸ
τῆς ΖΕΒ γραμμῆς καὶ τῆς ZB[4] εὐθείας περιεχόμενον σχῆμα, οὕτως ὁ ΑΓΔ κύκλος πρὸς τὸ ὑπὸ τῆς ΖΗΘ περιφερείας καὶ τῶν
ΖΒΘ εὐθειῶν περιεχόμενον σχῆμα, ἑκάτερον γὰρ ἑκατέρου τρίτον
ἐδείχθη μέρος, ὁ δὲ ΑΓΔ κύκλος πρὸς τὸ ὑπὸ τῶν ΖΒΘ εὐθειῶν καὶ
τῆς ΖΗΘ περιφερείας ἀπολαμβανόμενον[5] χωρίον τὸν συγκείμενον
ἔχει λόγον ἔκ τε τοῦ ὃν ἔχει ὁ ΑΓΔ κύκλος πρὸς τὸν ΖΗΘ κύκλον καὶ ἐξ οὗ
ὃν ἔχει ὁ ΖΗΘ κύκλος < πρὸς >[6] τὸ ὑπὸ τῶν ΖΒΘ εὐθειῶν καὶ τῆς ΖΗΘ περιφερείας ἀπολαμβανόμενον[7] χωρίον, ἀλλ' ὡς μὲν ὁ ΑΓΔ κύκλος πρὸς τὸν ΖΗΘ
κύκλον, οὕτως τὸ ἀπὸ τῆς ΑΒ πρὸς τὸ ἀπὸ τῆς ΒΖ, ὡς δὲ ὁ ΖΗΘ κύκλος
πρὸς τὸ εἰρημένον χωρίον, ἡ ὅλη αὐτοῦ περιφέρεια πρὸς τὴν ΖΗΘ, τουτέστιν ἡ τοῦ ΑΓΔ κύκλου περιφέρεια πρὸς τὴν ΓΔΑ, τουτέστιν διὰ τὸ
σύμπτωμα τῆς γραμμῆς ἡ ΑΒ εὐθεῖα πρὸς τὴν ΒΖ, καὶ τὸ μεταξὺ

---

[1] ἡ ἕλιξ αὐτὴ ἡ *AZ EB* AS ἡ ἕλιξ αὐτὴ ἡ *AZEB* Tr ἡ ἕλιξ ἡ αὐτὴ ἡ *AZEB* coni. Hu
[2] λέγω ὅτις A corr. BS Hu, Tr
[3] τῆς ΖΘ A corr. Hu, Tr
[4] ZB in rasura A
[5] ἀπολαμβάνον ABS corr. Hu, Tr
[6] πρὸς add. Hu, Tr
[7] ἀπολαμβάνον ABS corr. Hu, Tr

ἄρα τῆς ἕλικος καὶ τῆς ΑΒ εὐθείας σχῆμα πρὸς τὸ μεταξὺ τῆς ἕλικος καὶ
τῆς ΒΖ λόγον ἔχει τὸν συγκείμενον ἔκ τε τοῦ < ἀπὸ >[1] τῆς ΑΒ πρὸς τὸ ἀπὸ τῆς
ΖΒ καὶ ἔκ τε τοῦ[2] τῆς ΑΒ πρὸς ΒΖ. οὗτος δὲ ὁ λόγος ὁ αὐτός ἐστι τῷ

f. 45v (Prop. 22 and conchoid)

τοῦ ἀπὸ τῆς ΑΒ κύβου πρὸς τὸν[3] ἀπὸ τῆς ΒΖ κύβον.

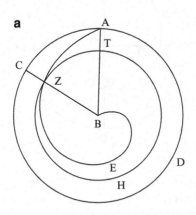

#25 ἐκ δὴ τούτου φανερὸν ὅτι, ἐὰν τῆς
ἕλικος ὑποκειμένης καὶ τοῦ
περὶ αὐτὴν κύκλου ἐκβληθῇ ἡ
ΑΒ ἐπὶ τὸ Δ καὶ πρὸς ὀρθὰς αὐτῇ ἀχθῇ ἡ ΓΖ ΕΚ[4], οἵου ἐστὶν ἑνὸς
τὸ μεταξὺ < τῆς >[5] ΒΛΕ γραμμῆς καὶ τῆς
ΒΕ εὐθείας χωρίον, τοιούτων ἐστὶν
τὸ μὲν μεταξὺ τῆς ΝΜΕ γραμ-
μῆς καὶ τῶν ΝΒΕ[6] εὐθειῶν χωρίον ἑπτά[7], τὸ δὲ μεταξὺ τῆς ΖΘΝ[8] γραμμῆς καὶ τῶν ΖΒΝ εὐθειῶν ιθ', τὸ δὲ μεταξὺ τῆς ΑΞΖ γραμμῆς καὶ τῶν
ΑΒΖ εὐθειῶν λζ', δῆλα γὰρ ταῦτα ἐκ [τε][9] τοῦ προδεδειγμένου θεωρήμα-

---

[1] ἀπὸ add. Tr, appendix Hu p. 1229

[2] ἐκ ///οῦ A

[3] τὸ A corr. BS Hu, Tr

[4] coniunx. B Hu, Tr

[5] τῆς add. Hu, Tr

[6] ΝΒ A corr. S Hu, Tr

[7] χωρίον ἔπα A corr. BS Hu, Tr

[8] ΖΘΗ ABS corr. Co, Hu, Tr

[9] τε del. S Hu, Tr

τος, καὶ ὅτι οἵων ἐστὶν ἡ ΑΒ δ[1], ἡ μὲν ΖΒ τριῶν, ἡ δὲ ΒΝ δύο, ἡ δὲ ΒΕ
ἑνός· καὶ γὰρ τοῦτο δῆλον ἔκ τε τοῦ τῆς γραμμῆς συμπτώματος καὶ
τοῦ τὰς ΑΓ ΓΔ ΔΚ ΚΑ περιφερείας ἴσας εἶναι.

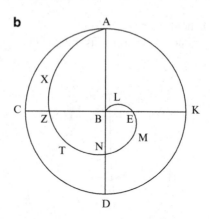

Conchoid

<u>#26</u> εἰς τὸν διπλασιασμὸν τοῦ κύβου
παράγεταί τις ὑπὸ Νικομή-
δους[2] γραμμὴ καὶ γένεσιν ἔχει
τοιαύτην. ἐκκείσθω εὐθεῖα ἡ
ΑΒ, καὶ αὐτῇ[3] πρὸς ὀρθὰς ἡ
ΓΔΖ, καὶ εἰλήφθω τι σημεῖον
ἐπὶ τῆς ΓΔΖ δοθὲν τὸ Ε, καὶ μέ-
νοντος τοῦ Ε σημείου ἐν ᾧ ἐστιν τόπῳ ἡ ΓΔΕΖ εὐθεῖα φερέσθω
κατὰ τῆς ΑΔΒ εὐθείας ἑλκομένη διὰ τοῦ Ε σημείου οὕτως ὥστε
διὰ παντὸς φέρεσθαι τὸ Δ ἐπὶ τῆς ΑΒ εὐθείας καὶ μὴ ἐκπίπτειν
ἑλκομένης τῆς *ΓΔ ΕΖ*[4] διὰ τοῦ Ε. τοιαύτης δὴ κινήσεως
γενομένης ἐφ᾽ ἑκάτερα, φανερὸν ὅτι τὸ Γ σημεῖον γράψει
γραμμὴν οἵα ἐστὶν ἡ ΛΓΜ, καὶ ἔστιν αὐτῆς τὸ σύμπτωμα τοιοῦτον.
ὡς ἂν εὐθεῖα προσπίπτῃ τις[5] ἀπὸ τοῦ Ε σημείου πρὸς τὴν γραμ-
μήν, τὴν ἀπολαμβανομένην μεταξὺ τῆς τε ΑΒ εὐθείας καὶ
τῆς ΛΓΜ γραμμῆς ἴσην εἶναι[6] τῇ ΓΔ εὐθείᾳ[7]· μενούσης γὰρ
τῆς ΑΒ καὶ μένοντος τοῦ Ε σημείου, ὅταν γένηται τὸ Δ ἐπὶ
τὸ Η, ἡ ΓΔ εὐθεῖα τῇ ΗΘ ἐφαρμόσει καὶ τὸ Γ σημεῖον ἐπὶ τὸ Θ[8]·

[1] *ΑΒΔ* ΑΒ *ΑΒ τεσσάρων Σ ΑΒ* δ᾽ Hu, Tr

[2] *νικομήδους* A corr. Hu, Tr

[3] *αὐτὴ* a A corr. S Hu, Tr

[4] coniunx. S Hu, Tr

[5] *τῆς* A corr. Hu, Tr

[6] *ποιεῖ* coni. Hu

[7] *τὴν ΓΔ εὐθεῖαν* A corr. S man. rec. Hu, Tr

[8] *ἐπὶ τὸ Θ πεσεῖται* Hu

f. 46 (Conchoid and Prop. 23)

ἴση ἄρα ἐστὶν ἡ ΓΔ τῇ ΗΘ. ὁμοίως καὶ ἐὰν ἑτέρα τις ἀπὸ τοῦ Ε
σημείου πρὸς τὴν γραμμὴν προσπέσῃ, τὴν ἀποτεμνομένην
ὑπὸ τῆς γραμμῆς καὶ τῆς ΑΒ εὐθείας ἴσην ποιήσει τῇ ΓΔ
[ἐπειδὴ ταύτῃ ἴσαι εἰσὶν αἱ προσπίπτουσαι][1]. καλείσθω δέ, φησίν, ἡ
μὲν ΑΒ εὐθεῖα κανών, τὸ δὲ σημεῖον πόλος, διάστημα δὲ ἡ
ΓΔ, ἐπειδὴ ταύτῃ ἴσαι εἰσὶν < αἱ >[2] προσπίπτουσαι πρὸς τὴν ΛΓΜ γραμ-
μήν, αὕτη[3] δὲ ἡ ΛΓΜ γραμμὴ κοχλοειδὴς[4] πρώτη[5], ἐπειδὴ
καὶ ἡ δευτέρα καὶ ἡ τρίτη καὶ ἡ τετάρτη ἐκτίθεται[6] εἰς ἄλλα θεωρήματα χρησιμεύουσαι.

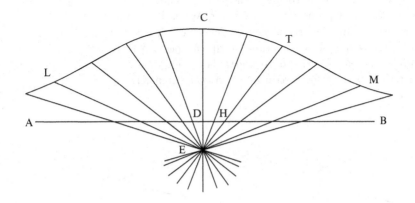

#27 ὅτι δὲ ὀργανικῶς δύναται
γράφεσθαι ἡ γραμμὴ καὶ ἐπ᾽[1]
ἔλαττον[7] ἀεὶ συμπορεύεσθαι[8]
τῷ κανόνι, τουτέστιν ὅτι πα-
σῶν τῶν ἀπό τινων σημεί-
ων τῆς ΛΓΘ γραμμῆς ἐπὶ
τὴν ΑΒ εὐθεῖαν καθέτων
μεγίστη ἐστὶν ἡ ΓΔ κάθετος, ἀεὶ δὲ ἡ ἔγγιον[9] τῆς ΓΔ ἀγομένη κάθετος τῆς ἀπώτερον μείζων ἐστίν, καὶ ὅτι, εἰς τὸν μεταξὺ τό-

---

[1] ἐπειδὴ...προσπίπτουσαι del. Hu, Tr

[2] αἱ add. B2 Hu, Tr

[3] αυτη sine spirit. et acc. A αὐτὴ Hu αὕτη Tr

[4] κογχοειδὴς A1 corr. A2 BS

[5] πρώτη A corr. Hu, Tr

[6] ἐκτίθεται A Hu ἐκτίθενται Tr, coniecit Hu

[7] ἐ-π᾽ ἔλαττον A

[8] συμπορεύεσθαι ABS Tr συμπορεύεται Hu

[9] η εγγειον sine spirit. et acc. A spirit. et acc. add. B corr. S Hu, Tr

πον τοῦ κανόνος καὶ τῆς κοχλοειδοῦς[1] ἐάν τις ᾖ εὐθεῖα, ἐκβαλλομένη τμηθήσεται ὑπὸ τῆς κοχλοειδοῦς[2], αὐτὸς ἀπέδειξεν ὁ Νικομήδης[3], καὶ ἡμεῖς ἐν τῷ εἰς τὸ ἀνάλημμα Διοδώρου[4], τρίχα τεμεῖν τὴν γωνίαν βουλόμενοι, κεχρήμεθα τῇ
προειρημένῃ γραμμῇ.

Prop. 23

διὰ δὴ τῶν εἰρημένων φανερὸν ὡς
δυνατόν ἐστιν γωνίας δοθείσης ὡς τῆς ὑπὸ ΗΑΒ καὶ σημείου
ἐκτὸς αὐτῆς τοῦ Γ διάγειν τὴν ΓΗ καὶ ποιεῖν τὴν ΚΗ μεταξὺ
τῆς γραμμῆς καὶ τῆς ΑΒ ἴσην τῇ δοθείσῃ. ἤχθω κάθετος ἀπὸ τοῦ Γ σημείου ἐπὶ τὴν ΑΒ ἡ ΓΘ καὶ ἐκβεβλήσθω, καὶ τῇ[5] δοθείσῃ ἴση ἔστω ἡ ΔΘ, καὶ πόλῳ μὲν
τῷ Γ, διαστήματι δὲ τῷ δοθέντι, τουτέστιν τῇ ΔΘ, κανόνι
δὲ τῷ ΑΒ γεγράφθω κοχλοειδὴς[6] γραμμὴ πρώτη ἡ ΕΔΗ[7]·
συμβάλλει ἄρα τῇ ΑΗ διὰ τὸ προλεχθέν. συμβαλλέτω
κατὰ τὸ Η, καὶ ἐπεζεύχθω ἡ ΓΗ· ἴση ἄρα καὶ ἡ ΚΗ τῇ
δοθείσῃ.

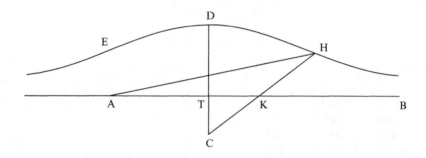

f. 46v (Prop. 24)
#28 τινὲς δὲ τῆς χρήσεως ἕνεκα παρα-
τιθέντες[8] κανόνα τῷ Γ κινοῦσιν
αὐτόν, ἕως ἂν ἐκ τῆς πείρας ἡ
μεταξὺ ἀπολαμβανομένη τῆς

[1] κοχλοειδοῦς A1B2S Hu γ superscriptum, λ expunctum A1

[2] κοχλοειδοῦς A1B2S Hu γ superscriptum, λ expunctum A1

[3] ὑπέδειξεν ὁ νικομήδης A corr. Hu, Tr

[4] διοδώρου A corr. Hu, Tr

[5] τῇ om. Tr

[6] κοχλοειδὴς A1B2S Hu γ superscriptum, λ expunctum A1

[7] πρώτη ἡ ΕΔΗ A corr. BS Hu, Tr

[8] παρατεθέντες ABS corr. Hu, Tr

ΑΒ εὐθείας καὶ τῆς ΕΔΗ γραμμῆς ἴση γένηται τῇ δοθείσῃ·
τούτου γὰρ ὄντος τὸ προκείμενον ἐξ ἀρχῆς δείκνυται· λέγω δὲ κύβος
κύβου διπλάσιος εὑρίσκεται. πρότερον δὲ δύο δοθεισῶν εὐθειῶν
δύο μέσαι κατὰ τὸ συνεχὲς[1] ἀνάλογον λαμβάνονται· ὧν ὁ Νικομήδης[2] τὴν κατασκευὴν ἐξέθετο μόνον[3], ἡμεῖς δὲ καὶ τὴν ἀπό-
δειξιν ἐφηρμόσαμεν τῇ κατασκευῇ τὸν τρόπον τοῦτον.

Prop. 24

δεδό-
σθωσαν γὰρ δύο εὐθεῖαι αἱ ΓΛ ΛΑ πρὸς ὀρθὰς ἀλλήλαις, ὧν δεῖ δύο
μέσας ἀνάλογον κατὰ < τὸ > συνεχὲς[4] εὑρεῖν, καὶ συμπεπληρώσθω τὸ
ΑΒΓΛ παραλληλόγραμμον, καὶ τετμήσθω δίχα ἑκατέρα τῶν ΑΒ
ΒΓ τοῖς ΔΕ[5] σημείοις, καὶ ἐπιζευχθεῖσα[6] μὲν ἡ ΔΛ ἐκβεβλήσθω
καὶ συμπιπτέτω[7] τῇ ΓΒ ἐκβληθείσῃ[8] κατὰ τὸ Η, τῇ δὲ ΒΓ
πρὸς ὀρθὰς ἡ ΕΖ, καὶ προσβεβλήσθω ἡ ΓΖ ἴση οὖσα τῇ ΑΔ, καὶ
ἐπεζεύχθω ἡ ΖΗ[9] καὶ αὐτῇ παράλληλος ἡ ΓΘ, < καὶ >[10] γωνίας οὔσης
τῆς ὑπὸ τῶν ΚΓΘ ἀπὸ δοθέντος τοῦ Ζ διήχθω ἡ ΖΘΚ ποιοῦσα
ἴσην τὴν ΘΚ τῇ ΑΔ ἢ τῇ ΓΖ, τοῦτο γὰρ ὡς δυνατὸν ἐδείχθη
διὰ τῆς κοχλοειδοῦς[11] γραμμῆς, καὶ ἐπιζευχθεῖσα ἡ ΚΛ ἐκβεβλήσθω καὶ συμπιπτέτω τῇ ΑΒ ἐκβληθείσῃ κατὰ τὸ Μ· λέγω ὅτι
ἐστὶν ὡς ἡ ΛΓ πρὸς τὴν[12] ΚΓ, ἡ ΚΓ πρὸς ΜΑ καὶ ἡ ΜΑ πρὸς τὴν ΑΛ. ἐπεὶ
ἡ ΒΓ τέτμηται δίχα τῷ Ε καὶ πρόσκειται αὐτῇ[13] ἡ ΚΓ, τὸ ἄρα ὑπὸ
ΒΚΓ[14] μετὰ τοῦ < ἀπὸ >[15] ΓΕ ἴσον ἐστὶ τῷ ἀπὸ ΕΚ. κοινὸν προσκείσθω τὸ ἀπὸ
ΕΖ· τὸ ἄρα ὑπὸ ΒΚΓ μετὰ τῶν ἀπὸ ΓΕΖ[16], τουτέστιν τοῦ ἀπὸ ΓΖ, ἴσον
ἐστὶν τοῖς ἀπὸ ΚΕΖ, τουτέστιν τῷ ἀπὸ ΚΖ. καὶ ἐπεὶ ὡς ἡ ΜΑ πρὸς
ΑΒ, ἡ ΜΛ πρὸς ΛΚ, ὡς δὲ ἡ ΜΛ πρὸς ΛΚ, οὕτως ἡ ΒΓ πρὸς ΓΚ, καὶ

---

[1] συνεχεῖ A corr. Hu, Tr

[2] νικομήδης A corr. Hu, Tr

[3] μόνην A μόνον Hu, Tr

[4] κατὰ συνεχὲς AS Tr τὸ add. B1 Eut. Hu

[5] distinx. S Hu, Tr

[6] ἐπιζευχθεῖσαν A corr. BS Hu, Tr

[7] συμπιπτέτω A corr. Hu, Tr

[8] ἐκβληθείσῃ A corr. Hu, Tr

[9] ἡ ΖΗ om. Tr

[10] καὶ add. Eut. Hu, Tr

[11] κοχλοειδοῦς A1B2S Hu γ superscriptum, λ expunctum A1

[12] τὴν om. Hu

[13] αυτη A corr. Hu, Tr

[14] ΒΓΚ ABS corr. Hu, Tr

[15] ἀπὸ add. Eut. Hu, Tr

[16] μετὰ τῶν ἀπὸ ΛΕΖ ABS corr. Hu, Tr μετὰ τῶν ἀπὸ ΓΕ ΕΖ Eut.

ὡς ἄρα ἡ ΜΑ πρὸς ΑΒ, οὕτως ἡ ΒΓ[1] πρὸς ΓΚ. καὶ ἔστι τῆς μὲν ΑΒ
ἡμίσεια ἡ ΑΔ, τῆς δὲ ΒΓ διπλῆ ἡ ΓΗ· ἔσται ἄρα καὶ ὡς ἡ
ΜΑ πρὸς ΑΔ, οὕτως ἡ ΗΓ πρὸς ΚΓ. ἀλλ' ὡς ἡ ΗΓ πρὸς ΓΚ, οὕτως
ἡ ΖΘ πρὸς ΘΚ διὰ τὰς παραλλήλους < τὰς > ΗΖ ΓΘ[2]· καὶ συνθέντι ἄρα
ὡς ἡ ΜΔ πρὸς ΔΑ, ἡ ΖΚ πρὸς ΚΘ. ἴση δὲ ὑπόκειται καὶ ἡ ΑΔ

f. 47 (Props. 24 and 25, and Quadratrix)
τῇ ΘΚ, ἐπεὶ καὶ τῇ ΓΖ ἴση ἐστὶν ἡ ΑΔ[3]· ἴση ἄρα καὶ ἡ ΜΔ τῇ
ΖΚ· ἴσον ἄρα καὶ τὸ ἀπὸ ΜΔ τῷ ἀπὸ ΖΚ. καὶ ἔστι τῷ μὲν ἀπὸ ΜΔ
ἴσον τὸ ὑπὸ ΒΜΑ[4] μετὰ τοῦ ἀπὸ ΔΑ, τῷ δὲ ἀπὸ ΖΚ ἴσον ἐδείχθη
τὸ ὑπὸ ΒΚΓ μετὰ τοῦ ἀπὸ ΖΓ, ὧν τὸ ἀπὸ ΑΔ ἴσον τῷ ἀπὸ ΓΖ. ἴση
γὰρ ὑπόκειται ἡ ΑΔ τῇ ΓΖ. ἴσον ἄρα καὶ τὸ ὑπὸ ΒΜΑ τῷ ὑπὸ
ΒΚΓ· ὡς ἄρα ἡ ΜΒ πρὸς ΒΚ, ἡ ΓΚ πρὸς ΜΑ[5]. ἀλλ' ὡς ἡ ΒΜ πρὸς ΒΚ,
ἡ ΛΓ πρὸς ΓΚ· ὡς ἄρα ἡ ΛΓ πρὸς ΓΚ, < ἡ ΓΚ πρὸς ΑΜ. ἔστι δὲ καὶ ὡς ἡ
ΜΒ πρὸς ΒΚ, >[6] ἡ ΜΑ πρὸς ΑΛ· καὶ ὡς ἄρα
ἡ ΛΓ πρὸς ΓΚ, ἡ ΓΚ πρὸς ΑΜ, καὶ ἡ ΑΜ πρὸς ΑΛ.

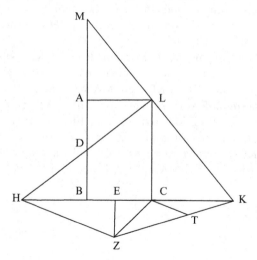

Prop. 25
#29 τούτου δειχθέντος πρόδηλον
ὅπως δεῖ κύβου δοθέντος
κύβον ἄλλον εὑρεῖν κατὰ τὸν
δοθέντα λόγον. ἔστω γὰρ ὁ

---

[1] ΒΓ A Hu ΒΚ Tr
[2] ἡ ΖΓΘ A corr. Hu, Tr τὰς ΗΖ ΓΘ Eut.
[3] ἐπεὶ καὶ τῇ ΓΖ ἴση ἐστὶν ἡ ΑΔ del. Hu
[4] τὸ ἀπὸ ΒΜΑ ABS corr. Hu, Tr τὸ ὑπὸ ΒΜΑ Eut.
[5] ἡ ΛΓ πρὸς ΓΚ ABS corr. Hu, Tr ἡ ΚΓ πρὸς ΑΜ Eut.
[6] ἡ ΓΚ...πρὸς ΒΚ om. ABS restit. Hu, Tr ἡ ΓΚ πρὸς ΑΜ. ἔστι δὲ καὶ ὡς ἡ ΛΓ πρὸς ΓΚ Eut.

δοθεὶς λόγος τῆς Α εὐθείας
πρὸς τὴν Β, καὶ τῶν ΑΒ¹ δύο
μέσαι ἀνάλογον κατὰ τὸ συνεχὲς εἰλήφθωσαν αἱ ΓΔ²· ἔσται
ἄρα ὡς ἡ Α πρὸς τὴν Β, οὕτως ὁ ἀπὸ τῆς Α κύβος πρὸς τὸν
ἀπὸ τῆς Γ κύβον, τοῦτο γὰρ δῆλον ἐκ τῶν στοιχείων.

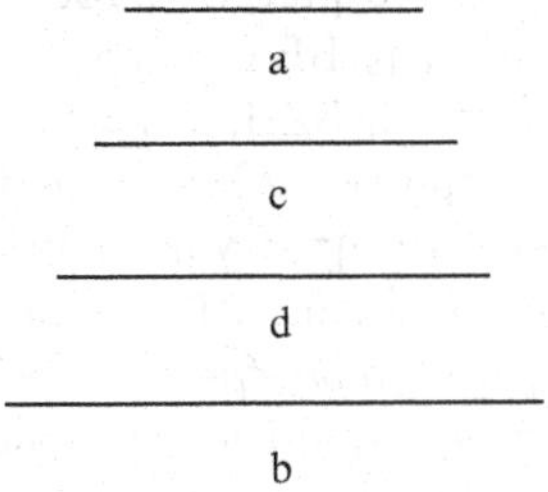

a

c

d

b

Quadratrix

 # 30 εἰς τὸν τετραγωνισμὸν τοῦ κύκλου παρελήφθη τις
ὑπὸ Δεινοστράτου καὶ Νικο-
μήδους³ γραμμὴ καί τινων
ἄλλων νεωτέρων ἀπὸ τοῦ
περὶ αὐτὴν συμπτώματος λαβοῦσα τοὔνομα, καλεῖται γὰρ
ὑπ' αὐτῶν τετραγωνίζουσα καὶ γένεσιν ἔχει τοιαύτην. ἐκκείσθω τετράγωνον τὸ ΑΒ ΓΔ⁴, < καὶ >⁵ περὶ κέντρον τὸ Α περιφέρεια
γεγράφθω < ἡ >⁶ ΒΕΔ, καὶ κινείσθω ἡ μὲν ΑΒ οὕτως ὥστε τὸ μὲν Α σημεῖον μένειν, τὸ δὲ Β φέρεσθαι κατὰ τὴν ΒΕΔ περιφέρειαν, ἡ δὲ
ΒΓ παράλληλος ἀεὶ διαμένουσα τῇ ΑΔ τῷ Β σημείῳ φε-
ρομένῳ⁷ κατὰ τῆς ΒΑ συνακολουθείτω⁸, < καὶ >⁹ ἐν ἴσῳ χρόνῳ ἥτε
ΑΒ κινουμένη¹⁰ ὁμαλῶς τὴν ὑπὸ ΒΑΔ γωνίαν, τουτέστιν

---

[1] distinx. B Hu, Tr

[2] distinx. BS Hu, Tr

[3] ὑπὸ δεινοστράτου καὶ νικοδήμου AB3 corr. S Hu, Tr ὑπὸ νικοστράτου B νικομήδου B1 To

[4] coniunx. B Hu, Tr

[5] καὶ add. To, Hu, Tr

[6] ἡ add. Hu, Tr

[7] τῷ Β σημεῖον φερον ἐν ᾧ ABS corr. To, Hu, Tr

[8] κατὰ τῆς Β συνακολουθεῖ τῷ AS corr. To, Hu, Tr συνακολουθεῖ τὸ Β

[9] καὶ add. To, Hu, Tr

[10] κινουμένης AB3S κινουμένη B1 Hu, Tr

τὸ Β σημεῖον τὴν ΒΕΔ < περιφέρειαν > [1], διαννύτω, καὶ ἡ ΒΓ τὴν ΒΑ εὐθεῖαν παροδευέτω, τουτέστιν τὸ Β σημεῖον κατὰ τῆς ΒΑ φερέσθω. συμ-

f. 47v (Quadratrix and Sporos)

βήσεται δῆλον[2] τῇ ΑΔ εὐθείᾳ ἅμα ἐφαρμόζειν ἑκατέραν[3] τήν τε ΑΒ καὶ τὴν ΒΓ. τοιαύτης δὴ γινομένης κινήσεως τεμοῦσιν ἀλλήλας ἐν τῇ φορᾷ αἱ ΒΓ ΒΑ εὐθεῖαι κατά τι σημεῖον αἰεὶ συμμεθιστάμενον αὐταῖς, ὑφ' οὗ σημείου γράφεταί τις ἐν τῷ μεταξὺ τόπῳ τῶν τε ΒΑΔ εὐθειῶν καὶ τῆς ΒΕΔ περιφερείας γραμ-μὴ ἐπὶ τὰ αὐτὰ κοίλη, οἷα ἐστὶν ἡ ΒΖΗ, < ἣ >[4] καὶ χρειώδης[5] εἶναι δοκεῖ πρὸς τὸ τῷ δοθέντι κύκλῳ τετράγωνον ἴσον εὑρεῖν. τὸ δὲ ἀρχικὸν αὐτῆς σύμπτωμα τοιοῦτόν ἐστιν· ἥτις γὰρ ἂν διαχθῇ τυχοῦσα < εὐθεῖα πρὸς τὴν περιφέρειαν, ὡς ἡ ΑΖΕ, ἔσται ὡς ἡ ὅλη >[6] περιφέρεια πρὸς τὴν ΕΔ, ἡ ΒΑ [περιφέρεια][7] εὐθεῖα πρὸς τὴν ΖΘ· τοῦτο γὰρ ἐκ τῆς γενέσεως τῆς γραμμῆς φανερόν ἐστιν.

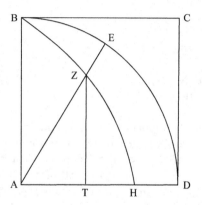

#31 δυσαρεστεῖται δὲ αὐτῇ[8] ὁ
Σπόρος[9] εὐλόγως διὰ ταῦτα. πρῶτον μὲν γὰρ

---

[1] περιφέρειαν add. To, Hu

[2] δῆλον A Hu δηλονότι vel δὴ coni. Hu δὴ[λον] Tr [δῆλον] Eberhard

[3] ἑκάτερα A ἑκατέρα B corr. S Hu, Tr

[4] ἣ add. Hu, Tr

[5] χρειῶδες ABS corr. To, Hu, Tr

[6] εὐθεῖα πρὸς τὴν περιφέρειαν, ὡς ἡ ΑΖΕ, ἔσται ὡς ἡ ὅλη add. Tr πρὸς τὴν περιφέρειαν, ὡς ἡ ΑΖΕ, ἔσται ὡς ὅλη ἡ add. Hu εὐθεῖα πρὸς τὴν περιφέρειαν, ὡς ἡ ΒΖΕ, ἔσται ὅλη ἡ ΒΕΔ add. To

[7] περιφέρεια del. S Hu, Tr

[8] αὐτῷ coni. Hu

[9] σπόρος A corr. Hu, Tr

πρὸς ὃ δοκεῖ χρειώδης
εἶναι πρᾶγμα, τοῦτ' ἐν
ὑποθέσει[1] λαμβάνει. πῶς
γὰρ δυνατόν, δύο σημείων ἀρξαμένων ἀπὸ τοῦ Β κινεῖσθαι,
τὸ μὲν κατ' εὐθείας ἐπὶ τὸ Α, τὸ δὲ κατὰ περιφερείας ἐπὶ
τὸ Δ ἐν[2] ἴσῳ χρόνῳ *συναποκαταστῆναι*[3] μὴ πρότερον τὸν λόγον[4]
τῆς ΑΒ εὐθείας πρὸς τὴν ΒΕΔ περιφέρειαν ἐπιστάμενον; ἐν
γὰρ τούτῳ τῷ λόγῳ[5] καὶ τὰ τάχη τῶν κινήσεων *ἀναγκαῖον*[6]. ἐπεὶ
πῶς *οἴονται*[7] συναποκαταστῆναι[8] τάχεσιν ἀκρίτοις *χρώμενα*[9],
πλὴν εἰ μὴ *ἂν*[10] κατὰ τύχην *ποτὲ*[11] *συμβῇ*[12]; τοῦτο δὲ πῶς οὐκ ἄλ-
ογον; ἔπειτα δὲ τὸ πέρας αὐτῆς ᾧ χρῶνται πρὸς τὸν τετραγωνισμὸν τοῦ κύκλου, τουτέστιν καθ' ὃ τέμνει σημεῖον τὴν ΑΔ
εὐθεῖαν, οὐχ εὑρίσκεται. νοείσθω δὲ ἐπὶ τῆς προκειμένης τὰ
λεγόμενα καταγραφῆς· ὁπόταν < *γὰρ* >[13] αἱ ΓΒ ΒΑ φερόμεναι συναπο-
κατασταθῶσιν, ἐφαρμόσουσιν *τὴν ΑΔ*[14] καὶ τομὴν οὐκέτι
ποιήσουσιν ἐν ἀλλήλαις· παύεται γὰρ ἡ τομὴ πρὸς *τῆς*[15] ἐπὶ
τὴν ΑΔ ἐφαρμογῆς, ἥπερ τομὴ πέρας *ἂν*[16] ἐγένετο τῆς
γραμμῆς καθ' ὃ τῇ ΑΔ εὐθείᾳ συνέπιπτεν. πλὴν εἰ μὴ λέγοι
τις ἐπινοεῖσθαι προσεκβαλλομένην τὴν γραμμὴν ὡς
ὑποτιθέμεθα τὰς εὐθείας ἕως τῆς ΑΔ· τοῦτο δ' οὐχ ἕπεται

f. 48 (Sporos and Prop. 26)
ταῖς ὑποκειμέναις ἀρχαῖς, ἀλλ' ὡς δ *ἂν*[17] ληφθείη τὸ Η σημεῖον
προειλημμένου τοῦ τῆς περιφερείας πρὸς τὴν εὐθεῖαν λόγου. χωρὶς

---

[1] *ὑ - ποθέσει* Α

[2] *τὸ ΔΕ Κ* Α corr. S Hu, Tr *τὸ δεη* Β

[3] *συναποκαταστῆσαι* Hu

[4] *τολον* Α *ὸν* superscriptum prima manu *τὸ ὅλον* Β Το *τὸν λόγον* S

[5] *ἐν γὰρ τῷ αὐτῷ λόγῳ* coni. Hu

[6] *ἀναγκαῖον* ABS Tr *ἀναγκαῖον εἶναι* (omisso posthac *ἐπεὶ*) Το *ἀνάγκη εἶναι* Hu

[7] *ἐπει πως οιονται* (sine acc.) Α *πῶς οἴονται γὰρ* Το *ἐπεὶ πῶς οἷόν τε* BS Hu, Tr quo pacto arbitrantur Co

[8] *συναποκαταστῆσαι* coni. Hu

[9] *χρώμενον* coni. Hu

[10] *ἂν* del. Το, probat et *συμβαίη* coni. Hu

[11] *τότε* Α corr. Το, Hu, Tr

[12] *συμβη* sine acc. Α

[13] *γὰρ* add. Hu

[14] *τῇ ΑΔ* Hu, Tr *ἐπὶ τὴν ΑΔ* Το

[15] *πρὸς τῆς* ABS *πρὸ τῆς* Το, Hu, Tr

[16] *ἂν* Hu

[17] *ἀλλ' ὡς δ ἂν* AB Το, Tr *ἄλλως δ ἂν* S *ἀλλ' ὡς ἂν* Hu

δὲ τοῦ δοθῆναι τὸν λόγον τοῦτον, οὐ¹ χρὴ τῇ² τῶν εὑρόντων ἀνδρῶν
δόξῃ³ πιστεύοντας παραδέχεσθαι τὴν γραμμὴν μηχανικω-
τέραν πῶς οὖσαν [καὶ εἰς πολλὰ προβλήματα χρησιμεύουσαν
τοῖς μηχανικοῖς]⁴, *πολὺ πρότερον παραδεκτέον ἐστὶ*⁵ τὸ δι' αὐτῆς δεικνύμενον πρόβλημα.

Prop. 26
τετραγώνου γὰρ ὄντος τοῦ ΑΒΓΔ⁶ καὶ
τῆς μὲν περὶ τὸ κέντρον τὸ Γ περιφερείας τῆς ΒΕΔ, τῆς
δὲ ΒΗΘ⁷ τετραγωνιζούσης γινομένης, ὡς προείρηται, δείκνυται,
ὡς ἡ ΔΕΒ περιφέρεια πρὸς τὴν ΒΓ εὐθεῖαν, οὕτως ἡ ΒΓ πρὸς
τὴν ΓΘ εὐθεῖαν. εἰ γὰρ μή ἐστιν⁸, ἤτοι πρὸς μείζονα ἔσται τῆς ΓΘ ἢ⁹
πρὸς ἐλάσσονα. ἔστω πρότερον, εἰ δυνατόν, πρὸς μείζονα τὴν ΓΚ,
καὶ περὶ κέντρον τὸ Γ περιφέρεια ἡ ΖΗΚ γεγράφθω τέμνουσα τὴν γραμμὴν κατὰ τὸ Η, καὶ κάθετος ἡ ΗΛ, καὶ ἐπιζευχθεῖσα ἡ ΓΗ ἐκβεβλήσθω ἐπὶ τὸ Ε. ἐπεὶ οὖν ἐστιν ὡς ἡ ΔΕΒ
περιφέρεια πρὸς τὴν ΒΓ εὐθεῖαν, οὕτως ἡ ΒΓ, τουτέστιν ἡ ΓΔ, πρὸς
τὴν ΓΚ, ὡς δὲ ἡ ΓΔ πρὸς τὴν ΓΚ, ἡ ΒΕΔ περιφέρεια πρὸς τὴν
ΖΗΚ περιφέρειαν, ὡς γὰρ ἡ διάμετρος τοῦ κύκλου πρὸς τὴν διάμετρον,
ἡ περιφέρεια τοῦ κύκλου πρὸς τὴν περιφέρειαν, φανερὸν ὅτι
ἴση ἐστὶν ἡ ΖΗΚ περιφέρεια τῇ ΒΓ εὐθείᾳ. καὶ ἐπειδὴ
διὰ τὸ σύμπτωμα τῆς γραμμῆς ἐστιν ὡς ἡ ΒΕΔ περιφέρεια
πρὸς τὴν ΕΔ, οὕτως ἡ ΒΓ πρὸς τὴν ΗΛ, καὶ ὡς ἄρα ἡ ΖΗΚ πρὸς
τὴν ΗΚ περιφέρειαν, οὕτως ἡ ΒΓ εὐθεῖα πρὸς τὴν ΗΛ. καὶ ἐδείχθη ἴση ἡ ΖΗΚ περιφέρεια τῇ ΒΓ εὐθείᾳ· ἴση ἄρα καὶ ἡ ΗΚ
περιφέρεια τῇ ΗΛ εὐθείᾳ, ὅπερ ἄτοπον. οὐκ ἄρα ἐστὶν ὡς ἡ
ΒΕΔ περιφέρεια πρὸς τὴν ΒΓ εὐθεῖαν, οὕτως ἡ ΒΓ πρὸς μείζονα τῆς ΓΘ.

#32 λέγω δὲ ὅτι οὐδὲ πρὸς ἐλάσσονα.
εἰ γὰρ δυνατόν, ἔστω πρὸς τὴν ΚΓ,
καὶ περὶ κέντρον τὸ Γ περιφέρεια γεγράφθω ἡ ΖΜΚ, καὶ πρὸς
ὀρθὰς τῇ ΓΔ ἡ ΚΗ τέμνου-

---

¹ η Α ἢ BS To, Tr οὐ Hu
² τη Α corr. Hu, Tr
³ δόξη Α corr. Hu, Tr
⁴ καὶ εἰς... μηχανικοῖς del. Hu, Tr
⁵ πολὺ πρότερον παραδεκτέον ἐστὶ ABS Tr ἀλλὰ πρότερον παραδεκτέον ἐστὶ Hu
παραδοτέον coni. Hu cf. versio Latina
⁶ ΑΒΓ ΑΒ1 corr. B2S Co, Hu, Tr
⁷ ΒΕΘ Α corr. To, Tr, Hu
⁸ μη εστιν sine acc. Α μή ἐστιν Tr μὴ ἔστιν Hu
⁹ τῆς ΓΘΗ ΑΒ1corr. B2S Hu, Tr

σα τὴν τετραγωνίζουσαν κατὰ
τὸ Η, καὶ ἐπιζευχθεῖσα ἡ ΓΗ

f. 48v (Prop. 26, 27, and 28)
ἐκβεβλήσθω ἐπὶ τὸ Ε. ὁμοίως δὲ τοῖς προγεγραμμένοις δείξομεν καὶ
τὴν ΖΜΚ περιφέρειαν τῇ ΒΓ εὐθείᾳ ἴσην, καὶ ὡς τὴν ΒΕΔ περιφέρειαν πρὸς τὴν ΕΔ, τουτέστιν[1] ὡς τὴν ΖΜΚ πρὸς τὴν ΜΚ, οὕτως
τὴν ΒΓ εὐθεῖαν [πρὸς τὴν *ΜΚ οὕτως τὴν ΒΓ εὐθεῖαν*][2] πρὸς τὴν ΗΚ.
ἐξ ὧν φανερὸν ὅτι ἴση ἔσται ἡ ΜΚ περιφέρεια τῇ ΚΗ εὐθείᾳ, ὅπερ
ἄτοπον. οὐκ ἄρα ἔσται ὡς ἡ ΒΕΔ περιφέρεια πρὸς τὴν ΒΓ εὐθεῖαν,
οὕτως ἡ ΒΓ πρὸς ἐλάσσονα τῆς ΓΘ. ἐδείχθη δὲ ὅτι οὐδὲ πρὸς μείζο-
να· πρὸς αὐτὴν ἄρα τὴν ΓΘ.

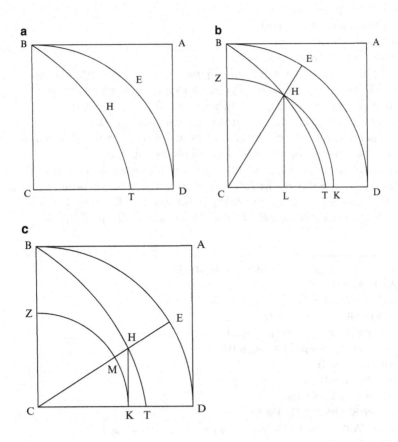

---

[1] *τοῦτό ἐστιν* ΑΒ *τουτέστιν* S Hu, Tr
[2] bis scripta del. S Hu, Tr

58 Part Ia Greek Text

Prop. 27
ἔστι δὲ καὶ τοῦτο φανερόν, ὅτι ἡ
τῶν ΘΓ ΓΒ εὐθειῶν τρίτη ἀνάλογον λαμβανομένη¹ εὐθεῖα ἴση
ἔσται τῇ ΒΕΔ περιφερείᾳ, καὶ ἡ τετραπλασίων αὐτῆς τῇ τοῦ
ὅλου κύκλου περιφερείᾳ. εὑρημένης δὲ τῇ τοῦ κύκλου περιφερείᾳ ἴσης εὐθείας, πρόδηλον ὡς δεῖ² καὶ αὐτῷ τῷ κύκλῳ ῥᾴδιον ἴσον τετράγωνον συστήσασθαι· τὸ γὰρ ὑπὸ τῆς περιμέτρου
τοῦ κύκλου καὶ τῆς ἐκ τοῦ κέντρου διπλάσιόν ἐστι τοῦ κύκλου,
ὡς Ἀρχιμήδης³ ἀπέδειξεν.

Prop. 28
33 αὕτη μὲν οὖν ἡ γένεσις τῆς
γραμμῆς ἐστιν, ὡς εἴρηται, μη-
χανικωτέρα· γεωμετρικῶς
δὲ διὰ τῶν πρὸς ἐπιφανείαις
τόπων ἀναλύεσθαι δύναται
τὸν τρόπον τοῦτον. θέσει κύ-
κλου τεταρτημόριον τὸ ΑΒΓ, καὶ διήχθω, ὡς ἔτυχεν⁴, ἡ ΒΔ, καὶ κάθετος ἐπὶ τὴν ΒΓ ἡ ΕΖ λόγον ἔχουσα δοθέντα πρὸς τὴν ΔΓ περι-
φέρειαν· ὅτι πρὸς γραμμὴν⁵ τὸ Ε. νοείσθω γὰρ ἀπὸ τῆς ΑΔΓ
περιφερείας ὀρθοῦ κυλίνδρου ἐπιφάνεια, καὶ ἐν αὐτῇ ἕλιξ
γεγραμμένη δεδομένη⁶ τῇ θέσει ἡ ΓΗΘ, καὶ πλευρά⁷ τοῦ κυλίνδρου ἡ
ΘΔ, καὶ τῷ τοῦ κύκλου ἐπιπέδῳ ὀρθαὶ ἤχθωσαν αἱ ΕΙΒΛ⁸ ἀ-
νεσταμέναι ὀρθαί⁹, διὰ δὲ τοῦ Θ¹⁰ τῇ ΒΔ παράλληλος ἡ ΘΛ. ἐπεὶ
< δοθεὶς μέν ἐστι >λόγος¹¹ τῆς ΕΖ εὐθείας¹² πρὸς τὴν ΔΓ περιφέρειαν, τῆς δὲ ΔΓ¹³
διὰ τὴν ἕλικα λόγος πρὸς τὴν ΔΘ¹⁴, ἔσται καὶ τῆς ΕΖ πρὸς ΕΙ¹⁵ λόγος δοθείς. καὶ εἰσὶν αἱ ΖΕ ΕΙ παρὰ θέσει· καὶ ἡ ΖΙ ἄρα ἐπιζευ-

¹τρίτη ἀνάλογον λαμβανομένη ABS corr. To, Hu, Tr
²δεῖ AB To δὴ S Hu, Tr
³ἀρχιμήδης A corr. Hu, Tr
⁴ὡς ἐτύχην AB1 corr. B3S Hu, Tr
⁵πρὸς γραμμὴν ABS πρὸς γραμμῇ Hu, Tr
⁶ἕλιξ γεγραμμένη δεδομένη A corr. B Hu, Tr
⁷ΠΛ AB corr. S Hu, Tr
⁸ΕΙ ΒΛ A ΕΙ ΛΒ B Hu, Tr
⁹ἀνεσταμέναι ορθαί del. Hu
¹⁰διὰ δὲ τοῦ Κ ABS corr. Co, Hu, Tr
¹¹ἐπι λόγος ABS ἐπίλογος To ἐπεὶ λόγος Hu δοθεὶς μέν ἐστι add. Tr
¹²τῆς ΕΙ ευθείας Hu
¹³περιφέρειαν, τῆς ΔΕ ΔΓ ABS corr. Tr περιφέρειάν ἐστιν δοθεὶς Hu
¹⁴τῆς δὲ ΔΓ διὰ τὴν ἕλικα λόγος πρὸς τὴν ΔΘ (ABS) Tr δοθεὶς δὲ καὶ ὁ τῆς ΕΖ λόγος πρὸς τὴν ΔΓ Hu
¹⁵Η ABS corr. Tr

χθεῖσα παρὰ θέσει. καὶ ἔστιν κάθετος ἐπὶ τὴν ΒΓ· ἐν ὀρθῷ[1] ἄρα
ἐπιπέδῳ ἡ ΖΙ, ὥστε καὶ τὸ Ι. ἔστιν δὲ καὶ ἐν κυλινδροειδεῖ ἐπι-

f. 49 (Props. 28, 29, and 30)
φανείᾳ[2], φέρεται γὰρ ἡ ΘΛ διά τε[3] τῆς ΘΗΓ ἕλικος καὶ τῆς ΛΒ εὐθείας καὶ αὐτῆς τῇ θέσει δεδομένης αἰεὶ παράλληλος οὖσα τῷ ὑποκειμένῳ ἐπι-
πέδῳ· πρὸς γραμμὴν[4] ἄρα τὸ Ι, ὥστε καὶ τὸ Ε. τοῦτο μέν οὖν ἀνελύθη καθόλου, ἂν δ' ὁ τῆς ΕΖ εὐθείας πρὸς τὴν ΔΓ περιφέρειαν[5] λόγος ὁ αὐτὸς ᾖ[6] τῷ τῆς
ΒΑ πρὸς τὴν ΑΔ Γ[7], ἡ προειρημένη τετραγωνίζουσα γίνεται γραμμή.

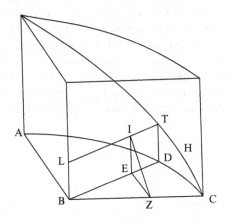

Prop. 29
34 δύναται δὲ καὶ διὰ τῆς ἐν ἐπιπέδῳ
γραφομένης ἕλικος ἀναλύεσθαι
τὸν ὅμοιον τρόπον. ἔστω γὰρ ὁ τῆς ΕΖ
πρὸς τὴν ΔΓ < περιφέρειαν >[8] λόγος ὁ αὐτὸς τῷ τῆς ΑΒ
πρὸς τὴν ΑΔΓ περιφέρειαν, καὶ ἐν ᾧ
ἡ ΑΒ εὐθεῖα περὶ τὸ Β κινουμένη[9]

---

[1] ὀρθῷ (vix legibile) A Tr τεμνόντι Hu

[2] ἐν κ////// ///φανείᾳ A.........ἐπιφανείᾳ BS ἐν κυλινδροειδεῖ ἐπιφανείᾳ Το ἐν κυλινδρικῇ
ἐπιφανείᾳ Hu ἐν κοχλοειδεῖ ἐπιφανείᾳ Tr ἐν πλεκτοειδεῖ ἐπιφανείᾳ coniecit Ver Eecke

[3] διά δε τῆς A corr. To, Hu, Tr

[4] πρὸς γραμμὴν ABS πρὸς γραμμῇ Hu, Tr

[5] τὴν ΔΘ περιφέρειαν ABS corr. Co, Hu, Tr

[6] η (sine spirit. et acc.) A corr. Hu, Tr

[7] coniunx. Hu, Tr

[8] περιφέρειαν add. Hu

[9] κινουμένη A corr. Hu, Tr

παροδεύει τὴν ΑΔΓ[1] περιφέρειαν, ση-
μεῖον ἐπ' αὐτῆς ἀρξάμενον ἀπὸ τοῦ Β ἐπὶ τὸ *Γ*[2] *παραγενέσθω*[3]
θέσιν λαβούσης τὴν < ΓΒ τῆς >[4] ΑΒ, καὶ ποιείτω τὴν *ΒΗ Α*[5] ἕλικα. ἔστιν
ἄρα ὡς ἡ ΑΒ πρὸς ΒΗ, ἡ ΑΔΓ περιφέρεια πρὸς τὴν ΓΔ, καὶ
ἐναλλάξ. ἀλλὰ καὶ ἡ ΕΖ πρὸς ΔΓ· ἴση ἄρα ἡ ΒΗ τῇ ΖΕ. ἤχθω
τῷ ἐπιπέδῳ ὀρθὴ ἡ ΚΗ ἴση τῇ ΒΗ· ἐν κυλινδροειδεῖ ἄρα
ἐπιφανείᾳ τῇ ἀπὸ τῆς ἕλικος τὸ Κ. ἀλλὰ καὶ ἐν κωνικῇ,
ἐπιζευχθεῖσα γὰρ ἡ ΒΚ ἐν κωνικῇ[6] γίνεται ἐπιφανείᾳ
ἡμίσειαν ὀρθῆς κεκλιμένη[7] πρὸς τὸ ὑποκείμενον καὶ
ἠγμένη διὰ δοθέντος τοῦ Β· πρὸς γραμμῇ[8] ἄρα τὸ Κ. ἤχθω διὰ
τοῦ Κ τῇ ΕΒ παράλληλος ἡ ΛΚΙ, καὶ ὀρθαὶ τῷ ἐπιπέδῳ
αἱ ΒΛ ΕΙ· ἐν πληκτοειδεῖ[9] ἄρα ἐπιφανείᾳ ἡ ΛΚΙ, φέρεται γὰρ
διά τε τῆς ΒΛ εὐθείας θέσει οὔσης καὶ διὰ θέσει γραμμῆς πρὸς ᾗ
τὸ Κ· καὶ τὸ Ι ἄρα < ἐν >ἐπιφανείᾳ[10]. ἀλλὰ καὶ ἐν ἐπιπέδῳ, ἴση γὰρ
ἡ ΖΕ τῇ ΕΙ, ἐπεὶ καὶ τῇ ΒΗ, καὶ γίνεται παρὰ θέσει ἡ ΖΙ κάθετος
οὖσα ἐπὶ τὴν ΒΓ· πρὸς γραμμῇ[11] ἄρα τὸ Ι [σ][12], ὥστε καὶ τὸ Ε. καὶ δῆλον ὅτι
ἂν ὀρθὴ < ᾖ >[13] ἡ ὑπὸ ΑΒΓ γωνία, ἡ προειρημένη τετραγωνίζουσα γραμμὴ γίνεται.

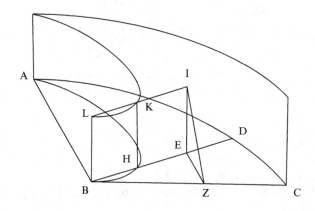

---

[1] *τὴν ΑΔ περιφέρειαν* A corr. Co, Hu, Tr
[2] *ἀπὸ τοῦ Β ἐπὶ τὸ Γ* A *ἀπὸ τοῦ Α ἐπὶ τὸ Β* Hu, Tr
[3] *παργενέσθω* A *παραγενέσθω* To, Tr *παραγινέσθω* Hu
[4] ΓΒ *τῆς* add. Hu, Tr
[5] coniunx. Hu, Tr
[6] *γωνικῇ* A corr. S Hu, Tr
[7] *κεκλιμένης* AB corr. S Hu, Tr
[8] *πρὸς γραμμῇ* A Tr *πρὸς γεγραμμένην* B1 corr. B2 vel B3 To, Hu *πρὸς γραμμὴν* S
[9] *πλέκτοειδεῖ* Hu
[10] *ἐπιφανεια* (sine acc.) A *ἐπιφάνεια* BS *ἐν ἐπιφανείᾳ* To, Hu, Tr
[11] *προσγραμμη* A *πρὸς γραμμὴ* B *προσγραμμὴ* S corr. To, Hu, Tr
[12] σ del. Hu, Tr
[13] *ᾖ* add. Hu, Tr

Prop. 30

#35 ὥσπερ < ἐν > ¹ ἐπιπέδῳ νοεῖται γι-
νομένη τις ἕλιξ φερομέ-
νου σημείου κατ' εὐθείας κύκλον περιγραφούσης, καὶ
ἐπὶ στερεῶν φερομένου ²

f. 49v (Prop. 30)
σημείου κατὰ μιᾶς πλευρᾶς τὴν³ ἐπιφάνειαν περιγραφούσης, οὕτω δὲ⁴ καὶ ἐπὶ σφαίρας ἕλικα νοεῖν ἀκόλουθόν ἐστι γραφο-
μένην τὸν τρόπον τοῦτον. ἔστω ἐν σφαίρᾳ μέγιστος κύκλος ὁ
ΚΛΜ περὶ πόλον τὸ⁵ Θ σημεῖον, καὶ ἀπὸ τοῦ Θ μεγίστου κύκλου τεταρτημόριον γεγράφθω τὸ ΘΝΚ, καὶ ἡ μὲν ΘΝΚ περιφέρεια, περὶ τὸ Θ μένον φερομένη κατὰ τῆς ἐπιφανείας
ὡς ἐπὶ τὰ Λ Μ⁶ μέρη, ἀποκαθιστάσθω πάλιν ἐπὶ τὸ αὐτό,
σημεῖον δέ τι φερόμενον ἐπ' αὐτῆς⁷ ἀπὸ τοῦ Θ ἐπὶ τὸ Κ πα-
ραγινέσθω· γράφει δή τινα ἐπὶ τῆς ἐπιφανείας ἕλικα,
οἵα ἐστὶν ἡ ΘΟΙΚ, καὶ ἥτις ἂν ἀπὸ τοῦ Θ γραφῇ μεγίστ < ου κύκλου περιφέρεια,
ἔστιν ὡς > ἡ τοῦ⁸
κύκλου περιφέρεια, πρὸς τὴν ΚΛ < περιφέρειαν λόγον ἔχει ὃν > ⁹ ἡ ΛΘ
πρὸς τὴν ΘΟ· λέγω δὴ
ὅτι, ἂν ἐκτεθῇ¹⁰ τεταρτημόριον τοῦ μεγίστου ἐν τῇ σφαίρᾳ κύκλου τ' οὗ ΑΒΓ περιφέρεια, κέντρον τὸ Δ¹¹, καὶ ἐπιζευχθῇ ἡ ΓΑ, γίνεται ὡς ἡ τοῦ ἡμισφαιρίου ἐπιφάνεια πρὸς
τὴν μεταξὺ τῆς ΘΟΙΚ ἕλικος καὶ τῆς ΚΝΘ περιφερείας ἀπολαμβανομένην¹² ἐπιφάνειαν, οὕτως ὁ ΑΒΓΔ
τομεὺς πρὸς τὸ ΑΒΓ τμῆμα. ἤχθω γὰρ ἐφαπτομένη τῆς
περιφερείας ἡ ΓΖ, καὶ περὶ κέντρον τὸ Γ διὰ τοῦ Α γε-
γράφθω¹³ περιφέρεια ἡ ΑΕΖ· ἴσος ἄρα ὁ ΑΒΓΔ τομεὺς

---

¹ ἐν add. Hu, Tr

² ε///////// φερομένου A ἔπειτα φερομένου voluit Co ἐπὶ στερεῶν φερομένου Hu ἐν ὀρθῷ
κώνου φερομένου Tr

³ τὴν A Tr τιν' Hu

⁴ οὕτω δὲ A οὕτως δὴ Hu, Tr

⁵ περὶ πόλον τὸν Θ σημεῖον A corr. Hu, Tr

⁶ τὰ ΛΑΜ μέρη A corr. Co, Hu, Tr

⁷ ἀπ' αὐτῆς A corr. Hu, Tr

⁸ μεγίστη τοῦ A –ου...ὡς add. Tr μεγίστου Hu

⁹ λόγον ἔχει ὃν add. Hu

¹⁰ ἐκτεθη A corr. Hu, Tr

¹¹ τοῦ ΑΒΓ περιφέρεια κέντρον A τὸ ΑΒΓ περὶ κέντρον Hu τοῦ ΑΒΓ Tr

¹² ἀπολαμβανομένης ABS corr. Hu, Tr

¹³ διὰ τοῦ ΑΓ εγράφθω A corr. Hu, Tr

τῷ ΑΕΖΓ, διπλασία μὲν γὰρ ἡ πρὸς τῷ Δ γωνία τῆς
ὑπὸ ΑΓΖ, ἥμισυ δὲ τὸ ἀπὸ ΔΑ τοῦ ἀπὸ ΑΓ. ὅτι ἄρα καὶ
ὡς αἱ εἰρημέναι ἐπιφάνειαι πρὸς ἀλλήλας, οὕτως ὁ
ΑΕΖΓ¹ τομεὺς πρὸς τὸ ΑΒΓ τμῆμα. ἔστω μέρος² ἡ ΚΛ περιφέρεια τῆς ὅλης τοῦ κύκλου περιφερείας, καὶ τὸ αὐτὸ μέρος
[ὅδε μέρος]³ ἡ ΖΕ τῆς ΖΑ, καὶ ἐπεζεύχθω ἡ ΕΓ· ἔσται δὴ καὶ ἡ
ΒΓ τῆς ΑΒΓ τὸ αὐτὸ μέρος. ὃ δὲ μέρος ἡ ΚΛ τῆς ὅλης
περιφερείας, τὸ αὐτὸ καὶ ἡ ΘΟ τῆς ΘΟΛ. καὶ ἔστιν ἴση
ἡ ΘΟΛ⁴ τῇ ΑΒΓ· ἴση ἄρα καὶ ἡ ΘΟ τῇ ΒΓ. γεγράφθω περὶ
πόλον τὸν Θ διὰ τοῦ Ο περιφέρεια ἡ ΟΝ, καὶ διὰ τοῦ Β
περὶ τὸ Γ κέντρον ἡ ΒΗ. ἐπεὶ οὖν ὡς ἡ ΛΚΘ σφαιρικὴ
ἐπιφάνεια πρὸς τὴν Ο Θ Ν⁵, ἡ ὅλη τοῦ ἡμισφαιρίου ἐπιφάνεια⁶
πρὸς τὴν τοῦ τμήματος⁷ ἐπιφάνειαν οὗ ἡ ἐκ⁸
τοῦ πόλου ἐστὶν ἡ ΘΟ, ὡς δ' ἡ⁹ τοῦ ἡμισφαιρίου ἐπιφάνεια
πρὸς τὴν τοῦ τμήματος ἐπιφάνειαν, οὕτως ἐστὶν τὸ ἀπὸ

f. 50 (Prop. 30 and metatheoretical passage)
τῆς τὰ ΘΛ¹⁰ ἐπιζευγνυούσης εὐθείας τετράγωνον πρὸς τὸ ἀπὸ τῆς
ἐπὶ τὰ ΘΟ¹¹, ἢ τὸ ἀπὸ τῆς ΕΓ τετράγωνον πρὸς τὸ ἀπὸ τῆς ΒΓ, ἔσται
ἄρα καὶ ὡς ὁ ΚΛΘ τομεὺς ἐν τῇ ἐπιφανείᾳ < πρὸς >¹² τὸν ΟΘΝ, οὕτως
ὁ ΕΖΓ τομεὺς πρὸς τὸν ΒΗΓ. ὁμοίως δείξομεν ὅτι καὶ < ὡς >¹³ πάντες οἱ
ἐν τῷ ἡμισφαιρίῳ τομεῖς οἱ ἴσοι τῷ ΚΛΘ, οἵ εἰσιν ἡ ὅλη τοῦ¹⁴
ἡμισφαιρίου ἐπιφάνεια πρὸς τοὺς περιγραφομένους περὶ
τὴν ἕλικα τομέας ὁμοταγεῖς τῷ ΟΘΝ, οὕτως < πάντες >¹⁵ οἱ ἐν τῷ ΑΖΓ
τομεῖς οἱ ἴσοι τῷ ΕΖΓ, τουτέστιν ὅλος ὁ ΑΖΓ τομεύς, πρὸς τοὺς
περιγραφομένους περὶ τὸ ΑΒΓ τμῆμα τοὺς¹⁶ ὁμοταγεῖς τῷ

---

¹ ΑΕΓΖ AS corr. B Hu, Tr
² ὃ μέρος Hu
³ ὁ δὲ μέρο ἡ A ὅδὲ μέρος ἡ B ὅδὲ μερη S ὁ δὲ μέρος del. Tr περιφέρεια Hu
⁴ ΘΟΑ AB2S corr. B1 Hu, Tr
⁵ coniunx. Hu, Tr
⁶ πρὸς...ἐπιφάνεια add. A2 in margine
⁷ τὴν τοῦ ἡμισφαιρίου ABS τὴν τοῦ τμήματος Co, Hu, Tr τὴν ἐντὸς τοῦ ἡμισφαιρίου coni. Hu
⁸ οὐκ ἐκ A corr. Co, Hu, Tr
⁹ δὴ A δ'ἡ Hu, Tr
¹⁰ distinx. B1S Hu, Tr
¹¹ distinx. B Hu, Tr Θ Ο, τουτέστιν Co
¹² πρὸς add. Hu, Tr
¹³ ὡς add. Hu, Tr
¹⁴ οιείσιν οι ολη A corr. S Hu
¹⁵ πάντες add. Hu, Tr
¹⁶ τοὺς A Hu τομέας Tr

ΓΒΗ. τῷ δ' αὐτῷ τρόπῳ δειχθήσεται καὶ ὡς ἡ τοῦ ἡμισφαιρί-
ου <ἐπιφάνεια >[1] πρὸς τοὺς ἐγγραφομένους τῇ ἕλικι τομέας, οὕτως ὁ ΑΖΓ τομεὺς πρὸς τοὺς ἐγγραφομένους τῷ ΑΒΓ τμήματι τομέας,
ὥστε καὶ ὡς ἡ τοῦ ἡμισφαιρίου ἐπιφάνεια πρὸς τὴν ὑπὸ
τῆς ἕλικος ἀπολαμβανομένην ἐπιφάνειαν, οὕτως ὁ ΑΖΓ
τομεύς, τουτέστιν [ὡς][2] τὸ ΑΒΓΔ τεταρτημόριον, πρὸς τὸ ΑΒΓ
τμῆμα. συνάγεται δὲ διὰ τούτου ἡ μὲν ἀπὸ τῆς ἕλικος
ἀπολαμβανομένη ἐπιφάνεια πρὸς τὴν ΘΝΚ περιφέρειαν ὀκταπλασία τοῦ ΑΒΓ τμήματος, ἐπεὶ καὶ ἡ τοῦ ἡμισφαιρίου ἐπιφάνεια τοῦ ΑΒΓΔ τομέως, ἡ δὲ μεταξὺ τῆς
ἕλικος καὶ τῆς βάσεως τοῦ ἡμισφαιρίου ἐπιφάνεια ὀκταπλασία τοῦ ΑΓΔ τριγώνου, τουτέστιν ἴση τῷ ἀπὸ τῆς διαμέτρου τῆς σφαίρας τετραγώνῳ.

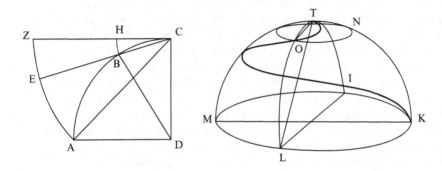

Metatheoretical passage
#36 τὴν δοθεῖσαν γωνίαν εὐθύγραμμον εἰς τρία ἴσα τεμεῖν οἱ παλαιοὶ
γεωμέτραι θελήσαντες ἠπόρησαν δι' αἰτίαν τοιαύτην. τρία
γένη φαμὲν εἶναι τῶν ἐν γεωμετρίᾳ προβλημάτων, καὶ τὰ μὲν
αὐτῶν ἐπίπεδα καλεῖσθαι, τὰ δὲ στερεά, τὰ δὲ γραμμικά.

f. 50v (metatheoretical passage)
τὰ μὲν οὖν δι' εὐθείας καὶ κύκλου περιφερείας δυνάμενα λύεσθαι λέγοιτο ἂν[3] εἰκότως ἐπίπεδα· καὶ γὰρ αἱ γραμμαὶ δι' ὧν εὑρίσκεται τὰ τοιαῦτα προβλήματα τὴν γένεσιν ἔχουσιν ἐν ἐπιπέδῳ.
ὅσα δὲ λύεται προβλήματα παραλαμβανομένης εἰς τὴν γένεσιν[4]
μιᾶς τῶν τοῦ κώνου τομῶν ἢ καὶ πλειόνων, στερεὰ ταῦτα κέκλη-

---

[1] ἐπιφάνεια add. Hu, Tr

[2] ὡς ABS del. Hu, Tr

[3] λέγοιτ'ἂν Hu

[4] εἰς τὴν γένεσιν ABS εἰς τὴν κατασκευὴν Co εἰς τὴν εὕρεσιν Hu

ται· πρὸς γὰρ τὴν κατασκευὴν χρήσασθαι στερεῶν σχημάτων
ἐπιφανείαις, λέγω δὲ ταῖς κωνικαῖς, ἀναγκαῖον. τρίτον δέ τι προβλημάτων ὑπολείπεται γένος, τὸ καλούμενον γραμμικόν· γραμμαὶ
γὰρ ἕτεραι παρὰ τὰς εἰρημένας εἰς τὴν κατασκευὴν λαμβάνονται ποικιλωτέραν ἔχουσαι τὴν γένεσιν καὶ βεβιασμένην μᾶλ-
λον, ἐξ ἀτακτοτέρων ἐπιφανειῶν καὶ κινήσεων ἐπιπεπλεγμέ-
νων γεννώμεναι. τοιαῦται δέ εἰσιν αἵτε ἐν τοῖς πρὸς ἐπιφανείαις
καλουμένοις τόποις εὑρισκόμεναι γραμμαί, ἕτεραί τε τούτων ποι-
κιλώτεραι καὶ πολλαὶ τὸ πλῆθος ὑπὸ Δημητρίου¹ τοῦ Ἀλεξαν-
δρέως² ἐν ταῖς γραμμικαῖς ἐπιστάσεσι καὶ Φίλωνος τοῦ Τυανέως³
ἐξ ἐπιπλοκῆς πληκτοειδῶν⁴ τε καὶ ἑτέρων παντοίων ἐπιφανειῶν εὑρισκόμεναι, πολλὰ καὶ θαυμαστὰ συμπτώματα περὶ
αὐτὰς⁵ ἔχουσαι. καί τινες αὐτῶν ὑπὸ τῶν νεωτέρων ἠξιώ-
θησαν λόγου πλείονος, μία δέ τις ἐξ αὐτῶν ἐστιν ἡ καὶ παράδοξος
ὑπὸ τοῦ Μενελάου⁶ κληθεῖσα γραμμή. τοῦ δὲ αὐτοῦ γένους ἕτεραι ἕλικές εἰσιν τετραγωνίζουσαί τε καὶ κοχλοειδεῖς⁷ καὶ κισ-
σοειδεῖς. δοκεῖ δέ πως ἁμάρτημα τὸ τοιοῦτον οὐ μικρὸν εἶναι τοῖς
γεωμέτραις, ὅταν ἐπίπεδον πρόβλημα διὰ τῶν κωνικῶῒ ἢ τῶν
γραμμικῶν ὑπό τινος εὑρίσκηται, καὶ τὸ σύνολον ὅταν ἐξ ἀνοι-
κείου λύηται γένους, οἷόν ἐστιν τὸ ἐν τῷ πέμπτῳ⁸ τῶν Ἀπολλωνίου⁹
κωνικῶν ἐπὶ τῆς παραβολῆς πρόβλημα καὶ < ἡ >¹⁰ ἐν τῷ περὶ τῆς
ἕλικος ὑπὸ Ἀρχιμήδους¹¹ λαμβανομένη στερεὰ νεῦσις¹² ἐπὶ κύ-
κλου¹³· μηδενὶ γὰρ προσχρώμενον στερεῷ δυνατὸν εὑρεῖν τὸ ὑπ᾽ αὐτοῦ
γραφόμενον θεώρημα, λέγω δὴ τὸ τὴν περιφέρειαν τοῦ ἐν τῇ
πρώτῃ περιφορᾷ κύκλου ἴσην ἀποδεῖξαι τῇ πρὸς ὀρθὰς
ἀγομένῃ εὐθείᾳ¹⁴ τῇ ἐκ τῆς γενέσεως < ἕως >¹⁵ τῆς ἐφαπτομένης τῆς
ἕλικος. τοιαύτης δὴ τῆς διαφορᾶς τῶν προβλημάτων ὑπαρχού-
σης οἱ πρότεροι γεωμέτραι τὸ προειρημένον ἐπὶ τῆς γωνίας

---

¹δημητρίου A corr. Hu, Tr

²ἀλεξανδρέως A corr. Hu, Tr

³φίλωνος το τυ*ανεως A corr. Hu, Tr

⁴πλεκτοειδῶν Hu

⁵περὶ αὐτὰς ABS corr. Hu, Tr

⁶μενελάου A corr. Hu, Tr

⁷κοχλοειδεῖς AB Hu, Tr κογχοειδεῖς S

⁸πρώτῳ voluit Hu

⁹ἀπολλωνίου A corr. Hu, Tr

¹⁰ἡ add. Hu, Tr

¹¹ἀρχιμήδους A corr. Hu, Tr

¹²στερεα νευσεις A στερεὰ νεῦσις B Tr στερεαὶ νεύσεις S στερεοῦ νεῦσις Hu

¹³ἐπὶ κύκλον Hu

¹⁴ἀγομένη εὐθεῖα A corr. BS Hu, Tr

¹⁵ἕως add. Hu, Tr

f. 51 (metatheoretical passage, Prop. 31 and 32)

πρόβλημα τῇ φύσει στερεὸν ὑπάρχον διὰ τῶν ἐπιπέδων ζητοῦντες
οὐχ οἷοί < τ' > [1] ἦσαν εὑρίσκειν, οὐδέπω γὰρ αἱ τοῦ κώνου τομαὶ συνήθεις ἦσαν
αὐτοῖς, καὶ διὰ τοῦτο ἠπόρησαν· ὕστερον μέντοι διὰ τῶν κωνικῶν
ἐτριχοτόμησαν[2] τὴν γωνίαν, εἰς τὴν εὕρησιν χρησάμενοι τῇ ὑπο-
γεγραμμένῃ νεύσει.

Prop. 31

παραλληλογράμμου δοθέντος ὀρθογωνίου τοῦ
ΑΒΓΔ καὶ ἐκβληθείσης τῆς ΒΓ, δέον ἔστω διαγαγόντα τὴν ΑΕ ποιεῖν
τὴν ΕΖ εὐθεῖαν ἴσην τῇ δοθείσῃ. γεγονέτω, καὶ ταῖς ΕΖ ΕΔ[3]
παράλληλοι ἤχθωσαν αἱ ΔΗ ΗΖ[4]. ἐπεὶ οὖν δοθεῖσά ἐστιν ἡ ΖΕ καὶ ἔστιν
ἴση τῇ ΔΗ, δοθεῖσα ἄρα καὶ ἡ ΔΗ. καὶ δοθὲν τὸ Δ· τὸ Η ἄρα πρὸς θέσει
κύκλου περιφερεία. καὶ ἐπεὶ τὸ ὑπὸ ΒΓΔ δοθὲν καὶ ἔστιν ἴσον τῷ
ὑπὸ ΒΖ ΕΔ[5], δοθὲν ἄρα καὶ τὸ ὑπὸ ΒΖ ΕΔ[6], τουτέστιν τὸ ὑπὸ ΒΖΗ[7]·
τὸ Η ἄρα πρὸς ὑπερβολήν[8]. ἀλλὰ καὶ πρὸς θέσει κύκλου περιφερείᾳ[9]· δοθὲν ἄρα τὸ Η.

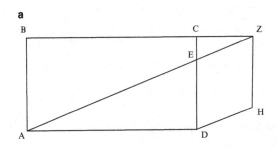

#37 συντεθήσεται δὴ τὸ πρόβλημα
οὕτως. ἔστω τὸ δοθὲν παραλληλόγραμμον τὸ ΑΒΓΔ, ἡ δὲ
δοθεῖσα εὐθεῖα τῷ μεγέθει ἡ

---

[1] τ' add. Hu, Tr

[2] ετριχατομησαν A corr. BS Hu, Tr

[3] ΖΔ ABS corr. Co, Hu, Tr

[4] ΗΘ ABS corr. Co, Hu, Tr

[5] ΒΕ ΖΔ A corr. Hu

[6] ΒΕ ΖΔ A corr. Hu

[7] ΒΘΗ A ΒΖΗ Hu ΒΕΗ Tr

[8] πρὸς ὑπερβολῇ Hu, Tr

[9] περιφέρεια ABS corr. Hu, Tr

Μ, καὶ ἴση αὐτῇ ἔστω ἡ ΔΚ, καὶ
γεγράφθω διὰ μὲν τοῦ Δ περὶ ἀσυμπτώτους τὰς ΑΒΓ ὑπερβολὴ ἡ
ΔΗΘ, τοῦτο γὰρ ἑξῆς ἀποδείξομεν, διὰ δὲ τοῦ Κ περὶ κέντρον τὸ Δ
κύκλου περιφέρεια ἡ ΚΗ τέμνουσα τὴν ὑπερβολὴν κατὰ
τὸ Η, καὶ τῇ[1] ΔΓ παραλλήλου ἀχθείσης τῆς ΗΖ ἐπεζεύχθω
ἡ ΖΑ· λέγω ὅτι ἡ ΕΖ ἴση ἐστὶν τῇ Μ. ἐπεζεύχθω γὰρ ἡ ΗΔ καὶ τῇ
ΚΑ παράλληλος ἤχθω ἡ ΗΛ· τὸ ἄρα ὑπὸ ΖΗΛ, τουτέστιν τὸ ὑπὸ ΒΖΗ,
ἴσον ἐστὶν τῷ[2] ὑπὸ ΓΔΑ, τουτέστιν τῷ ὑπὸ ΒΓ ΓΔ. ἔστιν ἄρα ὡς ἡ ΖΒ
πρὸς ΒΓ, τουτέστιν ὡς ἡ ΓΔ πρὸς ΔΕ, οὕτως ἡ ΓΔ πρὸς ΖΗ· ἡ ἄρα ΕΔ
ἴση τῇ ΖΗ. παραλληλόγραμμον ἄρα τὸ ΔΕ ΖΗ[3]· ἴση ἄρα ἡ ΕΖ τῇ
ΔΗ, τουτέστιν τῇ ΔΚ, τουτέστιν τῇ Μ.

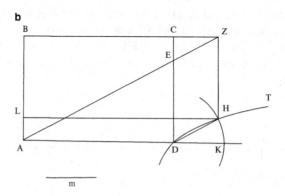

b

m

Prop. 32
#38 δεδειγμένου δὴ τούτου τρίχα
τέμνεται ἡ δοθεῖσα γωνία
εὐθύγραμμος οὕτως. ἔστω γὰρ
ὀξεῖα πρότερον ἡ ὑπὸ ΑΒΓ,
καὶ ἀπό τινος σημείου κά-

f. 51v (Prop. 32 and 33)
θετος ἡ ΑΓ, καὶ συμπληρωθέντος τοῦ ΓΖ παραλληλογράμμου ἡ ΖΑ
ἐκβεβλήσθω ἐπὶ τὸ Ε, καὶ παραλληλογράμμου ὄντος ὀρθογωνίου
τοῦ ΓΖ κείσθω μεταξὺ τῶν ΕΑΓ εὐθεῖα ἡ ΕΔ νεύουσα ἐπὶ τὸ Β ἴση
τῇ διπλασίᾳ τῆς ΑΒ, τοῦτο γὰρ ὡς δυνατὸν γενέσθαι προγέ-
γραπται· λέγω δὴ ὅτι τῆς δοθείσης γωνίας τῆς ὑπὸ ΑΒΓ τρίτον μέρος ἐστὶν ἡ ὑπὸ ΕΒΓ. τετμήσθω γὰρ ἡ ΕΔ δίχα τῷ Η, καὶ ἐπεζεύχθω ἡ ΑΗ[4]· αἱ τρεῖς ἄρα αἱ ΔΗ ΗΑ ΗΕ ἴσαι εἰσίν· διπλῆ ἄρα ἡ ΔΕ

---

[1] τῆς ΑΒ corr. S Hu, Tr
[2] τὸ Α corr. BS Hu, Tr
[3] coniunx. Hu, Tr
[4] ΑΕ Α corr. Co, Hu, Tr

τῆς ΑΗ. ἀλλὰ καὶ τῆς ΑΒ διπλῆ· ἴση ἄρα ἐστὶν ἡ ΒΑ τῇ ΑΗ, καὶ
ἡ ὑπὸ ΑΒΔ γωνία τῇ ὑπὸ ΑΗΔ. ἡ δὲ ὑπὸ ΑΗΔ διπλασία τῆς
ὑπὸ ΑΕΔ, τουτέστιν τῆς ὑπὸ ΔΒΓ· καὶ ἡ ὑπὸ ΑΒΔ ἄρα [διπλῆ]
διπλῆ[1] ἐστιν τῆς ὑπὸ ΔΒΓ. καὶ ἐὰν τὴν ὑπὸ ΑΒΔ δίχα τέμωμεν,
ἔσται [ἡ ὑπὸ ΑΒΔ δίχα τέμωμεν, ἔσται][2] ἡ ὑπὸ ΑΒΓ γωνία τρίχα
τετμημένη.

#39 ἐὰν δὲ ἡ δοθεῖσα γωνία ὀρθὴ
τυγχάνῃ, ἀπολαβόντες τινὰ
τὴν ΒΓ ἰσόπλευρον ἐπ᾿ αὐτῆς
γράψομεν τὸ ΒΔΓ, καὶ τὴν ὑπὸ
ΔΒΓ[3] γωνίαν δίχα τεμόντες ἕξομεν
τρίχα τετμημένην τὴν ὑπὸ ΑΒΓ γωνίαν.
#40 ἔστω δὲ ἀμβλεῖα ἡ γωνία καὶ τῇ

ΓΒ πρὸς ὀρθὰς ἡ ΒΔ, καὶ τῆς μὲν
ὑπὸ ΔΒΓ τρίτον ἀπειλήφθω
μέρος ἡ ὑπὸ ΔΒΖ, τῆς δὲ ὑπὸ
ΑΒΔ ὀξείας γωνίας τρίτον ἡ
ὑπὸ ΕΒΔ, ταῦτα γὰρ ἡμῖν προ-
δέδεικται· καὶ ὅλης[4] ἄρα τῆς ὑπὸ ΑΒΓ γωνίας τρίτον μέρος
ἐστὶν ἡ ὑπὸ ΕΒΖ. < ἐὰν δὲ τῇ ὑπὸ ΕΒΖ >[5] ἴσην συστησώμεθα πρὸς ἑκατέραν τῶν ΑΒΓ,
τρίχα τεμοῦμεν τὴν δοθεῖσαν γωνίαν.

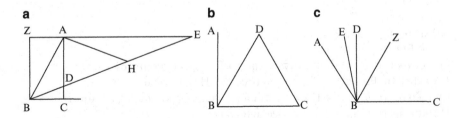

Prop. 33
#41 τὸ δὲ ὑπερτεθὲν πρόβλημα
νῦν ἀναλύσομεν. θέσει οὐσῶν δύο
εὐθειῶν τῶν ΑΒΓ καὶ δοθέντος
σημείου τοῦ Δ, γράψαι διὰ τοῦ Δ
περὶ ἀσυμπτώτους τὰς ΑΒΓ

[1]διπλῆ bis scriptum sed prius expunctum del. Hu, Tr
[2]ἡ ὑπὸ ΑΒΔ διχατεμωμεν ἔσται (sic) bis scripta corr. et alterum del. Hu, Tr
[3]ὑπὸ ΔΓ γωνίαν AB corr. S Hu, Tr
[4]ὅλη A corr. Hu, Tr
[5]ἐὰν· ΕΒΖ add. Hu, Tr

f. 52 (Prop. 33)

ὑπερβολήν. γεγονέτω, καὶ γεγράφθω ἡ ΕΔΖ, καὶ ἤχθω ἀπὸ τοῦ Δ ἐφαπτομένη αὐτῆς ἡ ΑΔΓ, καὶ διάμετρος ἡ ΗΒΔ, καὶ τῇ ΒΓ παράλληλος
ἡ ΔΘ. θέσει ἄρα αἱ ΗΔ ΔΘ¹, καὶ δοθὲν τὸ Θ. καὶ ἐπεὶ ἀσύμπτωτοί εἰσιν
αἱ ΑΒΓ τῆς ὑπερβολῆς, καὶ ἐφαπτομένη ἡ ΑΓ, ἴση ἄρα ἡ ΑΔ
τῇ ΔΓ, καὶ τὸ ἀφ' ἑκατέρας αὐτῶν τετράγωνον ἴσον ἐστὶν τῷ τετάρ-
τῳ τοῦ πρὸς τῇ ΗΔ εἴδους· ταῦτα γὰρ ἐν τῷ δευτέρῳ τῶν κωνικῶν
ἀποδέδεικται. ἐπεὶ οὖν ἴση ἡ ΓΔ τῇ ΔΑ, ἴση καὶ ἡ ΒΘ τῇ ΘΑ, καὶ
δοθεῖσα ἡ ΒΘ· δοθεῖσα ἄρα καὶ ἡ ΘΑ. καὶ δοθὲν τὸ Θ· δοθὲν ἄρα καὶ τὸ
Α· θέσει ἄρα ἡ ΑΔΓ < καὶ >² δοθεῖσα τῷ³ μεγέθει ἡ ΑΓ, ὥστε καὶ τὸ < ἀπὸ >⁴
ΑΓ⁵ δοθέν ἐστιν.
καὶ ἔστιν ἴσον τῷ πρὸς τῇ ΗΔ εἴδει· δοθὲν ἄρα καὶ τὸ πρὸς τῇ ΗΔ
εἶδος. καὶ δοθεῖσα ἡ ΗΔ, διπλῆ γάρ ἐστιν τῆς ΒΔ τῷ μεγέθει δεδομέ-
νης⁶ διὰ < τὸ >⁷ δοθὲν ἑκάτερον εἶναι τῶν ΒΔ⁸· δοθεῖσα ἄρα καὶ ἡ ὀρθία
τοῦ εἴδους πλευρά. γέγονεν δὴ πρόβλημα τοιοῦτον· θέσει καὶ μεγέθει
δύο δοθεισῶν εὐθειῶν τῆς τε ΗΔ καὶ τῆς ὀρθίας γράψαι περὶ
διάμετρον τὴν ΗΔ ὑπερβολήν, ἧς παρ' ἣν δύνανται⁹ ἔσται ἡ λοιπὴ
εὐθεῖα, καὶ αἱ καταγόμεναι τεταγμένως ἐπὶ τὴν ΗΔ παράλληλοι
ἔσονται θέσει τινὶ εὐθείᾳ¹⁰ τῇ ΑΓ. τοῦτο δὲ ἀναλέλυται ἐν τῷ
πρώτῳ τῶν κωνικῶν.

#42 συντεθήσεται δὴ οὕτως.
ἔστωσαν αἱ μὲν τῇ θέσει
δοθεῖσαι εὐθεῖαι αἱ ΑΒΓ,
τὸ δὲ δοθὲν σημεῖον τὸ
Δ, καὶ τῇ μὲν ΒΓ πα-
ράλληλος ἤχθω ἡ ΔΘ,
τῇ δὲ ΒΘ ἴση ἡ ΘΑ, καὶ
ἐπιζευχθεῖσα ἡ ΑΔ ἐκβεβλήσθω ἐπὶ τὸ Γ, ἐπιζευχθεῖσα δὲ καὶ ἡ
ΒΔ ἐκβεβλήσθω καὶ τῇ ΒΔ ἴση κείσθω ἡ ΒΗ, καὶ τῷ ἀπὸ τῆς ΑΓ
ἴσον ἔστω τὸ ὑπὸ τῆς ΗΔ καὶ ἑτέρας τινὸς τῆς κ, καὶ περὶ διάμετρον τὴν
ΗΔ καὶ ὀρθίαν τὴν κ¹¹ γεγράφθω ὑπερβο-
λὴ ἡ ΕΔΖ, ὥστε τὰς καταγομένας ἐπὶ τὴν ΗΔ παραλλήλους εἶναι
τῇ ΑΓ· ἡ ἄρα ΑΓ ἐφάπτεται τῆς τομῆς. καὶ ἔστιν ἡ ΑΔ τῇ ΔΓ ἴση,

---

¹ αἱ ΗΔ ΔΟ ABS corr. Co, Hu, Tr
² καὶ add. Hu, Tr
³ τῷ Α corr. Hu, Tr
⁴ ἀπὸ add. Co, Hu, Tr
⁵ τὸ ΔΓ ABS τὸ ἀπὸ ΑΓ Co, Hu, Tr
⁶ δεδομένη Α corr. Hu, Tr
⁷ τὸ add. Hu, Tr
⁸ distinx. Β Hu, Tr
⁹ παρην δύνανται Α corr. Hu, Tr παρ' ἣν δύναται S
¹⁰ εὐθεια (sine acc.) Α corr. Hu, Tr
¹¹ καὶ περὶ διάμετρον·τη* Κ add. Α2 in margine corr. Co, Hu, Tr

ἐπεὶ καὶ ἡ ΒΘ τῇ ΘΑ, καὶ φανερὸν ὅτι τὸ ἀφ᾽ ἑκατέρας τῶν ΑΔ ΔΓ τέταρτόν ἐστι
τοῦ πρὸς τῇ ΗΔ εἴδους· αἱ ἄρα ΑΒΓ ἀσύμπτωτοί εἰσι τῆς ΕΔΖ ὑπερβολῆς. γέγραπται
ἄρα διὰ τοῦ Δ περὶ τὰς δοθείσας εὐθείας ἀσυμπτώτους ὑπερβολή[1].

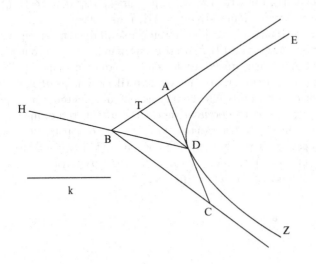

f. 52v (Prop. 33 and 34)
Prop. 34
#43 καὶ ἄλλως τῆς δοθείσης περι-
φερείας τὸ τρίτον ἀφαιρεῖ-
ται μέρος, χωρὶς τῆς νεύ-
σεως, διὰ στερεοῦ τόπου τοι-
ούτου. θέσει ἡ διὰ τῶν ΑΓ[2], καὶ
ἀπὸ δοθέντων τῶν[3] ἐπ᾽ αὐτῆς[4]
τῶν ΑΓ[5] κεκλάσθω ἡ ΑΒΓ διπλασίαν ποιοῦσα[6] τὴν
ὑπὸ ΑΓΒ[7] γωνίαν τῆς ὑπὸ
ΓΑΒ· ὅτι τὸ Β πρὸς ὑπερβολῇ[8]. ἤχθω κάθετος ἡ ΒΔ, καὶ τῇ[9] ΓΔ ἴση
ἀπειλήφθω ἡ ΔΕ· ἐπιζευχθεῖσα ἄρα ἡ ΒΕ ἴση ἔσται τῇ ΑΕ. κεί-

---

[1] ὑπερβολῇ A corr. Hu, Tr
[2] ΑΓ ΑΒ1 distinx. Β2S Hu, Tr
[3] τῶν om. Hu
[4] ἀπ᾽ αὐτῆς A ἐπ᾽ αὐτῆς Co, Hu, Tr
[5] ΑΓ ΑS distinx. B Hu, Tr
[6] ποιοῦσαν ΑS corr. B Hu, Tr
[7] ΑΒΓ A corr. Co, Hu, Tr
[8] προσυπερβολη A πρὸς ὑπερβολὴν B πρὸς ὑπερβολῇ S Hu, Tr
[9] τῇ A corr. Hu, Tr

σθω καὶ τῇ ΔΕ ἴση ἡ ΕΖ· τριπλασία ἄρα ἡ ΓΖ τῆς ΓΔ. ἔστω καὶ ἡ
ΑΓ τῆς ΓΗ τριπλασία· ἔσται δὴ δοθὲν τὸ Η, καὶ λοιπὴ ἡ ΑΖ τῆς
ΗΔ τριπλασία. καὶ ἐπεὶ τὸ ἀπὸ < ΒΔ τῶν ἀπὸ > ΒΕ ΕΖ ὑπεροχή ἐστιν[1],
ἔστιν δὲ καὶ
τὸ ὑπὸ ΔΑ ΑΖ τῶν αὐτῶν ὑπεροχή, ἔσται [ἴσον][2] τὸ ὑπὸ ΔΑΖ, τουτέστιν
τὸ τρὶς ὑπὸ ΑΔΗ[3], ἴσον τῷ ἀπὸ ΒΔ· πρὸς ὑπερβολῆ[4] ἄρα τὸ Β, ἧς
πλαγία μὲν τοῦ πρὸς ἄξονι εἴδους ἡ ΑΗ, ἡ δὲ ὀρθία τριπλασία
τῆς ΑΗ. καὶ φανερὸν ὅτι τὸ Γ σημεῖον ἀπολαμβάνει πρὸς τῇ
Η κορυφῇ τῆς τομῆς τὴν ΓΗ ἡμίσειαν τῆς πλαγίας τοῦ εἴδους
πλευρᾶς τῆς ΑΗ. καὶ ἡ σύνθεσις φανερά· δεήσει γὰρ τὴν ΑΓ
τεμεῖν ὥστε διπλασίαν εἶναι τὴν ΑΗ τῆς ΗΓ, καὶ περὶ ἄξονα
τὸν ΑΗ γράψαι διὰ τοῦ Η ὑπερβολήν, ἧς ὀρθία τοῦ εἴδους πλευρὰ
τριπλασία τῆς ΑΗ. καὶ δείκνυται[5] ποιοῦσαν αὐτὴν τὸν εἰρημένον διπλάσιον λόγον τῶν γωνιῶν. καὶ ὅτι τῆς δοθείσης κύκλου περιφερείας τὸ γ" ἀποτέμνει[6] μέρος ἡ τοῦτον γραφομένη
τὸν τρόπον ὑπερβολὴ συνιδεῖν[7] ῥᾴδιον τῶν ΑΓ[8] σημείων
περάτων τῆς περιφερείας ὑποκειμένων.

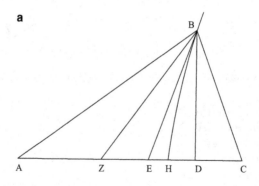

#44 ἑτέρως δὲ τὴν ἀνάλυσιν τοῦ τρί-
χα τεμεῖν τὴν γωνίαν ἢ περιφέρειαν ἐξέθεντό τινες ἄνευ
τῆς νεύσεως. ἔστω δὲ ἐπὶ περι-

---

[1] ἐπεὶ τὸ ἀπὸ ΒΕ ΕΖ ὑπεροχή ἐστιν Α ΒΔ τῶν ἀπὸ add. Tr ἐπεὶ τῶν ἀπὸ ΒΕ ΕΖ ὑπεροχή
ἐστιν τὸ ἀπὸ ΒΔ Co, Hu
[2] ἴσον del. Tr ἄρα Hu
[3] ΔΑΗ Α corr. Co, Hu, Tr
[4] προσυπερβολὴ ΑΒ πρὸς ὑπερβολῇ S Hu, Tr
[5] δεικνύναι Hu
[6] τὸ Γ ἀποτέμνειν ΑΒS τὸ γ" ἀποτέμνει Hu τὸ τρίτον ἀποτέμνει Tr
[7] συνειδεῖν Α corr. BS Hu, Tr
[8] distinx. BS Hu, Tr

φερείας ὁ λόγος, οὐδὲ¹ γὰρ δια-
φέρει γωνίαν ἢ περιφέρειαν

f. 53 (Props. 34 and 35)

τεμεῖν. γεγονέτω δή, καὶ τῆς ΑΒΓ περιφερείας τρίτον ἀπειλήφθω
μέρος ἡ ΒΓ, καὶ ἐπεζεύχθωσαν αἱ ΑΒ ΒΓ [μέρος ἡ ΒΓ]² ΓΑ· διπλασίων³
ἄρα ἡ < ὑπὸ >⁴ ΑΓΒ⁵ τῆς ὑπὸ ΒΑΓ. τετμήσθω δίχα ἡ ὑπὸ ΑΓΒ τῇ ΓΔ, καὶ
κάθετοι αἱ ΔΕΖΒ⁶· ἴση ἄρα ἡ ΑΔ τῇ ΔΓ⁷, ὥστε καὶ ἡ ΑΕ τῇ ΕΓ· δοθὲν
ἄρα τὸ Ε. ἐπεὶ οὖν ἐστιν ὡς ἡ ΑΓ πρὸς ΓΒ, οὕτως ἡ ΑΔ πρὸς ΔΒ,
τουτέστιν
ἡ ΑΕ πρὸς ΕΖ, καὶ ἐναλλὰξ ἄρα ἐστὶν ὡς ἡ ΓΑ πρὸς ΑΕ, ἡ ΒΓ πρὸς
ΕΖ. διπλῆ δὲ ἡ ΓΑ τῆς ΑΕ⁸· διπλῆ ἄρα καὶ ἡ ΒΓ⁹ τῆς ΕΖ. τετραπλά-
σιον ἄρα τὸ ἀπὸ ΒΓ, τουτέστιν τὰ ἀπὸ τῶν ΒΖΓ, τοῦ ἀπὸ τῆς ΕΖ.
ἐπεὶ οὖν δύο δοθέντα ἐστὶν τὰ ΕΓ¹⁰, καὶ ὀρθὴ ἡ ΒΖ, καὶ λόγος ἐστὶν
τοῦ ἀπὸ ΕΖ πρὸς τὰ ἀπὸ τῶν ΒΖΓ, τὸ Β ἄρα πρὸς ὑπερβολῇ¹¹. ἀλλὰ
καὶ [τὰ] πρὸς θέσει περιφερείᾳ¹²· δοθὲν ἄρα τὸ Β. καὶ ἡ σύνθεσις
φανερά.

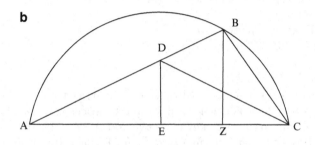

Prop. 35
#45 τὸ μὲν οὖν τὴν δοθεῖσαν γωνίαν
ἢ περιφέρειαν τρίχα τεμεῖν

---

¹οὐδὲν Hu
²μέρος ἡ ΒΓ del. Co, Hu, Tr
³διπλάσιον A corr. Hu, Tr
⁴ὑπὸ add. Hu, Tr
⁵ΑΙΒ A corr. S Hu, Tr
⁶ΔΕ ΖΒ BS Hu, Tr
⁷τῆς ΔΓ A corr. Hu, Tr
⁸τῇ ΑΕ A Hu τῆς Tr
⁹ἡ Β Α ΒΓ Co, Hu, Tr
¹⁰ΕΓ ΑΒ1S distinx. Β3 (vel Β2) Hu, Tr
¹¹προσυπερβολη A πρὸς ὑπερβολῇ S Hu, Tr
¹²καὶ τὰ προσθέσει περιφερείαις ABS corr. Hu, Tr

στερεόν ἐστιν, ὡς προδέδεικται·
τὸ δὲ τὴν δοθεῖσαν γωνίαν ἢ
περιφέρειαν εἰς τὸν δοθέντα λόγον τεμεῖν γραμμικόν ἐστιν, καὶ
δέδεικται μὲν ὑπὸ τῶν νεωτέρων, γραφήσεται δὲ καὶ ὑφ" ἡμῶν δι-
χῶς. ἔστω γὰρ κύκλου τοῦ ΚΛΘ περιφέρεια ἡ ΛΘ, καὶ δέον ἔστω τεμεῖν
αὐτὴν εἰς δοθέντα λόγον. ἐπὶ τὸ κέντρον αἱ ΛΒΘ, καὶ τῇ ΒΘ
πρὸς ὀρθὰς ἡ ΒΚ, καὶ διὰ τοῦ Κ γεγράφθω τετραγωνίζουσα γραμμὴ ἡ ΚΑ ΔΓ[1], καὶ κάθετος ἀχθεῖσα ἡ ΑΕ τετμήσθω κατὰ τὸ Ζ, ὥστε
εἶναι ὡς τὴν ΑΖ πρὸς ΖΕ, οὕτως τὸν δοθέντα λόγον εἰς ὃν διελεῖν θέ-
λομεν τὴν γωνίαν, καὶ τῇ μὲν ΒΓ παράλληλος ἡ ΖΔ. ἐπεζεύχθω δὲ
ἡ ΒΔ, καὶ κάθετος ἡ ΔΗ. ἐπεὶ οὖν διὰ τὸ σύμπτωμα τῆς γραμμῆς ἐστιν
ὡς ἡ ΑΕ πρὸς ΔΗ, τουτέστιν πρὸς ΖΕ, ἡ ὑπὸ ΑΒΓ γωνία πρὸς τὴν ὑπὸ
ΔΒΓ, διελόντι ἄρα ἐστὶν ὡς ἡ ΑΖ πρὸς ΖΕ, τουτέστιν ὡς ὁ δοθεὶς λόγος,
οὕτως ἡ ὑπὸ[2] ΑΒΔ γωνία πρὸς τὴν[3] ὑπὸ ΔΒΓ, τουτέστιν ἡ ΛΜ περιφέρεια
πρὸς ΜΘ.

#46 ἑτέρως δὲ τέμνεται < κύκλου >[4] τοῦ ΑΗΓ ἡ ΑΓ
περιφέρεια. ὁμοίως ἐπὶ τὸ κέν-
τρον αἱ ΑΒΓ, καὶ γεγράφθω διὰ
τοῦ Β ἡ ἕλιξ ἡ ΒΖ ΔΓ[5] ἧς ἡ ἐν τῇ

f. 53v (Props. 35, 36, and 37)
γενέσει εὐθεῖα ἡ ΓΒ, καὶ τῷ δοθέντι λόγῳ ὁ αὐτὸς ἔστω[6] ὁ τῆς ΔΕ πρὸς
ΕΒ, καὶ διὰ τοῦ Ε περὶ κέντρον τὸ Β κύκλου περιφέρεια ἡ ΕΖ τέμνουσα τὴν ἕλικα κατὰ τὸ Ζ, καὶ ἐπιζευχθεῖσα ἡ ΒΖ ἐκβεβλήσθω ἐπὶ τὸ Η· ἔστιν ἄρα διὰ τὴν ἕλικα ὡς ἡ ΔΒ[7] πρὸς ΒΖ, τουτέστιν
πρὸς ΒΕ, οὕτως ἡ ΑΗΓ[8] περιφέρεια πρὸς ΓΗ, καὶ διελόντι ὡς ἡ ΔΕ πρὸς
ΕΒ, οὕτως ἡ ΑΗ περιφέρεια πρὸς ΗΓ. ὁ δὲ τῆς ΔΕ πρὸς ΕΒ λόγος
ἐστὶν ὁ αὐτὸς τῷ δοθέντι· καὶ ὁ τῆς ΑΗ ἄρα περιφερείας πρὸς τὴν
ΗΓ λόγος ὁ αὐτός ἐστιν τῷ δοθέντι. τέτμηται ἄρα. ※

---

[1] coniunx. BS Hu, Tr
[2] ἀπὸ Hu
[3] πρὸς τῇ ABS corr. Hu, Tr
[4] κύκλου add. Hu
[5] διὰ τοῦ Β ἕλιξ ἡ ΒΖΔΓ Hu
[6] ἔσται AB corr. S Hu, Tr
[7] AB A corr. S Hu, Tr
[8] ΑΓ A corr. Co, Hu, Tr

a

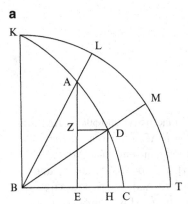

b

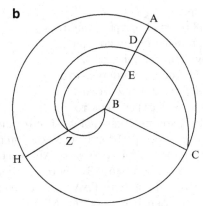

Prop. 36

#47 ἐκ δὴ τούτου φανερὸν ὡς δυνατόν
ἐστιν ἀπὸ δύο κύκλων ἀνίσων
ἴσας[1] περιφερείας ἀφελεῖν. γεγονέτω γάρ, καὶ ἀφηρήσθωσαν ἴσαι
αἱ ΑΗΒ ΓΘΔ, ἔστω δὲ μείζων ὁ περὶ
κέντρον τὸ Ε· μείζων ἄρα ἡ ὁμοία < τῇ >[2] ΓΘΔ
τῆς ΑΗΒ. ἔστω οὖν τῇ[3] ΑΗΒ ὁμοία ἡ
ΓΘ· λόγος ἄρα ὁ τῆς ΑΗΒ πρὸς ΓΘ < δοθείς >[4]· ὁ γὰρ αὐτός ἐστιν ταῖς
ὅλαις τῶν κύκλων
περιφερείαις ἢ ταῖς τῶν κύκλων διαμέτροις. ἴση δὲ ἡ ΑΗΒ τῇ ΓΘΔ·
λόγος ἄρα δοθεὶς καὶ τῆς ΓΘΔ πρὸς τὴν ΓΘ. καὶ διελόντι γέγονεν
οὖν τέμνειν τὴν ΓΘΔ περιφέρειαν εἰς δοθέντα λόγον κατὰ τὸ Θ·
τοῦτο δὲ προγέγραπται.

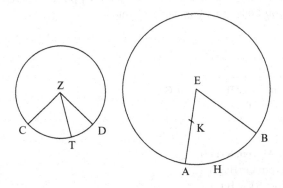

---

[1] ἴ- σας Α

[2] τῇ add. Hu, Tr

[3] συντῇ Α σὺν τῇ Β τῇ S Co οὖν τῇ Hu, Tr

[4] δοθείς add. Hu, Tr

Prop. 37

#48 ἰσοσκελὲς τρίγωνον συσ-
τήσασθαι[1] ἔχον ἑκατέραν
τῶν πρὸς τῇ βάσει γωνι-
ῶν λόγον ἔχουσαν δοθέντα
πρὸς τὴν λοιπήν. γεγο-
νέτω, καὶ συνεστάτω τὸ
ΑΒΓ[2], καὶ περὶ κέντρον τὸ Β διὰ τῶν ΑΓ[3] κύκλος γεγράφθω ὁ ΑΔΓ, καὶ
ἐκβεβλήσθω ἡ ΑΒ[4] ἐπὶ τὸ Δ, καὶ ἐπεζεύχθω ἡ ΔΓ. ἐπεὶ οὖν λόγος
ἐστὶν δοθεὶς τῆς ὑπὸ τῶν ΓΑΒ[5] γωνίας πρὸς τὴν ὑπὸ τῶν ΑΒΓ[6].
καὶ ἔστιν τῆς ὑπὸ ΑΒΓ[7] ἡμίσεια [ἡμίσεια][8] ἡ πρὸς τῷ Δ, λόγος ἄρα
δοθεὶς καὶ τῆς ὑπὸ ΓΑΔ γωνίας πρὸς τὴν ὑπὸ ΑΔΓ, ὥστε καὶ τῆς
ΔΓ περιφερείας πρὸς τὴν ΑΓ λόγος. ἐπεὶ οὖν ἡ ΑΓΔ περιφέρεια
τοῦ ἡμικυκλίου εἰς δοθέντα λόγον τέτμηται, δοθέν ἐστιν τὸ Γ,

f. 54 (Props. 37, 38, 39, and 40)
καὶ δοθὲν τῷ εἴδει τὸ ΑΒΓ τρίγωνον. συντεθήσεται δὲ οὕτως. ἔστω γὰρ ὁ
δοθεὶς λόγος, ὃν ἔδει ἔχειν ἑκατέραν τῶν πρὸς τῇ βάσει γωνιῶν πρὸς τὴν
λοιπήν, ὁ τῆς ΕΖ πρὸς ΖΗ, καὶ τετμήσθω ἡ ΖΗ δίχα τῷ[9] Θ, καὶ ἐκκείσθω κύκλος ὁ ΑΔΓ περὶ κέντρον τὸ Β καὶ διάμετρον τὴν ΑΔ, καὶ
τετμήσθω ἡ ΑΓΔ περιφέρεια κατὰ τὸ Γ, ὥστε εἶναι ὡς τὴν ΔΓ περιφέρειαν πρὸς τὴν ΓΑ, οὕτως τὴν ΕΖ πρὸς ΖΘ, τοῦτο γὰρ προγέγραπται,
καὶ καθόλου πῶς ἡ δοθεῖσα περιφέρεια εἰς δοθέντα λόγον τέμνεται, καὶ ἐπεζεύχθωσαν αἱ ΒΓ ΓΑ ΓΔ. ἐπεὶ οὖν ἐστιν ὡς ἡ ΔΓ περιφέρεια πρὸς τὴν ΓΑ, τουτέστιν ὡς < ἡ >[10] ὑπὸ ΔΑΓ γωνία πρὸς τὴν ὑπὸ ΑΔΓ,
οὕτως ἡ ΕΖ < πρὸς ΖΘ >[11], καὶ τὰ διπλάσια τῶν ἑπομένων, ὡς ἄρα ἡ ὑπὸ ΓΑΒ
πρὸς τὴν ὑπὸ ΑΒΓ[12], οὕτως ἡ ΕΖ πρὸς ΖΗ. ἰσοσκελὲς ἄρα τὸ[13] τρίγωνον
συνέσταται τὸ ΑΒΓ ἔχον ἑκατέραν τῶν πρὸς τῇ βάσει γωνιῶν λόγον ἔχουσαν τὸν δοθέντα πρὸς τὴν λοιπήν.

---

[1] συστήσας ΑΒ corr. S Hu, Tr συστῆσαι Co
[2] ὁ ΑΒΓ Α corr. BS Hu, Tr
[3] distinx. BS Hu, Tr
[4] ΑΔ ΑΒ corr. S Hu, Tr
[5] τῆς ὑπὸ τῶν ΑΒ ΑΒ corr. S Hu, Tr
[6] ὑπὸ τὴν ΑΒΓ Α corr. Co, Hu, Tr
[7] τῆς ὑπὸ ΑΒ Α corr. S Co, Hu, Tr
[8] bis scriptum del. Hu, Tr
[9] τω Α corr. Hu, Tr
[10] ἡ add. Β Hu, Tr
[11] πρὸς ΖΘ add. S Hu, Tr
[12] πρὸς τὸ ὑπὸ ΑΒΓ Α corr. Hu, Tr
[13] τὸ τρίγωνον Α τὸ om. Hu

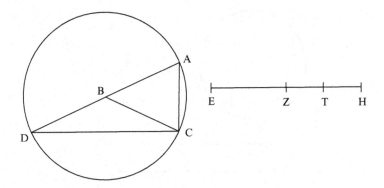

Prop. 38

#49 δεδειγμένου δὴ τούτου φανερὸν
ὡς δυνατὸν ἐγγράψαι πολύγω-
νον εἰς κύκλον ἰσόπλευρον καὶ
ἰσογώνιον πλευρὰς ἔχον ὅσας
ἄν τις ἐπιτάξῃ.

Prop. 39

πῶς δ' εὑρίσκεται
κύκλος οὗ ἡ περιφέρεια ἴση[1]
ἐστὶν τῇ δοθείσῃ εὐθείᾳ[2] συν-
ιδεῖν[3] εὔκολον. εὑρήσθω γὰρ τῇ Γ εὐθείᾳ ἴση ἡ τοῦ Α κύκλου περιφέρεια, καὶ ἐκκείσθω κύκλος τυχὼν ὁ Β, καὶ τῇ περιφερείᾳ αὐτοῦ
ἴση διὰ τῆς τετραγωνιζούσης[4] εὑρήσθω ἡ Δ εὐθεῖα. ἔστιν ἄρα ὡς ἡ
Γ πρὸς τὴν Δ, οὕτως ἡ ἐκ τοῦ κέντρου τοῦ Α κύκλου πρὸς τὴν ἐκ
τοῦ κέντρου τοῦ Β. λόγος δὲ τῆς Δ πρὸς Γ· λόγος ἄρα καὶ τῶν
ἐκ τοῦ κέντρου πρὸς ἀλλήλας. καὶ ἔστιν δοθεῖσα ἡ ἐκ τοῦ κέντρου
τοῦ Β· δοθεῖσα ἄρα καὶ ἡ ἐκ τοῦ κέντρου τοῦ Α, ὥστε καὶ αὐτὸς ὁ Α.
καὶ φανερὰ ἡ σύνθεσις.

---

[1] ι - ση Α
[2] εὐθεία Α corr. Hu, Tr
[3] συ - νειδειν Α corr. Hu, Tr
[4] τετραγονιζούσης Α corr. BS Hu, Tr

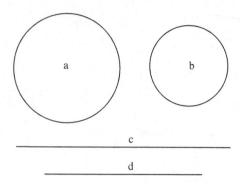

Prop. 40
#50 εὐθείας τῇ θέσει καὶ τῷ
μεγέθει δεδομένης τῆς
ΑΒ γράψαι¹ διὰ τῶν *ΑΒ²*
κύκλου περιφέρειαν
λόγον ἔχουσαν τὸν δοθέντα

f. 54v (Prop. 40 and 41)
πρὸς τὴν ΑΒ εὐθεῖαν. γέγραφθω ἡ ΑΓΒ, καὶ ἐκκείσθω τεταρτημόριον κύκλου θέσει δεδομένον τὸ ΖΗΕ³, καὶ γεγράφθω τετραγωνίζουσα
ἡ ΖΘΚ, καὶ τῇ βεβηκυίᾳ γωνίᾳ ἐπὶ τῆς ΑΓ περιφερείας πρὸς
τῇ λοιπῇ περιφερείᾳ⁴ ἴση συνεστάτω < ἡ >⁵ ὑπὸ ΕΗΛ, καὶ ἤχθωσαν
κάθετοι αἱ ΛΜ ΘΝ. ἔσται οὖν διὰ τὸ ἰδίωμα τῆς γραμμῆς ὡς ἡ
ΕΛΖ περιφέρεια πρὸς τὴν ΖΗ⁶ εὐθεῖαν, τουτέστιν ὡς ἡ ΛΗ πρὸς ΗΚ, οὕτως ἡ ΛΕ⁷ περιφέρεια πρὸς τὴν ΘΝ εὐθεῖαν. ἀλλὰ καὶ ὡς ἡ ΘΗ⁸
πρὸς τὴν ΗΛ, ἡ ΘΝ⁹ πρὸς ΛΜ· καὶ ὡς ἄρα ἡ ΘΗ πρὸς ΗΚ, οὕτως ἡ ΕΛ
περιφέρεια πρὸς τὴν ΛΜ εὐθεῖαν. εἰλήφθω δὴ τὸ κέντρον τῆς
ΑΓΒ¹⁰ περιφερείας τὸ Ξ, καὶ κάθετος ἐπὶ τὴν ΑΒ ἡ ΞΡΓ· ἴση ἄρα

¹*ΑΕ γράψαι* ΑS corr. Β Hu, Tr
²distinx. ΒS Hu, Tr
³*ΖΗΘ* Α1Β corr. Α2S Hu, Tr
⁴*τῇ λοιπῇ περιφερείᾳ* Α Tr *τῇ ΖΕ περιφερείᾳ* Hu *πρὸς τῇ λοιπῇ περιφερείᾳ* delenda esse
censuit Co, Hu
⁵*ἡ* add. ΒS Hu, Tr
⁶*ΖΝ* ΑΒS corr. Hu, Tr
⁷*ΛΒ* ΑΒS corr. Co, Hu, Tr
⁸*ΘΒ* ΑΒS corr. Co, Hu, Tr
⁹*ΘΚ* ΑΒS corr. Co, Hu, Tr
¹⁰*ΑΒΓ* ΑΒS corr. Co, Hu, Tr

ἡ ὑπὸ ΓΞΑ¹ γωνία τῇ ὑπὸ ΕΗΛ². καὶ ἔστι³ κέντρα τὰ ΞΗ⁴· ὡς ἄρα
ἡ ΑΓ περιφέρεια πρὸς τὴν ΑΡ ευθεῖαν, τουτέστιν ἡ ΘΗ πρὸς τὴν ΗΚ,
< οὕτως ἡ ΑΓΒ περιφέρεια πρὸς τὴν ΑΒ εὐθεῖαν >⁵
καὶ λόγος τῆς *ΑΓΒ*⁶ πρὸς τὴν ΑΒ· λόγος ἄρα καὶ ὁ τῆς ΘΗ
πρὸς ΗΚ. καὶ δοθεῖσα ἡ ΗΚ· δοθεῖσα ἄρα καὶ ἡ ΗΘ· πρὸς περιφερείᾳ
ἄρα τὸ Θ⁷. ἀλλὰ καὶ πρὸς τῇ ΖΘΚ γραμμῇ⁸. δοθὲν ἄρα τὸ Θ. θέσει
ἡ ΗΘΛ⁹· δοθεῖσα ἄρα ἡ ὑπὸ ΕΗΛ γωνία. καὶ ἔστιν ἴση τῇ ὑπὸ ΓΞΑ,
καὶ ἔστιν θέσει ἡ ΓΞ, καὶ δοθὲν τὸ Α· θέσει ἄρα ἡ ΑΞ, ὥστε καὶ ἡ ΑΓΒ¹⁰
περιφέρεια. καὶ ἡ σύνθεσις φανερά. δεῖ γὰρ τῷ δοθέντι λόγῳ
τὸν αὐτὸν ποιῆσαι¹¹ τὸν τῆς ΔΗ πρὸς ΗΚ¹², καὶ περὶ κέντρον τὸ Η
διὰ τοῦ Δ γράψαι περιφέρειαν, καὶ λαβεῖν τὸ Θ, καθ' ὃ τέμνει¹³ τὴν
τετραγωνίζουσαν, καὶ ἐπιζεῦξαι¹⁴ τὴν ΘΗ, καὶ δίχα τεμόντα¹⁵
τὴν ΑΒ καὶ ὀρθὴν ἀναστήσαντα τὴν ΡΞ *κατάγειν*¹⁶ τὴν ΑΞ περι-
έχουσαν μετὰ τῆς ΞΡ γωνίαν¹⁷ ἴσην τῇ ὑπὸ ΚΗΘ, καὶ περὶ κέντρον
τὸ Ξ διὰ τοῦ Α γράψαι κύκλου περιφέρειαν τὴν ΑΓΒ ἔχειν αὑτῆς¹⁸
λόγον πρὸς τὴν ΑΒ βάσιν τὸν αὐτὸν τῷ δοθέντι.

---

¹ *ΓΞΗ* AS corr. B Hu, Tr

² *τῇ ὑπὸ* /// A *τῇ ὑπὸ εηλ* B *τῇ ὑπὸ θηλ* S

³ *ἔστι* BS Tr scriptura evanida in A *ἔστιν* Hu

⁴ *κέντρα τὰ* Hu, Tr *Ξ Η* distinx. S Hu, Tr

⁵ *οὕτως...εὐθεῖαν* add. Co, Hu, Tr

⁶ *ΑΒΓ* A Hu corr. Tr

⁷ *ἡ ΖΚ* A corr. Co, Hu, Tr

⁸ *γραμμη* A corr. BS Hu, Tr

⁹ *ΘΛ* A corr. Co, Hu, Tr

¹⁰ *ΑΒΓ* A corr. Co, Hu, Tr

¹¹ *προῆκται* A2B *προῆχθαι* S constituere Co *ποιῆσαι* Hu, Tr

¹² *τῆς ΖΗ πρὸς τὴν ΗΔ* ABS corr. Co, Hu, Tr

¹³ *τεμνόντα* A1 *τέμνο* A2 *τέμνει* B Hu, Tr

¹⁴ *ἐπιζευχθῇ* A *ἐπιζεῦξαι* Hu, Tr

¹⁵ *τέμνονται* A corr. Hu, Tr

¹⁶ *κατάγειν* AS Tr *καταγεῖν* B *καταγαγεῖν* Hu

¹⁷ *γωνίας* ABS corr. Hu, Tr

¹⁸ *ἔχειν αὑτῆς* A Tr *ἔχουσαν* Hu

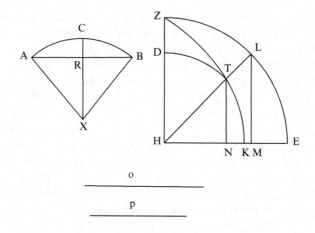

Prop. 41
#51 οὐκ ἀπίθανον δὲ οὐδὲ
τὸ γωνίας¹ ἀσυμμέτρους
εὑρεῖν, διὰ τούτου γὰρ καὶ
[διὰ]² τοῦ αὐτοῦ κύκλου
ἀσύμμετροι ληφθήσον-
ται περιφέρειαι, κἂν
ῥητὴν ὑποστησώμεθα τὴν μίαν γωνίαν ἢ περιφέρειαν, ἄλογος³
ἡ λοιπὴ γενήσεται. ἐκκείσθω τὸ ΑΒΓ τεταρτημόριον, καὶ ἡ

f. 55 (Props. 41, 42, and 43)
ἐν αὐτῷ⁴ τετραγωνίζουσα ἡ ΑΕΔΖ, καὶ διήχθω ἡ ΒΕ, καὶ τῇ ΒΓ
παράλληλος ἡ ΕΗ, καὶ ἀπειλήφθω ἡ ΒΘ ἀσύμμετρος μήκει τῇ ΒΗ,
καὶ ἤχθω παράλληλος ἡ ΔΘ, καὶ ἐπεζεύχθω ἡ ΔΒ· λέγω ὅτι ἀσύμμε-
τρός ἐστιν ἡ ὑπὸ⁵ ΕΒΖ γωνία τῇ ὑπὸ ΔΒΖ. ἤχθω κάθετος ἡ ΔΝ⁶·
ἔστιν ἄρα διὰ τὴν γραμμὴν ὡς ἡ ΕΚ πρὸς ΔΝ⁷, οὕτως ἡ ὑπὸ ΕΒΖ γωνία πρὸς τὴν ὑπὸ ΔΒΖ. ἀσύμμετρος δὲ ἡ ΕΚ⁸ τῇ ΔΝ, ἐπεὶ καὶ ἡ ΗΒ⁹
τῇ ΒΘ· ἀσύμμετρος ἄρα καὶ ἡ γωνία τῇ γωνίᾳ, κἂν ῥητὴν ὑπο-

¹ἀγωνίας AS ex γωνίας γωνίας Β τὸ γωνίας Hu
²διὰ del. Hu
³////// Α ἄλογος coni. Co, Hu, Tr
⁴τεταρτημόριον καὶ ἐν αὐτῇ Σ αὐτῷ corr. Co τεταρτημόριον καὶ ἐν αὐτῷ Ηu τεταρτημόριον καὶ ἡ ἐν αὐτῷ Tr
⁵ἀπὸ Hu
⁶ΔΗ ΑΒS corr. Co, Hu, Tr
⁷ὡς ΗΝ τῇ ΔΝ Α ὡς ἡ ΕΚ πρὸς ΔΝ Co, Hu
⁸ΕΓ Α corr. Co, Hu
⁹ΗΘ Α corr. Co, Hu

στησώμεθα τὴν ὑπὸ ΕΒΖ γωνίαν, [κἂν ἡμίσειαν ὀρθῆς¹], ἄλογος ἔσται
ἡ ὑπὸ ΔΒΖ.

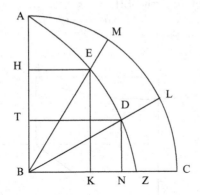

Prop. 42
#52 τῆς ὑπὸ Ἀρχιμήδους² ἐν τῷ
περὶ ἑλίκων βιβλίῳ λαμβα-
νομένης νεύσεως τὴν ἀνάλυ-
σίν σοι³ κατέταξα, ἵνα τὸ βιβλίον
διερχόμενος περὶ τῶν [περὶ
τῶν]⁴ ἑλίκων μὴ διαπορῇς.
λαμβάνονται δὲ εἰς αὐτὴν οἱ ὑπογεγραμμένοι τόποι καὶ πρὸς ἄλλα
πολλὰ τῶν στερεῶν προβλημάτων χρήσιμοι. θέσει εὐθεῖα
ἡ ΑΒ, καὶ ἀπὸ δοθέντος σημείου τοῦ Γ προσπιπτέτω τις ἡ ΓΔ⁵, καὶ
πρὸς ὀρθὰς τῇ ΑΒ ἡ ΔΕ. ἔστω δὲ λόγος τῆς ΓΔ⁶ πρὸς ΔΕ· ὅτι τὸ Ε
πρὸς ὑπερβολήν⁷. ἤχθω διὰ τοῦ Γ τῇ ΔΕ παράλληλος < ἡ ΓΖΗ >⁸·
δοθὲν ἄρα τὸ Ζ. καὶ τῇ ΑΒ παράλληλος ἡ ΕΗ, καὶ τῷ τῆς ΓΔ⁹
πρὸς ΔΕ λόγῳ¹⁰ ὁ αὐτὸς ἔστω τῆς ΓΖ¹¹ πρὸς ἑκατέραν τῶν ΖΘ ΖΚ.
δοθὲν ἄρα ἑκάτερον τῶν ΘΚ¹². ἐπεὶ οὖν ἐστιν ὡς τὸ ἀπὸ τῆς ΓΔ

---

¹κἂν ἡμίσειαν ὀρθῆς del. Hu, Tr

²ἀρχιμήδους A corr. Hu, Tr

³οὐ ABS corr. Hu, Tr

⁴bis scripta del. Hu, Tr

⁵ΓΒ A corr. Co, Hu, Tr

⁶ΓΛ A corr. Hu, Tr

⁷πρὸς ὑπερβολῇ Hu, Tr

⁸τῇ πρὸς ΔΕ παράλληλος A τῇ πρὸς ὀρθὰς παράλληλος ἡ ΓΖ Hu πρὸς del., ἡ ΓΖΗ add.
Tr

⁹ΓΑ A corr. Hu

¹⁰λόγος A Tr λόγῳ Hu

¹¹ΓΞ A corr. Hu

¹²distinx. BS Hu, Tr

< πρὸς τὸ ἀπὸ τῆς ΔΕ, οὕτως > [1]
τὸ ἀπὸ τῆς ΓΖ πρὸς τὸ ἀπὸ τῆς ΖΘ, καὶ λοιποῦ ἄρα τοῦ ἀπὸ
τῆς ΖΔ, τουτέστιν τοῦ ἀπὸ τῆς ΕΗ[2], πρὸς λοιπὸν τὸ ὑπὸ τῶν ΚΗΘ
λόγος ἐστὶν δοθείς. καὶ ἔστι δοθέντα τὰ *ΚΘ*[3]· τὸ Ε[4] ἄρα πρὸς ὑπερβολῇ ἐρχομένῃ[5] διὰ τῶν Θ Ε[6].

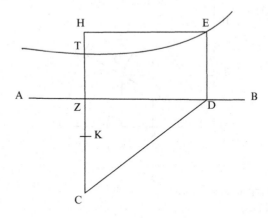

Prop. 43
#53 ἔστω θέσει καὶ μεγέθει δοθεῖσα ἡ
ΑΒ, καὶ πρὸς ὀρθὰς ἡ ΔΓ[7]· ἔστω δὲ
τὸ ὑπὸ τῶν ΑΓΒ *ἴσον* τῷ ὑπὸ
δοθείσης καὶ τῆς ΓΔ· ὅτι τὸ Δ
σημεῖον ἅπτεται θέσει παραβολῆς.
τετμήσθω γὰρ ἡ ΑΒ δίχα τῷ Ε,

f. 55v (Props. 43 and 44)
καὶ πρὸς ὀρθὰς ἡ ΕΖ, καὶ τῷ ἀπὸ τῆς ΕΒ *ἴσον* ἔστω τὸ ὑπὸ τῆς
δοθείσης καὶ τῆς ΕΖ· δοθὲν ἄρα τὸ Ζ. καὶ τῇ ΑΒ παράλληλος
ἡ ΔΗ[8]· λοιπὸν ἄρα τὸ ἀπὸ τῆς ΕΓ, τουτέστιν τὸ ἀπὸ τῆς ΔΗ, *ἴσον* ἐστὶ
τῷ ὑπὸ τῆς δοθείσης καὶ τῆς ΖΗ. καὶ ἔστι δοθὲν τὸ Ζ· τὸ ἄρα Δ
σημεῖον ἅπτεται παραβολῆς ἐρχομένης[9] διὰ τῶν Α Ζ Β[10], ἧς ἄξων
ἐστὶν ὁ ΕΖ.

---

[1] πρὸς...οὕτως add. Hu, Tr

[2] *ΒΗ* A corr. Co, Hu, Tr

[3] distinx. BS Hu, Tr

[4] *τὰ Θ* A corr. Co, Hu, Tr

[5] πρὸς ὑπερβολῃ ἐρχομένης A πρὸς ὑπερβολὴν ἐρχομένης S πρὸς ὑπερβολῇ ἐρχομένη Hu Tr

[6] διὰ τοῦ Θν A corr. Co, Hu διὰ τοῦ Θ[ν] Tr

[7] *ΔΖ* A corr. Co, Hu, Tr

[8] *ΔΓ* A *ΔΗ* Co, Hu

[9] κατερχομένης Hu

[10] διὰ τῆς *ΑΖΒ* A corr. Hu

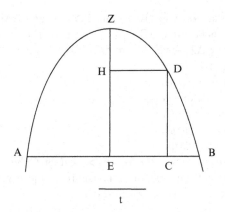

Prop. 44

#54 τούτων προγεγραμμένων < ἡ > προκειμένη
< νεῦσις ἀναλύεται > προγινομένη¹ τὸν
τρόπον τοῦτον. θέσει ὄντος κύκλου τοῦ²
ΑΒΓ, καὶ θέσει ἐν αὐτῷ εὐθείας τῆς
ΒΓ³, καὶ δοθέντος⁴ ἐπὶ τῆς περιφε-
ρείας τοῦ Α, θεῖναι μεταξὺ τῆς τε⁵ ΒΓ εὐθείας καὶ τῆς ΒΕΓ περιφερείας
εὐθεῖαν
ἴσην τῇ τεθείσῃ νεύουσαν ⁶ πρὸς τὸ Α⁷. γεγονέτω γάρ, καὶ κείσθω τῇ
ΕΔ ἴση, καὶ τῇ ΒΓ πρὸς ὀρθὰς ἤχθω ἡ ΔΖ ἴση τῇ ΑΔ. ἐπεὶ οὖν
πρὸς θέσει τὴν ΒΓ⁸ ἀπὸ δοθέντος τοῦ Α⁹ προσβέβληται¹⁰ ἡ ΑΔ, καὶ
ἴση¹¹ τῇ πρὸς ὀρθὰς ἐφέστηκεν [ἡ ἀπὸ]¹², τὸ < Ζ σημεῖον ἄρα πρὸς > ¹³

---

¹ ἡ et νεῦσις ἀναλύεται add. Tr προγεγραμμένων προκειμένη..... (lacuna) προγενομένη Hu
ἡ προκειμένη ἀνάλυσις δείκνυται γινομένη Hu/Baltzer

² θέσει ὄντος κύκλου τοῦ restit. S

³ ΕΓ ABS corr. Hu, Tr

⁴ δοθέντος Hu pro δοθὲν ὡς

⁵ τε om. Hu

⁶ τῆς ΒΖΓ περιφέρειαν ἴσην τῇ τε θείσῃ νεύουσαν Α τῆς ΒΖΓ περιφέρειαν ἴσην τῇ θέσει
νεύουσαν Β τῆς ΒΖΓ περιφερείας ἴσην τῇ τεθείσῃ Hu τῆς ΒΕΓ περιφερείας ἴσην τῇ ΗΘ
δοθείσῃ νεύουσαν Hu/Baltzer τῆς ΒΕΓ περιφερείας εὐθεῖαν ἴσην τῇ τεθείσῃ νεύουσαν
Tr

⁷ πρὸς τὸ Γ Α Hu πρὸς τὸ Α Hu/Baltzer, Tr

⁸ τῇ ΒΓ ABS Tr τὴν ΒΓ Hu

⁹ τοῦ Δ ABS τοῦ Α Hu, Tr

¹⁰ πρ///ληται Α προβέβληται S restit. Hu, Tr

¹¹ ἴση/ Α ἴσην BS Tr ἴση Hu

¹² ἡ ἀπὸ Α ἡ ΔΖ coni. Hu ἡ ἀπὸ τοῦ Δ Hu/Baltzer

¹³ Ζ...πρὸς add. Tr τὸ Ζ ἄρα ἔστιν πρὸς add. Hu/Baltzer

ὑπερβολῇ. <καὶ >[1] ἐπεὶ ἴσον ἐστὶν τὸ ὑπὸ ΒΔΓ τῷ ὑπὸ ΑΔΕ[2], τουτέστιν τῷ
ὑπὸ ΖΔΕ, καὶ ἔστιν δοθεῖσα ἡ ΔΕ[3], τὸ ἄρα ὑπὸ ΒΔΓ ἴσον ἐστὶν τῷ
ὑπὸ δοθείσης καὶ τῆς ΔΖ. <τὸ Ζ ἄρα πρὸς παραβολῇ >[4]. δοθὲν ἄρα τὸ
Ζ. < ἀναλύεται ἄρα. τούτῳ τῷ >[5]
προβλήματι χρῆται ὁ Ἀρχιμήδης[6] πρὸς τὸ δεῖξαι κύκλου περιφερείᾳ ἴσην εὐθεῖαν, αἰτιῶνται
δὲ αὐτοῦ τινες ὡς οὐ δεόντως
χρησαμένου[7] στερεῷ προβλή-
ματι, <δυνατὸν γάρ, ὡς ἀποδεικνύουσιν >[8] καὶ διὰ τῶν ἐπιπέδων
εὑρεῖν εὐθεῖαν ἴσην τῇ[9] τοῦ κύκλου περιφερείᾳ, χρησάμενον
τοῖς ἐπὶ τῆς ἕλικος εἰρημένοις θεωρήμασιν.

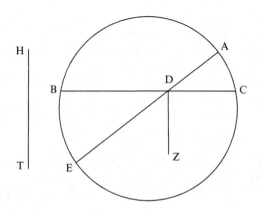

Πάππου συναγωγῆς Δ
ὅπερ ἐστὶν ἀνθηρῶν
θεωρημάτων ἐπιπέδων
καὶ στερεῶν καὶ γραμμικῶν.[10]

---

[1] καὶ add. Tr πάλιν add. Hu/Baltzer

[2] τὸ ὑπὸ ΒΑΓ τῷ ὑπὸ ΑΔΓ A τὸ ὑπὸ ΒΔΓ τῷ ὑπὸ ΑΔΕ Hu/Baltzer

[3] ΑΕ A corr. Hu

[4] το Ζ...παραβολῇ add. Hu, Tr

[5] ἀναλυ*** τῷ A ἀναλύεται ἄρα. τούτῳ τῷ Tr τούτῳ τῷ coni. Hu

[6] ἀρχιμήδης A corr. Hu, Tr

[7] χρησάμενον AB3S χρησάμενου B1 Hu, Tr

[8] προβλήματι***δεικνύουσιν A προβλήματι. δυνατὸν γάρ, ὡς ἀποδεικνύουσιν Tr προβλήματι *** δεικνύουσιν ὡς Hu

[9] εὑρεῖν ἔστιν εὐθεῖαν ἴσην τῇ Hu εὑρεῖν εὐθεῖαν ἴσην τῇ Tr

[10] πάππου·γραμμικῶν A τέλος τοῦ τῆς πάππου τοῦ ἀλεξανδρέως συναγωγῆς τετάρτου
ὅπερ ἐστὶν etc. B τοῦ δ¹ βιβλίου τέλος S

Part Ib
Annotated Translation of *Collectio* IV

Props. 1–3: Euclidean Plane Geometry: Synthetic Style

Prop. 1: Generalization of the Pythagorean Theorem

<u>#</u>1 When ABC is a triangle, and over AB and BC any parallelograms ABED and BCZH[1] are described, and DE and ZH are produced to T, and TB is joined, then the parallelograms ABED and BCZH <taken together> turn out to be equal to the parallelogram comprised by AC/TB, with an angle <at A> that is equal to the sum of the angles BAC and DTB.

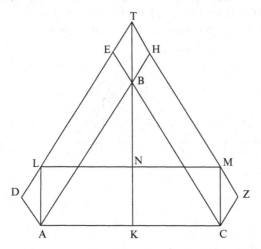

For:
    Produce TB to K, and through A and C draw the parallels AL and CM to TK, and join LM.

---

[1] The text of Prop. 1 shows a number of idiosyncrasies in labeling parallelograms. I have followed Hu in standardizing.

H. Sefrin-Weis, *Pappus of Alexandria: Book 4 of the Collection*,
Sources and Studies in the History of Mathematics and Physical Sciences,
DOI 10.1007/978-1-84996-005-2, © Springer-Verlag London Limited 2010

Since ALTB is a parallelogram,[1] AL and TB are equal and parallel. Similarly, MC and TB are both equal and parallel as well, so that LA and MC, also, are both equal and parallel. Therefore, LM and AC are both equal and parallel as well. Therefore, ALMC is a parallelogram with angle LAC, i.e.: with an angle that is the sum of angle BAC and angle DTB. For the angle DTB is equal to the angle LAB.[2]

And since the parallelogram DABE is equal to the parallelogram LABT (for they are both (erected) over the same base AB, and (lie) within the same parallels AB and DT[3]), but LABT is equal to LAKN (for they are both (erected) over the same base LA, and <lie> within the same parallels LA and TK[4]), ADEB is therefore equal to LAKN as well.

For the same reason, BHZC is equal to NKCM as well. Therefore, the parallelograms DABE and BHZC <taken together> are equal to LACM, i.e.: to the <parallelogram spanned by> AC/TB, with the angle LAC, which is equal to the sum of the angles BAC and BTD.

And this is much more general than what was proved in the Elements about right-angled <triangles> concerning the squares.[5]

Prop. 2: Construction of a Minor[6]

#2 <Let there be given> a semicircle over AB that has a *rational*[7] diameter, and let BC be on AB produced and equal to the radius, and CD a tangent <to the semicircle>, and let the arc BD be bisected in [the] point E, and CE joined.

<I claim> that[8] CE is an *irrational*,[9] the so-called *Minor*.

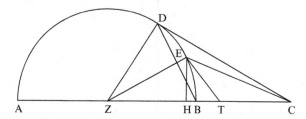

---

[1] By construction.

[2] I, 29.

[3] I, 35.

[4] I, 35.

[5] Reference to I, 47; Hu 178, 13 + app. Hu notes that a later manuscript has added a phrase that establishes a connection to VI, 31, which is not envisaged in Prop. 1; see the apparatus to the Greek text and the commentary.

[6] For the definition of a *"Minor"* see X, 76, for its classification see X, 82, and for its construction (used here in Prop. 2) see X, 94.

[7] For the meaning of *"rational"* see the commentary.

[8] The manuscript A has "οὐτῶς," Hultsch changes to the standard ὅτι (cf. also Co p. 58 A). Treweek follows him. The use of the differing conjunction is an idiosyncrasy of Props. 2–6. In the Greek text, I have kept the reading of A in all instances.

[9] For the meaning of the term *"irrational"* in this context see the commentary.

Take Z as the center of the circle, and join ZD and ZE. Since the angle ZDC is a right angle, it lies in the semicircle over ZC,[1] the center of which is B. And when BD is joined, the triangle BZD turns out to be equilateral, so that the angle DZB is two thirds <of a right angle>, and the angle EZB one third <of a right angle>.

Draw the perpendicular HE from E onto the diameter AB. Then the triangle CZD is equiangular to the triangle EZH, and EZ is to ZH as ZC is to CD.[2] However, the square over ZC is 4/3 of the square over CD.[3] Therefore, the square over EZ is 4/3 of the square over ZH, also.[4] Therefore, the ratio of the square over EZ to the square over ZH is the one that 16 <has> to 12, whereas the ratio of the square over ZC to the square over EZ is the one that 64 has to 16,[5] and therefore the ratio of the square over ZC to the square over ZH is the one that 64 has to 12.

However, let ZB be 4BT.[6] ZC is 2BZ, also. Therefore, the ratio of ZC to ZT is the one that 8 has to 5, and the ratio of ZT to TC is the one that 5 has to 3. Therefore, the ratio of the square over ZC to the square over ZT is the one that 64 has to 25, also.[7]

It has, however, been shown that the ratio of the square over CZ to the square over ZH is the one that 64 has to 12. Therefore, the ratio of the square over TZ to the square over ZH is as 25 to 12, also.[8] Therefore, TZ and ZH are *rationals*, commensurable in square only, and TZ in square exceeds ZH in square by a square whose side is incommensurable with it.[9] And the whole <line> ZT is commensurable with the *rational* <line> AB[10]. Therefore, TH is a *fourth Apotome*.[11]

However, ZC is *rational*, and its double is so, also. Therefore, the line the square of which is two times the rectangle ZC/HT is an *irrational*, the so-called *Minor*.[12] And the square of CE is double the rectangle CZ/HT[13]; therefore, CE is a *Minor*.

That, however, the square of CE is two times the rectangle CZ/HT will be clear in the following way: Join ET. Since the square over EC is equal to the <sum of the> squares over ET and TC, plus two times the rectangle CTH,[14] whereas the <sum of the> squares over ET and TZ is equal to the square over EZ plus two times

---

[1] III, 31.

[2] VI, 4.

[3] I, 47: $CD^2 = 3ZB^2$, and $ZC^2 = 4ZB^2$.

[4] VI, 23.

[5] ZC = 2ZB, ZE = ZB.

[6] Choose T on ZB, Z – B – T, with TB = 1/4ZB.

[7] ZT = 3BT; ZC = 2ZB = 8BT; TC = 5BT.

[8] V, 23 with V, 16.

[9] $ZT^2:(ZT^2 - ZH^2) = 25:13$; X, 9 with X, 5/6.

[10] ZT = 3BT; AB = 2ZB = 8BT; X, 9.

[11] X, 73; X, 84 a 4. The *Apotome* is introduced in X, 73, divided into subtypes in X, 84, with geometrical constructions in X, 85–90.

[12] X, 94.

[13] This will be shown below.

[14] II, 12.

the rectangle ZTH, also,[1] the <sum of the> squares over CE, ET, and TZ is therefore
equal to the <sum of the> squares over ET, TC, and EZ, plus two times the rectangle
CTH, together with two times the rectangle ZTH, i.e.: with two times the rectangle
CZ/HT. Take the common square over ET away. Then the remaining <sum of the>
squares over EC and ZT is equal to the <sum of the> squares over EZ and TC,
together with two times the rectangle CZ/HT. Of these, the square over ZT is equal
to <the sum of> the squares over EZ and TC (for the square over ZT is 25,[2] whereas
the square over TC is 9, and the square over EZ is 16). Therefore, the remaining
square over CE is equal to two times the rectangle ZC/HT.

Prop. 3: Construction of an Irrational Beyond Euclid

#3 <Let there be given> a semicircle over AC that has a rational diameter, and let
CD be equal to the radius, and DB tangent <to the semicircle>, and let the angle[3]
CDB be bisected by DZ.

<I claim> that DZ is the excess by which a *Binomial*[4] exceeds a *Line that
produces with a rational area a medial whole.*[5]

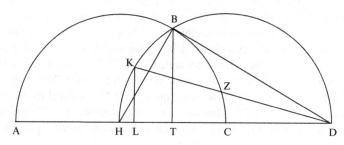

---

[1] II, 13 generalized. The proof and theorem of II, 13 in the *Elements* specifies acute-angled triangles,
but it can easily be extended (cf. Heath 1926 I, pp. 406–409). Within the present translation and
commentary, I will refer to "II,12/13 generalized," assuming that Pappus expects familiarity with
the generalized versions. A theorem much like II, 12/13 generalized seems to be invoked *inter alia*
in Prop. 7 and Prop. 8. The following piece of text is bracketed by Hultsch as a later addition:
Therefore, proportion holds. As the square over CE is to the (sum of the) squares over ET and TC,
together with two times the rectangle, so is the sum of the squares over ET and TZ to the square
over EZ, together with two times the rectangle ZTH (in A:ZHΘ). And as one to one, so are all (to
all, add. Hu). And the square over CE is equal to the sum of the squares over ET and TC, plus two
times the rectangle CTH (in A:ΓΕΘΗ).

[2] I.e., 25 BH². BH² appears as a unit of measure. The areas of squares are directly identified with
numbers. This is unusual.

[3] Note the connection to Prop. 2. There the arc between point of touch and base was bisected, here
it is the angle between the tangent and the base.

[4] The *Binomial* is introduced as a sum in X, 36, shown to be uniquely determined this way in X, 42,
split up into six types in X, 47–53, and geometrically constructed and characterized in X, 54.

[5] The *Line that produces with a rational area a medial whole* is introduced as a difference of lines
in X, 77, the uniqueness of this determination is proved in X, 83, and the line is constructed and
characterized geometrically in X, 95.

For:

Take H as the center of the semicircle, and join BH, and describe over HD the semicircle HBD,[1] and produce DZ to K. Then the arc BK is equal to the arc KH. Draw the perpendicular KL onto AC. And since BH is the side of a hexagon,[2] whereas KL is half of the side of a hexagon (for when it is produced, it subtends two times the arc KH), BH is therefore two times KL, i.e.: CK is two times KL. And the angle KLC is a right angle. Therefore, the square over KC is 4/3 of the square over CL,[3] i.e.: the square over DC is 4/3 of the square over CL. Therefore, DC and CL are *rationals*, commensurable in square only,[4] and the square of DC exceeds the square of CL by the square over a line that is commensurable with it,[5] and the larger <line> DC is commensurable with the *Rational* AC. Therefore, LD is a First *Binomial*,[6] whereas HD is *rational*. Therefore, the line the square of which is equal to the area of the rectangle between HD/DL is an *irrational*, the so-called *Binomial*.[7] However, the square of DK is equal to this <area> (for on account of the fact that the triangle HDK is equiangular to the triangle DLK,[8] KD is to DL as HD is to DK[9]). Therefore, DK is a *Binomial*.

And since the angle BHC is two thirds <of a right angle>, and HB is equal to HC, the triangle BHC is therefore equilateral. Now, draw the perpendicular BT; then HC, i.e.: DC, is two times CT.[10] And it has been shown that the square over DC is 4/3 of the square over CL. Therefore, the square over LC is three times the square over CT. Therefore, LC and CT are *rationals*, commensurable in square only, and the square of LC exceeds the square of CT by the square over a line that is incommensurable with it, and the smaller item CT is commensurable with the *Rational* AC.[11] Therefore, LT is a *fifth Apotome*.[12] And since the rectangle DHT is equal to the square over BH on account of the fact that the triangles BHT and BHD are equiangular,[13] whereas the rectangle DHL is equal to the square over KH, on account of the fact that the triangles KHL and KHD are equiangular,[14] the rectangle DHL is, therefore, to the square over KH as the rectangle DHT is to the square over BH.

---

[1] B on the semicircle, because ∠HBD = $\pi/2$.

[2] I.e., of a regular hexagon inscribed in the circle with diameter HD.

[3] See Prop. 2 for this intermediate step.

[4] X, 9.

[5] X, 9.

[6] X, 47 a 1.

[7] X, 54.

[8] VI, 8.

[9] VI, 4; VI, 17.

[10] Equilateral triangle HBC.

[11] X, 9.

[12] X, 84 a 5.

[13] VI, 4; VI, 17.

[14] VI, 4; VI, 17.

And alternate.[1] As the rectangle DHT is to the rectangle DHL, however, so is TH to HL.[2] And therefore, as HT is to HL, so is the square over BH, i.e.: the square over ZH, to the square over HK. Separando, therefore, as TL is to LH, so is the square over KZ to the square over HK.[3] And it has been shown that the rectangle between DH/HL is equal to the square over HK. Therefore, the rectangle between DH/LT is equal to the square over KZ, also. And LT is a *fifth Apotome*, whereas DH is *rational*. Therefore, KZ is a *Line that produces with a rational area a medial whole*.[4]

However, it has been shown also that DK is a *Binomial*. Therefore, (the remaining) DZ is the excess by which a *Binomial* exceeds a *Line that produces with a rational area a medial whole*.

Props. 4–6: Plane Analysis Within Euclidean Elementary Geometry

Prop. 4: Structure of Analysis-Synthesis

#4 Let ABC be a circle with center E and diameter BC, and AD a tangent intersecting BC in D, and let DZ[5] be drawn, and AE produced, after it has been joined, to H, and let ZKH and HLT be joined;
<I claim> that EK is equal to EL.

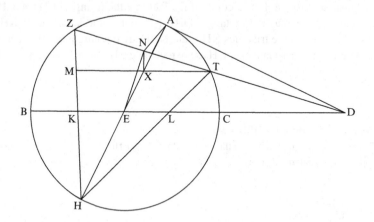

---

[1] V, 16; rectangle DHT:rectangle DHL = BH²:KH².
[2] The following phrase was bracketed by Hultsch; it translates to: For the height is equal.
[3] V, 17.
[4] X, 95.
[5] DZ secant to the circle, chosen at liberty between DC and DA.

Assume that it has turned out to be so,[1] and draw the parallel TXM to KL. Then MX is equal to XT, also.[2] Draw the perpendicular EN from E onto ZT. Then ZN is equal to NT.[3] However, MX was equal to XT, also. Therefore, NX is parallel to MZ. Thus,[4] the angle between TN/NX is equal to the angle between NZ/ZM,[5] i.e.: to the angle between TA/AX.[6] Thus, the points A, N, X, and T lie on a circle.[7] Thus, the angle between AN/NT is equal to the angle between AX/XT, i.e.: to the angle between AE/EL.[8] Thus, the points A, N, E, and D lie on a circle.[9]

They do, however <lie on a circle>. For the angles between EA/AD and EN/ND are both right angles.[10]

Now, the argument will be put together[11] in the following way. Since the angles between EA/AD and between EN/ND are both right angles, the points A, D, E, and N lie on a circle.[12] Therefore, the angle AND is equal to the angle AED. But the angle AED is equal to the angle AXT, on account of the parallels ED and XT.[13] Therefore, the points A, N, X, and T lie on a circle.[14] Therefore, the angle TAX is equal to the angle TNX. But the angle TAX is equal to the angle TZM.[15] Therefore, ZM is parallel to NX. And ZN is equal to NT.[16] Therefore, MX is equal to XT, also. And as XH to HE, so is, on the one hand, XM to EK, and, on the other hand, TX to LE; and therefore: as XM to EK, so <is> TX to LE. And <equation holds after> alternation.[17] MX is equal to XT as well. Therefore, KE is equal to LE, also.

---

[1] Analysis – assumption: EK = EL.

[2] ΔHKE ~ ΔHMX; ΔHEL ~ ΔHXT; VI, 4 and V, 16; KE = EL by assumption in the analysis.

[3] III, 3.

[4] οὕτως ἄρα; the occurrence of this phrase is a peculiarity of the analysis in Prop. 4. It will be translated as "thus."

[5] I, 29.

[6] III, 21.

[7] Converse of III, 21; see the commentary.

[8] III, 21; I, 29.

[9] ∠AND = ∠AED. The analysis proper ends here.

[10] III, 31. Note that this observation constitutes the *resolutio* for the analysis in Prop. 4. The position of E, N, A, and D on a circle is independent from the analysis-assumption.

[11] Technical term: συνθετήσεται. The synthesis begins here.

[12] III, 31; compare the above *resolutio*. See the commentary.

[13] By construction, I, 29.

[14] See the commentary. An appeal to the converse of III, 21 is not permissible in the synthesis.

[15] III, 21.

[16] III, 3.

[17] V, 16 yields MX:TX = EK:LE.

Props. 5/6: Reciprocity in Plane Geometry

Prop. 5

#5 Let ABC be a circle, and AD and DC tangents, and let AC be joined, and EZ drawn through the interior <of the angle>. EH, however, should be equal to HZ.
   <I claim> that TH is equal to HK, also.

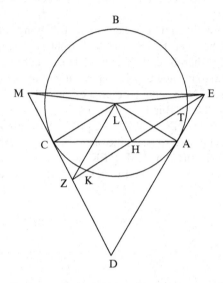

Draw the parallel EM to AC, and take L as the center of the circle, and join LA, LZ, LC, LM, LE, and LH. Since EH is equal to HZ, MC is equal to CZ, also.[1] And it is perpendicular to CL.[2] Therefore, LZ is equal to LM.[3] And since AD is equal to DC,[4] AE is equal to MC.[5] However, AL is equal to LC, also, and the right angle between EA/AL is equal to the right angle between MC/CL. Therefore, EL is equal to LM, also, i.e.: to LZ.[6] But EH is equal to HZ, also. Therefore, HL is a perpendicular onto EZ.[7] Therefore, TH is equal to HK.[8]

---

[1] ΔMZE ~ ΔCZH, VI, 2.

[2] III, 18.

[3] I, 4.

[4] Triangle LAC is isosceles; therefore, ∠LAC = ∠LCA (I, 5); therefore, ∠CAD = ∠ACD, and triangle ACD is isosceles (I, 6). Co refers to III, 36, with corollaries; cf. 191, * Hu.

[5] AC ‖ EM; therefore, ΔEDM is isosceles (I, 29; I, 6). Co refers to VI, 4; Hu's Latin paraphrase suggests using VI, 4, also.

[6] I, 4.

[7] ΔEHL ≅ ΔZHL; therefore, the neighboring angles at H are equal.

[8] III, 3.

Prop. 6

#6 Let ABC be a circle, and AD and DC tangents, and let AC be joined, and EZ be drawn through <the interior of the angle>; HT, however, should be equal to HK.

<I claim> that EH is equal to HZ, also.

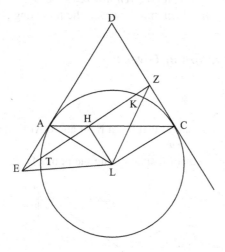

Take L as the center of the circle, and join EL, LA, LH, LZ, and LC. Since the angles between EA/AL and between EH/HL are both right angles, the points E, A, H, and L lie on a circle.[1] Therefore,[2] the angle between HA/AL is equal to the angle between HE/EL.[3] Again, since the angles between LH/HZ and between LC/CZ are both right angles, the points L, H, Z, and C lie on a circle.[4] Therefore, the angle between HC/CL, i.e.: the angle between HA/AL,[5] i.e.: the angle between HE/EL, is equal to the angle between HZ/ZL.[6] Therefore, EL is equal to LZ, also.[7] And LH is a perpendicular <onto EZ[8]>. Therefore, EH is equal to HZ.

---

[1] III, 18; III, 3.

[2] III, 31.

[3] III, 21.

[4] III, 31; compare above.

[5] Isosceles triangle ACL; I, 5.

[6] III, 21.

[7] Isosceles triangle ELZ; I, 5.

[8] III, 3.

Props. 7–10: Analysis, Apollonian Style (Focus: Resolutio)

Theorem (cf. Prop. 10)

Whenever there are three circles, *given*[1] in position and size, touching each other, the circle comprising them[2] will be *given* in size as well.

Before <discussing this theorem>, however, the following is written down.

Prop. 7: Determination of Givens

7a: First Example for Prop. 7

#7 Let ABCD be a quadrilateral that has a right angle ABC, and each of the lines AB, BC, CD, and DA are *given*.

<The task is> to show that the line joining the points D and B is *given*.

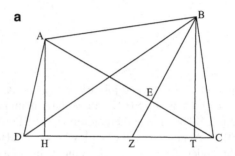

Join AC and draw the perpendiculars AH onto CD, on the one hand, and BE onto AC, on the other hand.[3] Now, since AB and BC are both *given*, and the angle ABC is a right angle, and BE is a perpendicular <onto AC>, each of the <lines> AE, EC, AC, and BE will therefore be *given*, also. For the rectangle ACE turns out to be *given*, because it is equal to the square over BC,[4] and AC is *given*,[5] so that each of the

---

[1] For information on the technical term "*given*" (Latin:datum, Greek:δοθέν) see the introduction to Prop. 7 in the commentary. Determining *givens* (data) is the central task of the *resolutio* stage of Greek geometrical analysis (see introduction to Props. 4–12). The terminology will also be employed in Props. 28 and 29, 35–41, 42–44, and 31–34. In the latter cases, Pappus is operating outside the scope of plane geometry, and the analysis serves very different functions.

[2] I.e., the circumscribed circle, touching each of the three given ones.

[3] E is on AC. Its position must be A – E – C because of the right angle at B. H is on DC. Its position D – H – C implies a special configuration for Prop. 7. See the commentary. Z is taken to lie on DC and BE.

[4] I.e.: it is *given* in size. *Elem. VI, 8, Porisma*, and VI, 17: $BC^2 = AC \times EC$. *Data 52*: with BC *given*, BC^2 is *given*.

[5] I, 47: $AC^2 = AB^2 + BC^2$, *Data 52*: AB, BC *given* ⇒ AB^2, BC^2 *given*; *Data 3*: $AB^2 + BC^2$ *given*, i.e., AC^2 *given*; *Data 55*: AC *given*.

<lines> AE, EC, and BE will be *given*.[1] Again, since each of the lines AC, CD, and DA is *given*, and AH is a perpendicular <onto DC>, each of the <lines> DH, HC, and AH is *given* as well. For the excess of the square over AC over the square over DA, when it is applied to CD, makes the excess of CD over HD a *given* one, as is stated in a lemma, so that each of the <lines> DH, HC, and AH will have been *given*, also.[2] And since the triangle AHC is equiangular to the triangle CEZ, as HC <is> to CE, so are both AC to CZ and AH to EZ.[3] And the ratio of HC to CE is *given*.[4] Therefore, both CZ and ZE will be *given*.[5] But so are both EB and BC. Therefore, each of the <lines> ZB, BC, and CZ is *given* as well.[6] Now, draw the perpendicular BT onto CZ.[7] Then each of the <lines> ZT, TC, and BT is *given*,[8] so that both DT and TB are *given*, also.[9] And the angle BTD is a right angle. Therefore, BD is *given*.[10]

7b: Second Example for Prop. 7

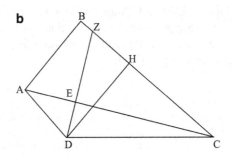

---

[1] *Data* 57: AC × EC *given*, AC *given* ⇒ **EC** *given*. Data 4: AC, EC *given* ⇒ **AC − EC**, i.e., **AE** *given*. I, 47: $BC^2 = BE^2 + EC^2$, i.e.: $BC^2 − EC^2 = BE^2$; *Data* 52: BC, EC *given* ⇒ BC^2, EC^2 *given*. *Data* 4: $BC^2 − EC^2$ *given*; *Data* 55: **BE** *given*.

[2] According to Hultsch, no such lemma is still extant. He provides a proof for "CD − HD is *given*" at Hu p. 193, # 4. II, 12/13 generalized: $AC^2 = AD^2 + DC^2 − 2DC × DH$; therefore: $AC^2 − AD^2 = DC × (DC − 2DH)$. *Data* 52: AC^2, AD^2 *given*; *Data* 4: $AC^2 − AD^2$ given, i.e., DC × (DC − 2DH) given; *Data* 57: DC, DC × (DC − 2DH) *given* ⇒ DC − 2DH *given*; *Data* 4: DC, DC − 2DH *given* ⇒ 2DH *given*; *Data* 2: **DH** *given*. Data 4: DC, DH *given* ⇒ **HC** *given*. I, 47: $AD^2 = AH^2 + DH^2$, i.e., $AH^2 = AD^2 − DH^2$; *Data* 52: AH^2, DH^2 *given*; *Data* 4: $AD^2 − DH^2$ *given*, i.e., AH^2 *given*; *Data* 55: **AH** *given*.

[3] VI, 4, V, 16.

[4] *Data* 1 (HC and CE are *given*).

[5] *Data* 2 (AC, AH, and HC:CE are *given*).

[6] *Data* 3 (ZB = ZE + BE).

[7] H is assumed to lie on DC, with D − Z − T − C. This means that, again, only one of several possible sub-cases is discussed here; see the commentary on the purpose of Prop. 7.

[8] ΔZBT ~ ΔZEC implies BZ:ZT = CZ:ZE; *Data* 2: BZ, BZ:ZT *given* implies **ZT** *given*. *Data* 4: **CZ − ZT = TC** *given*. I, 47: $BT^2 = BZ^2 − ZT^2$. *Data* 52, *Data* 4: BT^2 *given*; *Data* 55: **BT** *given*.

[9] *Data* 4: DC, CT *given* implies DT *given*. BT was shown to be *given* already.

[10] I, 47: $BT^2 + DT^2 = BD^2$; *Data* 52: BT^2, DT^2 *given*; *Data* 3: BD^2 *given*; *Data* 55: BD *given*.

#8 Draw the perpendicular DE onto AC and produce it to Z.[1]

Since each of the <lines> AD, DC, and CA is *given*,[2] and DE is a perpendicular, the <lines> AE and EC will both be *given*.[3] And since the triangle ABC is equiangular to the triangle CEZ, CB is to BA as CE <is> to EZ.[4] The ratio of CB to BA is, however, *given*. Therefore, the ratio of CE to EZ is *given* as well. And CE is *given*. Therefore, EZ is *given* as well.[5] However, DE was *given*, also.[6] Therefore, the whole DZ will be *given* as well.[7] For the same reason, BZ and ZC will both be *given*, also. For as AC <is> to BC, so <is> ZC to CE; and the ratio of AC to CB is *given*.[8] Now again, draw the perpendicular DH from D <onto BC[9]>. Then ZH and HC are both *given*,[10] so that BH and HD are both *given* as well.[11] And the angle at H is a right angle. Therefore, BD is *given* as well.[12]

Prop. 8: Analysis, Apollonian Style

(Apollonius, *Tangencies*, cf. *Coll.* VII, Props. 102–107)

#9 Let there be *given* <two> equal circles in position and size, with centers A and B, and let the point C be *given*,[13] and through C the circle CEZ touching the circles with centers A and B should be described.

---

[1] E is on AC, and Z is on BC. Again, the position A – E – C constitutes one of several possible cases. Furthermore, the argument will assume B– Z – H – C for the relative position of the intersection points of DE and BC, and the perpendicular. As remarked above, Pappus covers only two of a number of possible cases. The proof is completely analogous in all cases (cf. 195, * Hu). See the commentary on Prop. 7.

[2] As in Prop. 7a: I, 47: $AC^2 = AB^2 + BC^2$; with AB, BC *given*, *Data 52, Data 4*: AC^2 *given*; *Data 55*: AC *given*.

[3] As in 7a, for DH, HC, AH; II, 12/13, generalized: $AC^2 = AD^2 + DC^2 - 2CE \times AC$, i.e., $AC(AC + 2CE) = AD^2 + DC^2$; with AD, DC *given*, *Data 52, 3*: AC (AC + 2CE) *given*; AC, AC(AC + 2EC) *given* $\Rightarrow$ AC + 2EC *given* (*Data 57*); *Data 4, Data 2*: **EC** *given*. With *Data 4*: **AE** *given*.

[4] VI, 4.

[5] *Data 1*: CB:BA *given*, i.e., CE:EZ *given*. *Data 2*: CE, CE:CZ *given* $\Rightarrow$ EZ *given*.

[6] I, 47: $DE^2 = DC^2 - EC^2$; *Data 52, 4*: DE^2 *given*; *Data 55*: DE *given*.

[7] *Data 3* (DZ = DE + EZ, and DE, EZ are *given*).

[8] Apply VI, 4, for similar triangles ABC, CEZ:AC: BC = CZ:EZ; *Data 1*: AC:BC (thus: CZ:EZ) *given*; *Data 2*: **CZ** *given* (EZ, CZ:EZ are), *Data 4*: **BZ** *given* (BZ = BC – ZC).

[9] H on BC; the relative position B – Z – H – C constitutes one of several possible cases. See the commentary.

[10] II, 12/13, generalized: $ZC^2 = DZ^2 + DC^2 - 2DC \times ZH$; *Data 52, 4*: DC × (DC – 2ZH) *given*; *Data 57*: DC – 2ZH *given Data 4, 2*: **ZH** *given*; *Data 4*: ZC – ZH, i.e., **HC** *given*. Compare 7a for ∆ABC with lines DH, HC.

[11] *Data 4*: **BH** *given* (BC – HC), I, 47: $ZH^2 + HD^2 = DZ^2$, i.e., $HD^2 = DZ^2 - ZH^2$. *Data 52*: DC^2, HC^2 *given*, *Data 4*, HD^2 (= $DC^2 - HC^2$) *given*; *Data 55*: **HD** *given*; compare the argument for AH in 7a.

[12] I, 47, $BD^2 = BH^2 + DH^2$; with BH, HD given: *Data 52, Data 3*: BD^2 *given*; *Data 55*: BD *given*. Compare the last step in 7a.

[13] Co p. 66, B points out that the argument in Prop. 8 implies that CA, CB are *given* in position and size.

<I claim> that its diameter is *given*.[1]

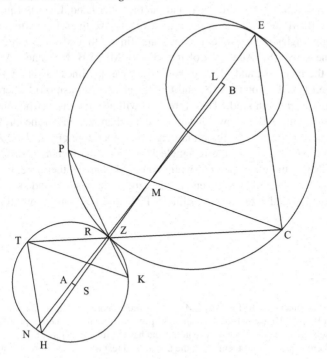

Join EZH, CZT, CMP, AB, CE, PZK, TK, and TH.[2]

Now, HT turns out to be parallel to CE on account of the fact that the vertex angles EZC and HZT are equal, and the arcs EPZ and HKZ similar,[3] and the triangle ECZ is equiangular to the triangle ZHT. For the same reasons TK is parallel to PC, also. And the circles with centers A and B are equal. Therefore, ZH is equal to DE.[4]

---

[1] Prop. 8 gives the *resolutio* for a special configuration in one of the cases treated in Apollonius, *Tangencies*. The *Tangencies* are lost, but Pappus' commentary on it can be found in *Coll.* VII. *Coll.* VII, Props. 102–107, are directly relevant for Prop. 8, as can be seen in the footnotes below. It is quite possible that Prop. 8 is in fact a (so far overlooked) testimony for a fragment from Apollonius' lost work. The connection of Prop. 8 to Apollonius' *Tangencies*, specifically the case of three touching circles, is noted also in Heath (1921, II, pp. 182–184).

[2] Extension of the configuration. Inconsistencies of labeling occur throughout Prop. 8 (compare 194/196 + app. Hu). Some of them probably go back to Pappus. For Prop. 8 shows clear signs of a not quite complete revision of a source text, after the insertion of additional material. See the commentary on this issue.

[3] For a proof, compare *Coll.* VII, Prop. 102, p. 826 Hu. (Jones 1986a, Vol. 1 p. 234, # 164). The proposition in *Coll.* VII is Pappus' commentary on Apollonius, *Tangencies*, I, 16. Also, compare Hultsch, p. 197, #2; Co p. 66/67, Lemma in E for a different explanation via similar arcs.

[4] For a proof, compare *Coll.* VII, 106, p. 833/834 Hu (Jones 1986a, Vol. 1. p. 238, # 169). The proposition in *Coll.* VII is a lemma by Pappus on Apollonius, *Tangencies* I, 17. Also, compare p. 197, #3 Hu and Co p. 67, G for a different explanation via similar arcs.

Draw the perpendiculars AS and BL <onto AB>. Then AS is equal to BL,[1] so that, on the one hand, BM is equal to MA, and, on the other hand, LM <is equal> to MS as well. For BLM and ASM are two triangles that have the same vertex angles <at M> and right angles at the points L and S, and finally they also have one side, BL, equal to one side, AS.[2] And each of the <lines> ML, LB, MS, and SA is *given*.[3] Therefore, the lines BM and MA are both *given*. But the lines AC and CB are both *given* as well (for the points A, B, and C <are *given*> in position).[4] Therefore, the triangle ABC is *given* in kind.[5] Therefore, CM will also be *given* (when the perpendicular from C onto AB is drawn).[6] And since the diameter NR of the circle HTK is *given*,[7] but MA is *given*, also, the remaining MR is therefore *given*, also.[8] And since the rectangle NMR is *given*, the rectangle HMZ, i.e.: the rectangle EMZ, i.e.: the rectangle CMP is therefore *given* as well.[9] And CM is *given*, therefore, CP is *given*, also.[10] Now, since the circle with center A is <*given*> in position and size, and CP is *given* in position and size, and the <lines> PZK and CZT are drawn through the

---

[1] III, 14.

[2] The explicit argument for BM = MA, LM = MS is much more elementary than the rest of the inferences in Prop. 8. Hultsch (196, 9–16 app.) suspects interpolation. Another possibility is that Pappus himself inserted this elementary material and has not fully integrated his resulting overall argument. There are further problems with the transmitted text and its line of reasoning (see 196, 17–198, 18 + app. Hu).

[3] Hultsch deletes the following here: "and in the same way both ZH and DE and BL and LS" (196, 18/19 + app. Hu). The phrase does not fit the context of the argument as given. Perhaps it is a leftover from a version of the text that was replaced by the suspected lines discussing BM, MA, LM, LS. Compare the preceding footnote. The implicit argument given for the status of ML, LB, MS, SA as *givens* – which the reader is perhaps meant to supply – shows strong affinities to Prop. 7. Compare p. 197/199, #4 Hu, including a reference to notes #2 and #3 on Prop. 7. A shorter route, avoiding the connection with Prop. 7, would have been to infer B, A *given* ⇒AB *given* (*Data* 26) ⇒ BM, MA *given* (*Data* 7).

[4] *Data* 26.

[5] *Data* 39. Indeed, the triangle is then *given* in position and size as well. The ensuing argument does not take advantage of these facts, and this may be yet another sign that Pappus has introduced material (from the *Data*, this time) into an argument that perhaps did not use the *Data*.

[6] Appeal to Prop. 7; compare p. 193, #3 Hu. Hultsch brackets the reference to the drawing of a perpendicular (thus, the reference to Prop. 7 is eliminated) and offers alternative arguments for "CM *given*" at 199, # 5 Hu. Evidently, Hultsch viewed the reference to Prop. 7 as something that is not of one piece with the main body of the argument in Prop. 8. We have yet another indication for Pappus' introduction and incomplete integration of material into Prop. 8.

[7] *Data*, def. 5: AR *given*; Then NR (its double) is *given*, also. The argument will use AR.

[8] *Data* 4; compare 199, # 6 Hu (covering NR, AR, MR, MN).

[9] III, 35; III, 36.

[10] *Data* 57. Note that CP is in fact *given* in position and size.

interior in such a way that KT is parallel to CP, the diameter of the circle <circum-scribed> around the triangle CZP is *given*[1] i.e.: the <diameter> of the circle CEZ.

Prop. 9: Lemma for Prop. 10

#10 Let ABC be a triangle that has each of its sides <as> *given*, and let D be an internal point, and let the difference of AD and CD be equal to the difference of CD and DB.[2] and let this difference[3] be *given*[4]

<I claim> that each of the <lines> AD, DC, and DB is *given*.

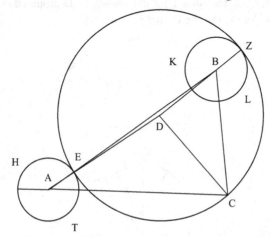

Since the difference of AD and DC is *given*, let AE and BZ both be equal to this difference.[5] Then the three <lines> ED, DC, and DZ, are equal to one another.[6] Describe the circle CEZ with center D. Now, on account of what has been

---

[1] That Z (and therefore all sides of the triangle CZP) is *given*, can be shown by *Coll.* VII, 105, pp. 830–831 Hu (Jones 1986a, Vol. 1, p. 236, # 168). That it is the point of touch for the sought circle with the circle around A, can be derived via *Coll.* VII, 104, pp. 828–829 Hu (Jones 1986a, Vol. 1, p. 234, # 166). Both lemmata are taken from Pappus' commentary on Apollonius, *Tangencies* I, 16. The latter lemma is the converse of *Coll.* VII, 102, quoted above.

[2] AD > CD > DB. A more literal translation of the sentence is: (let) that by which CD exceeds DB be equal to that by which AD exceeds CD.

[3] The text has "excess" (ὑπεροχή). In Prop. 10, Pappus will use the word "difference" (διαφορά).

[4] $d = AD - CD = CD - DB$, and d is *given*.

[5] E lies on AD, A – E – D; Z lies on DB, D – B – Z.

[6] $ED = AD - d$, $DZ = DB + d = DB + (AD - CD) = AD - d$. $DC = AD - d$.

written down above, DZ is *given*.[1] Of it, BZ is *given*.[2] Therefore, the remaining BD is *given*.[3] But AD and DC are both *given*,[4] also. Therefore, each of the lines AD, DC, and DB is *given*.

Prop. 10: Resolutio for a Sub-case of the Apollonian Problem

#11 Now these are the lemmata, whereas the following is the initial <problem>[5]:

<*Given* are> three unequal circles with centers A, B, and C, with *given* diameters, touching each other, and the circle DEZ <circumscribed> around them and touching them <is sought>; let the task be to find its diameter.

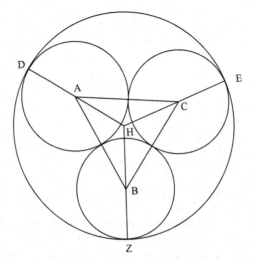

Let its center be H,[6] then, and join AB, AC, and CB toward the centers A, B, and C, and in addition <join the lines> HAD, HBZ, and HCE.[7]

Now, since the diameters of the circles with centers A, B, and C, are *given*, each of the <lines> AB, BC, and CA will turn out to be *given*, also. The differences[8] of

---

[1] Prop. 8: the diameter of the circle CEZ is *given*. Then its radius is *given*, also.

[2] BZ = d.

[3] *Data* 4.

[4] AD = DZ + BZ; *Data* 3; DC = DZ.

[5] Prop. 10 (in a much more general version) was announced before Prop. 7. Prop. 10 is essentially the *resolutio* of an analysis for a single very specific case out of several possible cases for the Apollonian problem. Construction and *apodeixis* are not offered. See the commentary.

[6] Whereas A labels the center of the (sought for) comprising circle with "H" here, the accompanying diagram, and parts of the text further down take the center to be N.

[7] D, E, Z will be the points of touch with the sought circle: III, 11 and 12.

[8] Here, the word used in A is διαφορά, whereas in Prop. 9, the word ὑπεροχή was used.

<the lines> AH, HC, and HB are *given* as well. Therefore, on account of what has been written down above, AH is *given*.[1] But AD is *given*, also, so that the diameter of the circle DEZ is *given*.

And this (issue) has an end for me here, whereas I will write down the rest later on.[2]

Props. 11 and 12: Analysis: Extension of Configuration[3]/*Apagoge*

Prop. 11: Chords, Perpendicular, and Diameter in a Circle

#12 Let ABC be a semicircle; let CBA be bent, and CD be drawn through the interior, and let BC be equal to the sum of AB and CD,[4] and let the perpendiculars BE and DZ <onto AC> be drawn.[5]

<I claim> that AZ is two times BE.

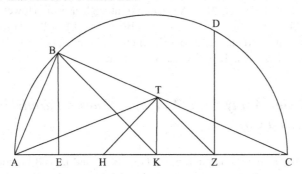

For:

Position EH, on the one hand, equal to AE, and BT, on the other hand, equal to AB,[6] and join AT, TH, and TZ, and draw the perpendicular TK <onto AC>, and join BK.[7]

---

[1] Pappus appeals to Prop. 9. However, in Prop. 9, the additional assumption was made that d = AD − DC = DC − CB, and this is not stated in Prop. 10. Pappus would have had to furnish an extension of Prop. 9, or else formulate an appropriate restriction on the configuration for Prop. 10. Hultsch p. 201, #3, supplies part of an argument, via Prop. 9, to establish that AH is *given*. On the issue of the gap in Prop. 10 see also appendix Hu p. 1227, and the commentary.

[2] The issue is not picked up again in *Coll.* IV. Perhaps Pappus intended to revise Props. 7–10.

[3] But see the commentary on this tentative interpretation of Prop. 11.

[4] D is chosen on the circumference so that CD = BC − AD.

[5] I have translated the text as read/reconstructed by Treweek, treating the phrase "ἐπὶ τὴν ΑΓ" as an explanatory addition (cf. 202, 3 + app. Hu).

[6] H on AC, A − E − H, AE = EH; T on BC, B − T − C, BT = BA.

[7] This passage contains the extension of the configuration (five auxiliary lines and points). From here, the *symperasma* can be directly deduced using the resulting triangles. See the commentary for a conjecture on how this might be indicative for the purpose of Prop. 11 within a group of propositions on analysis-synthesis.

Since CB is equal to the sum of AB and DC, of which BT is equal to BA, the remaining TC is therefore equal to the remaining CD. Therefore, the square over CD is equal to the square over CT, also. However, the rectangle between AC/CZ is equal to the square over DC.[1] Therefore, the rectangle between AC/CZ is equal to the square over CT, also. Therefore, the angle between ZT/TC is equal to the angle between TA/AH.[2]

Again, since the rectangle between CA/AE is equal to the square over AB,[3] two times the rectangle between CA/AE, i.e.: the rectangle between CA/AH,[4] is therefore equal to two times the square over AB, i.e.: to the square over AT,[5] also. Therefore, the angle between AT/TH is equal to the angle between TC/CZ.[6] However, the angle between TA/AH is equal to the angle between ZT/TC, also. Therefore, the remaining angle between AH/HT is equal to the remaining angle between TZ/ZC. Therefore, the angle THZ is equal to the angle TZH, also.[7] And TK has been set forth as a perpendicular.[8] Therefore, ZK is equal to KH.[9] And since the angles between AB/BT and AK/KT are both right angles, the quadrilateral ABTK lies on a circle.[10] The angle between BT/TA is therefore equal to the angle between BK/KA.[11] However, the angle between BT/TA is half a right angle.[12] Therefore, the angle between BK/KA is half a right angle as well. However, the angle between BE/EK is a right angle. Therefore, BE is equal to EK.[13] However, AZ is two times EK (since AE is equal to EH, whereas ZK is equal to KH). Therefore, AZ is two times EB, also. This is what was required to prove.

Prop. 12: *Plane Analysis via Apagoge; Chords, Parallels, and Angles in a Circle*

#13 Let ABC be a semicircle, and let ABD be bent, and let AB be equal to BD,[14] and DE drawn at a right angle <to BD>, and let BE be joined, and EZ drawn at right

---

[1] △ADC ~ △DZC; VI, 8, VI, 4, VI, 17.

[2] △ATC ~ △TZC; VI, 17, VI, 6.

[3] △ABC ~△AEB: VI, 8; VI, 4, VI, 17.

[4] AH = 2AE by construction.

[5] I, 47.

[6] △ATC ~ △ATH (VI, 17, VI, 6), and △ATH ~ △TZC has been shown.

[7] Complementary angles; △HTZ is therefore isosceles (I, 6), and TZ = TH. In the manuscript A, TH = TZ is claimed directly (202, 19 f. + app. Hu). Perhaps the manuscript reading would have been preferable.

[8] TK is perpendicular to AC by construction.

[9] I, 26 for △TKH, △TKZ.

[10] Circle with diameter AT; III, 31.

[11] III, 21.

[12] △ABT is isosceles, and the angle at B is a right angle by construction.

[13] △BEK has a right angle at E, half a right angle at K; it is isosceles (I, 6).

[14] Note the similarity of the starting configuration to the one in Prop. 11. An implicit assumption in Prop. 12 is arc ABC < arc of quadrant.

angles to it, and let H be the center <of the semicircle>, and let DT be to TZ as AH is to HD,[1] and let TE be joined;

<I claim> that the angle between BE/ED is equal to the angle between DE/ET.

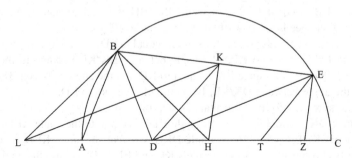

Draw the perpendicular HK from H onto BE. Then BK is equal to KE.[2] And the angle between BD/DE is a right angle. Therefore, the three <lines> BK, KD, and KE are equal to each other.[3] And HK is parallel to EZ.[4] And since one sought <to show> that the angle between KE/ED is equal to the angle between DE/ET, and <one knows that> DK is equal to KE, therefore <one knows> that the angle KED is equal to the angle KDE, therefore <one needs to show[5]> that the angle KDE is equal to the angle DET, also. Therefore <one needs to show> that DK is parallel to ET.[6]

Draw the parallel KL to DE, also, and produce CD to L, and join BL. Now, since KL is parallel to DE, whereas KH <is parallel> to EZ, and it is sought, however, that KD is parallel to ET,[7] therefore (on account of the fact that the triangle KLH is equiangular to the triangle EDZ, whereas <the triangle> DKH <is equiangular> to <the triangle> ETZ), <it is sought> that DZ is to ZE as LH is to HK, whereas EZ is to ZT, as KH <is> to HD.[8] Therefore, <one needs to show,> also, that DZ is to ZT as LH <is> to HD (namely, ex aequali[9]). Therefore, <one needs to show,> also, that DT is to TZ as LD <is> to DH (namely, separando[10]).

---

[1] Choose T on AC, A – D – H – T – Z, with DT:TZ = AH:HD.

[2] III, 3.

[3] III, 31.

[4] I, 29.

[5] Prop. 12 contains several series of phrases starting with "therefore, that X", all dependent on some single "one sought to show that Y". Within *Coll.* IV, this stylistic feature is unique. To facilitate reading, I have added the implicit phrases in brackets.

[6] I, 29. The analysis in Prop. 12 is predominantly reductive and deductive (with minimal input by extension of configuration). All steps are also convertible, and the synthesis will therefore mirror the analysis exactly. See the commentary on Prop. 12, and the introduction to Props. 4–12 on analysis-synthesis for this feature of the analysis in Prop. 12 in the context of plane geometry.

[7] Above, the claim in Prop. 12 was reduced to this statement.

[8] If KD is parallel to ET, ΔDKH ~ ΔETZ (ΔKLH ~ ΔEDZ by construction; I, 29); then the above mentioned proportions hold. The claim of the statement has been reduced to yet another condition that must be fulfilled.

[9] V, 22.

[10] V, 17. At this point, the initial claim has been reduced to: LD:HD = DT:ZT.

It was, however, assumed, also, that as DT <is> to TZ, so <is> AH to HD.[1] Therefore, <one needs to show> that DT is to TZ, i.e.: AH to HD, as LD <is> to DH.[2] Therefore, <one needs to show> that LD is equal to AH[3]; therefore, that LA is equal to DH, also.[4] But AB is equal to BD as well.[5] Therefore, <one needs to show> that LB is equal to BH as well.[6] But BH is equal to both LD and AH. Therefore, <one needs to show> that BL is equal to LD, also.[7]

However, this is the case <i.e.: BL is in fact equal to LD>.[8] For since KL is parallel to DE, and DK is equal to KE,[9] the angle between BK/KL is equal to the angle between LK/KD.[10] Now, since BK is equal to KD and the angle between BK/KL is equal to the angle between DK/KL, BL is therefore equal to LD, also.[11]

And the synthesis follows the analysis step by step.[12]

For since DK is equal to KE,[13] the angle KDE is equal to the angle KED, also.[14] But the angle KDE is equal to the angle DKL, whereas the angle KED is equal to the angle BKL on account of the parallels KL and ED.[15] Therefore, the angle BKL is equal to the angle DKL as well. However, the straight line BK is equal to <the straight line> KD,[16] also. Therefore, the base BL is equal to the base LD as well,[17] so that the angle between LB/BD <is equal> to the angle BDA, also, i.e.: to the

---

[1] Hypothesis of Prop. 12.

[2] Using the result of the first sequence of reductions.

[3] V, 9.

[4] LA = LD − AD; DH = AH − AD.

[5] Hypothesis of Prop. 12.

[6] Since AB = BD, the reduced claim LA = DH implies that ΔLAB ~ ΔHDB must hold (I, 4).

[7] BH = AH: radii of initial semicircle; AH = LD needs to be shown (see above); therefore, the claim of Prop. 12 has been reduced to BL (= BH = AH) = LD.

[8] Beginning of the *resolutio*. BL = LD holds independently of the analysis-assumption.

[9] By construction.

[10] I, 29, I, 5, I, 29.

[11] BK = KD by construction; I, 4 for ΔLBK, ΔLDK. The *resolutio* ends here.

[12] Greek word: ἀκολούθως (translation: following step by step). This term was subject to considerable debate in the discussion about the interpretation of Greek geometrical analysis and its logical structure. Some authors hold that it must mean "logically derived", and maintain that analysis is deductive, since it proceeds "akolouthos." I agree with Hintikka and others that it does not have to be interpreted so narrowly, and that it rather means "follows in sequence, in an orderly fashion". Co p. 70 translates "compositio vero resolutioni congruens erit." See the excursus on analysis-synthesis in the introduction to Props. 4–12 in the commentary. The synthesis is not a direct logical deduction from the analysis. Further occurrences of this word and its derivatives in *Coll*. IV, where regularly it does not carry the force of "logical derivation" will be noted ad locum.

[13] III, 3; III, 31.

[14] I, 5.

[15] I, 29.

[16] III, 31; III, 3.

[17] I, 4.

angle DAB, i.e.: to the angle ABH.[1] Take away the common angle ABD. Then the remaining angle LBA is equal to the remaining angle DBH. But the angle BDH is equal to the angle BAL, also.[2] Now, BDH and BAL are two triangles that have two angles equal to two <corresponding> angles and one side, AB, equal to, <one side,> BD. Therefore, BH is equal to BL, whereas DH is equal to LA,[3] so that LD is equal to AH, also.[4] Now, since it was assumed that, as AH is to HD, <so is> DT to TZ,[5] whereas AH is equal to LD, DT is therefore to TZ as LD is to DH. Componendo, then: as LH <is> to HD, so <is> DZ to ZT.[6] However, as LH <is> to HK, so is DZ to ZE, also[7]; and as KH <is> to HD, so <is> EZ to ZT.[8]

And the angle EZT is equal to the angle KHD on account of the fact that EZ and KH are parallels.[9] Therefore, the angle ETZ is equal to the angle KDH, also.[10] Therefore, KD is parallel to ET as well.[11] Therefore, the <angle> KDE, i.e.: the angle KED, is equal to the angle DET.[12]

Props. 13–18: Arbelos Treatise: Plane Geometry, Archimedean Style

#14 In certain <books> an ancient proposition of the following sort is reported.

Posit three semicircles ABC, ADE, and EZC, touching each other, and into the space between their circumferences, which is in fact called "arbelos,"[13] describe any number of circles, touching both the semicircles and each other, like the ones around the centers H, T, K, and L.[14]

---

[1] Isosceles triangles, I, 5.

[2] Isosceles triangle ABD, I, 6.

[3] I, 26.

[4] Add AD.

[5] Hypothesis of Prop. 12.

[6] V, 18.

[7] ΔLKD ~ ΔDEZ; VI, 4.

[8] V, 22.

[9] I, 29.

[10] VI, 6.

[11] I, 27; the corresponding step in the analysis (converse) rests on I, 29.

[12] I, 29; I, 6.

[13] The meaning of the term "arbelos" is not quite clear. One of the possible meanings is "shoemaker's knife". Apparently, ancient shoemakers used a tool with a shape that was similar to the one formed in the figure.

[14] Only the first of the inscribed circles touches all three initial semicircles; all others touch two of the semicircles and their own predecessor and successor. Note the motivic connection to Props. 7–10. Each inscribed circle in the arbelos sequence is a solution to the Apollonian problem. Only the starting configuration, however, is directly related to the special case treated in Prop. 10; cf. Jones (1986a, p. 539). See also Hofmann (1990) II, pp. 146–164, and the notes in the commentary on Props. 13–18.

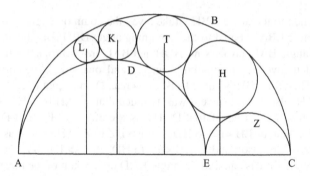

<The task is> to show that the perpendicular from the center H onto AC is equal to the diameter of the circle around H, whereas the perpendicular from T is double the diameter of the circle around T, and the perpendicular from K three times <the diameter of its circle>, and the perpendiculars in sequence are multiples of their respective diameters according to the sequence of numbers exceeding one another by a unit, when the inscription of circles continues indefinitely. However, the lemmata[1] will be proved before.

Prop. 13: Preparatory Lemma: Points of Similarity and Touching Circles

#15 Let there be <*given*> two circles ZB and BM with centers A and C, touching each other in B, and let <the circle> BM be the larger one. In addition, let there be given some other circle KL with center H, touching them in K and L <respectively>, and join CH and AH (they will in fact pass through K and L[2]), and the straight line joining K and L, when produced, will, on the one hand, intersect the circle ZB, and, on the other hand, it meets[3] the straight line through the centers A and C, when it is produced (on account of the fact that the side AK of the trapezoid AKDC is larger than the <side> CD).[4]

---

[1] τὰ λαμβανόμενα; a certain preference for this word, as a label for preliminary lemmata that are presented before the main body of a treatise, is attested for Archimedes, though he is not completely consistent in his usage of the word. In Prop. 17, the word λῆμμα will be used.

[2] III, 12.

[3] Note the change of tense. This could be an indication that the statement about KL intersecting AC was originally not part of the *ekthesis,* but of the proposition. As indicated by the way I set up the paragraphs above, I think that the whole text from "and the straight line" to "AB (is) to BC" is the proposition. Prop. 13's claim thus has two parts: (i) AKDC is a trapezoid, i.e., KL and AC meet (in E), (ii) AE:EC = AB:BC. See the commentary for a defense of my decision. It has consequences for the converse of Prop. 13 as well. For the converse can then assume both AK ∥ CD and AE:EC = AB:BC (even if the former condition is not explicitly mentioned), and derive K–L–D–E from there. The converse will be used in Props. 15 and 17.

[4] That AH ∥ CD and that therefore AKDC is a trapezoid will be shown in the first part of the *apodeixis.*

Now, let it meet <AC> in E, intersecting the circle <ZB> in D. <The task is> to show that AE is to EC as AB <is> to BC.[1]

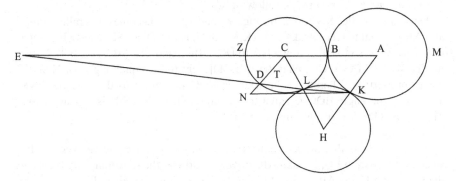

However, this is obvious when CD is joined.[2] For the triangles CDL and LKH turn out to be equiangular, since they have equal vertex angles at L,[3] and the sides adjacent to the angles <at> C and H are proportional,[4] so that the alternate angles DCH and CHA are equal, and CD is <thus> parallel to AH,[5] and as AE <is> to EC, so <is> AK to CD, i.e.: AB to BC.[6]

<Converse:>

However, the counterpart is obvious as well. For whenever AE is to EC, as AB <is> to BC, KD turns out to be on a straight line with DE.

For AK is parallel to CD,[7] and also, AK is to CD as AB <is> to BC,[8] i.e.: as AE <is> to EC.[9] Therefore, KD is on a straight line with DE. For if the <straight line> through K and E does not pass through D, also, but through T,[10] it turns out that AK <is> to CT as AE <is> to EC,[11] which is impossible. Similarly, it will not reach and

---

[1] This means that E is a point of similarity.

[2] Hultsch brackets the phrase "when CD is joined" as an interpolation (210, 8 app. Hu). For on his reading, the line CD is mentioned already in the *ekthesis*, and should not occur in the proof. See the above footnote and the commentary for the reconstruction of the overall argument in Prop. 13.

[3] Here Hultsch (perhaps unnecessarily) brackets the words "at L" (210, 10 + app. Hu), and in the following line "ἔχοντα"(210, 11 + app. Hu).

[4] VI, 7.

[5] I, 27. Now we have shown that CD is parallel to AH. Therefore, we indeed have a trapezoid AKDC, with AH > CD; therefore, AC and KD (=KL) meet, and we call the point of intersection E. See the commentary.

[6] VI, 4, V, 16.

[7] Hultsch 211, # 1, claims that AH ∥ CD can be shown exactly as above, from ΔLHK ∼ ΔDCL. However, that similarity rested on the assumption that D, K, and L lie on a straight line, and this is exactly what the converse is about to prove. In my opinion, we rather have to assume that DC is a parallel to AH in the converse. See the commentary.

[8] AK = AB and CD = BC, as radii of the respective circles.

[9] By assumption.

[10] With D – T – C.

[11] AE:AK = EC:CT (VI, 4); apply V, 16.

intersect CD produced beyond D, for example, in N.[1] For again AK will be to CN as AE <is> to EC,[2] which is impossible. For it is <so> to CD.

Or <it can be shown> in the following way.

Draw the parallel KN to AE through K, and ACNK becomes a parallelogram,[3] and AK is equal to CN.[4] And since AK, i.e.: CN, is to CD, as AE <is> to EC, separando, as AC <is> to CE, <so is> ND to DC.[5] Alternando,[6] as AC, i.e.: as KN, <is> to ND, so <is> EC to CD. And the sides adjacent to the equal angles at N and C are in proportion. Therefore, the triangle EDC is similar to the triangle DNK. Therefore, the angle EDC is equal to the angle NDK. And CN is a straight line. Therefore, KDE is a straight line as well.[7]

<Addition:>

Moreover, I say that the rectangle KEL is also equal to the square over EB. For since as AE <is> to EC, so <is> AB to BC, i.e.: to CZ, the remaining BE will be to the remaining EZ as AE <is> to EC, i.e.: as KE <is> to ED, also.[8] But as KE <is> to ED, so is the rectangle KEL to the rectangle LE/ED,[9] whereas as BE <is> to EZ, so <is> the square over BE to the rectangle BEZ,[10] and the rectangle LE/ED is equal to the rectangle BE/EZ.[11] Therefore, the rectangle KEL is equal to the square over EB, also.

Prop. 14: Technical Lemma. Perpendiculars and Diameters in Configurations with Three Touching Circles

#16 <Let there be given> two semicircles BHC and BED, and the circle EZHT touching them,[12] and let the perpendicular AM from its center A onto the base BC of the semicircles be drawn.

---

[1] N – D – C.

[2] VI, 4 and V, 16, as above.

[3] Again, it seems apparent that one must assume that CD ‖ AK (compare the above footnote).

[4] I, 34.

[5] CN:CD = AE:EC by assumption. CN = CD + DN, and AE = EC + AC; therefore, DN:CD = AC:EC (V, 17).

[6] V, 16.

[7] CDN is a straight line ⇒ ∠CDK + ∠KDN = π. It was shown above that ∠EDC = ∠NDK; therefore, ∠CDK + ∠EDC = π.

[8] AE:EC = (AB + BE):(EZ + ZC) = AB:CZ (assumption in the converse) ⇒ BE:EZ = AE:EC (V, 19); AC:EC = DK:DE (VI, 2), and thus AE:EC = KE:ED (VI, 1).

[9] VI, 1, height EL.

[10] VI, 1, height BE.

[11] III, 36.

[12] The diameters BC and BD of the semicircles are assumed to be in line. E and H are the points of touch with the third circle. There are three possibilities for the relative position of the semicircles and the circle involved (configurations 1–3, cf. figures a–c). Even though only two of these configurations are needed for the arbelos theorem, the author of the little treatise gives a complete account of the lemmata involved. See the commentary on the Archimedean features of Props. 13–18.

<I claim> that as MB <is> to the radius of the circle EZHT, so is, in the first configuration, the sum of CB and BD to their difference,[1] CD, whereas in the second and third configuration the difference between CB and BD <is> so to the sum of CB and BD, i.e.: to CD.[2]

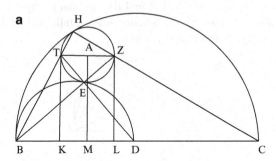

Draw the parallel TZ to BC through A. Now, since the two circles BHC and EZHT touch each other in H, and the diameters BC and ZT in them are parallel, the lines through H, T, and B and through H, Z, and C will both be straight lines.[3] Again, since the two circles BED and EZHT touch each other in E, and the diameters TZ and BD in them are parallel, the lines through Z, E, and B and through T, E, and D will both be straight lines.[4]

*Draw the perpendiculars TK and ZL from the points T and Z <onto BD>, also. Now, on account of the similarity of the triangles BHC and BTK,[5] BT <is> to BK as BC <is> to BH, and the area comprised by CB and BK[6] is equal to the

---

[1] The Greek word is "ὑπεροχή" (excess); as in Prop. 9, it is translated as "difference".

[2] The technical Prop. 14 yields the central result needed for establishing the arbelos theorem. Specifically, it is the intermediate step labeled as "*" below that is most important for the following theorems. See the commentary on Archimedean features of Props. 13–18, and compare the footnote on "*."

[3] Hultsch (p. 215, # 1 Hu) provides a proof involving an auxiliary construction, and reference to Prop. 13. Instead, one could simply assume implicit appeal to an elementary step of inference, capturing the same content as the group of theorems in *Coll.* VII, 102–106 mentioned in the footnotes to Prop. 8: Whenever one has a configuration with parallel chords in tangent circles, the lines connecting the endpoints "crosswise" also go through the point of touch. Another possibility for this step in Prop. 14, though valid only for configurations 1 and 2, would have been to appeal to a theorem as in *Lib. ass.* I. See also Co p. 74, A.

[4] Again, we may have an appeal to a theorem about parallel chords in tangent circles (cf., e.g., *Coll.* VII, 102–106; compare the preceding footnote).

[5] Both triangles have a right angle, and they have the angle at B in common.

[6] I.e., the area of the rectangle with sides CB and BK. The Greek text has "τὸ ὑπὸ ΓΒ ΒΚ περιεχόμενον χωρίον." The fact that we are dealing with areas is explicitly emphasized, and this seems to be a peculiarity of the text in Props. 13–18. It may very well go back to the style of the original author of the treatise.

<area comprised> by HB and BT,[1] whereas on account of the similarity of the triangles BZL and BED, BZ <is> to BL as DB <is> to BE, and the <area comprised by> DB and BL is equal to the <area comprised> by ZB and BE.[2] The <area comprised> by HB and BT is equal to the <area comprised> by ZB and BE as well.[3] Therefore, the <area comprised> by CB and BK is equal to the <area comprised> by DB and BL, also.[4] When the perpendicular from Z falls onto D, however, <this area CB/BK is equal> to the square over BD.[5]*

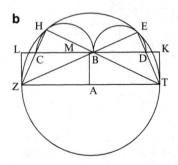

Therefore, in the first configuration, as CB <is> to BD, so <is> LB to BK,[6] so that <as> the sum <of> CB and BD <is> to their difference CD, so <is> the sum <of> LB and BK to their difference KL as well.[7] And BM is half of the sum <of> LB and BK (on account of the fact that KM is equal to ML[8]), whereas MK is half of LK. Therefore, BM <is> to MK, i.e.: to the radius of the circle EZTH, as the sum <of> CB and BD is to CD as well.

---

[1] The Greek phrasing is "τὸ ὑπὸ HB BΘ." This is also different from the abbreviations "τὸ ὑπὸ HBΘ" and "τὸ ὑπὸ τῶν HBΘ", which are taken, in this translation, as technical formulae for the rectangle HBT. The expression "τὸ ὑπὸ HB BΘ" is elliptic for "τὸ ὑπὸ HB BΘ περιεχόμενον χωρίον." The translation will keep track of this differentiation by adding the phrase "area comprised" in brackets throughout Props. 13–18. Compare the preceding footnote. BT:BK = BC:BH (VI, 4) ⇒ BT × BH = BC × BK (VI,16).

[2] The argument is completely analogous to the one in the preceding step. The triangles are similar because of the right angles and the common angle at B; similarity implies the stated proportion (VI, 4), and thus (VI, 16) the equality of the rectangles.

[3] III, 36 for configurations 1 and 3; III, 35 for configuration 2.

[4] We have shown: CB × BK = HB × BT, DB × BL = ZB × BE, and HB × BT = ZB × BE. Therefore, CB × BK = DB × BL.

[5] In that case, DB = BL. The limit case will be used in Prop. 17 and may have been inserted here precisely for that purpose (cf. 214, 20–216, 1 app. Hu).The passage framed by "*" is the core of Prop. 14. Its result will be quoted several times in what follows, independently of Prop. 14 itself.

[6] VI, 16.

[7] (BC + BD):BD = (BL + BK):BK (V, 18); (BC − BD):BD = (BL − BK):BK (V, 17); therefore: (BC + BD):(BC − BD) = (BL + BK):(BL − BK) (V, 22).

[8] In numbers, BM is the arithmetic mean of BK and BL; the author of the arbelos treatise avoids using terms coined for numbers to label properties of magnitudes.

In the second and third configuration, on the other hand, since the rectangle CBK has been shown to be equal to the rectangle DBL,[1] LB <is> therefore to BK as CB <is> to BD.[2] Componendo, KL <is> to KB as CD (is) to DB; so that KL <is> to the difference of LB and BK as CD <is> to the difference of CB and BD, also.[3] And the radius of the circle EZHT is half of KL, whereas BM is half of the difference of LB and BK (on account of the fact that LM is equal to MK[4]), so that as MB <is> to the radius of the circle EZHT, so <is> in the first configuration the sum <of> CB and BD to their difference CD, whereas in the second and third configuration the difference of CB and BD <is so> to the sum CBD, i.e.: to CD.

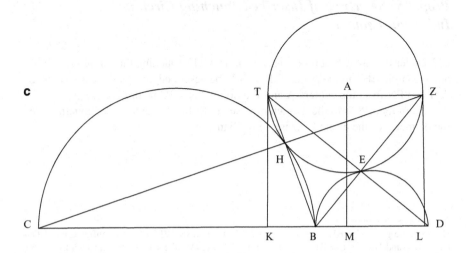

<Addition 1:>

At the same time, however, it is established by investigation that the <area comprised> by BK and LC is equal to the square over AM as well. For on account of the similarity of the triangles BTK and ZLC[5] ZL <is> to LC as BK <is> to KT, and the <area comprised> by BK and LC is equal to the <area comprised> by TK and ZL,[6] i.e.: to the square over AM.[7]

---

[1] The equality of the rectangles mentioned was shown in * above.

[2] VI, 16.

[3] (LB − BK):BK = (CB − BD):BD (V,17); from KL:KB = CD:DB we therefore get KL:(LB − BK) = CD:(CB − BD) (V, 22).

[4] LB = LM + MB; BK = MK − MB = LM − MB. From here, it follows immediately that BM:radius = CD:(CB − CD). Instead of giving the result for configurations 2 and 3 explicitly, and then restating it in the summary of what has been shown, the text proceeds directly to the summary.

[5] The triangles are both similar to the triangle BHC.

[6] VI, 4 and VI, 16.

[7] By construction: TK = ZL = AM, because TZ is a diameter parallel to BC.

<Addition 2:>

However, on account of the fact that as BC <is> to CD, so <is> BL to KL,[1] the <area comprised> by BC and by the <straight line> KL, i.e. <by BC and> the diameter of the circle, turns out to be equal to the <area comprised> by BL and DC,[2] also, whereas on account of the fact that as BD <is> to CD, so <is> BK to KL,[3] the <area comprised> by BD and KL, i.e.: <by BD and> the diameter of the circle <turns out to be> equal to the <area comprised> by BK and DC.

Prop. 15: Sequence of Inscribed Touching Circles: Induction Lemma

#17 Under the same conditions, let the circle TRT', touching the initial semicircles and the circle EHT in the points T, R, and T' be described, and let the perpendiculars AM and PN from the centers A and P onto the base BC be drawn.[4]

I claim that PN is to the diameter of the circle TRT' as AM, taken together with the diameter of the circle EH <is> to its diameter.[5]

---

[1] From *, we get: BC × BK = BD × BL ⇒ BC:BD = BL:BK (VI, 16); first configuration: BD + CD = BC, and BK + KL = BL; thus: CD:BD = KL:BK (V, 17); second and third configuration: BC + BD = DC, and BL + BK = KL; thus: CD:BD = KL:BK (V, 18). From these equations, it follows in all three cases that BC:CD = BL:KL (V, 16 and V, 22).

[2] VI, 16.

[3] The argument is analogous to the preceding one. From *, we get: BC × BK = BD × BL ⇒ BC:BD = BL:BK (VI, 16); first configuration: BC = BD + CD, and BL = BK + KL; thus: CD:BD = KL:BK (V, 17) second and third configuration: DC = BC + BD, and KL = BL + BK, and DC:BD = KL:BK (V, 18), also. In both cases, V, 16 yields BD:CD = BK:KL.

[4] R, T', and T are the points of touch with the semicircles over BD and BC and the first added circle EHT respectively. M and N lie on BC.

[5] Again we get three possible configurations, on the basis of the configurations in Prop. 14. Each of them leads to exactly one possibility for the second circle to be inscribed into the respective configuration. Note that in Hultsch's edition, configuration 1 from Prop. 14 leads to configuration 1 in Prop. 15, whereas configuration 2 leads to configuration 3, and configuration 3 to configuration 2. I have numbered the figures in concurrence with Prop. 14. In A, the second diagram for Prop. 15 concerns a limit case that is not treated in the text, but relevant for Prop. 17 and Addition 2 to Prop. 16. For a correct diagram and a reconstruction of the proof for the limit case see appendix Hu p. 1227 f.; cf. also Co p. 78 P. The figure for the limit case given in A is reproduced in an appendix to this edition. The figure for the configuration that results from building on configuration 3 has been added here; it is modeled on Hu, since it is missing in A. The occurrence of the figure for the limit case indicates a loss of text that was originally part of the source at some stage in the transmission.

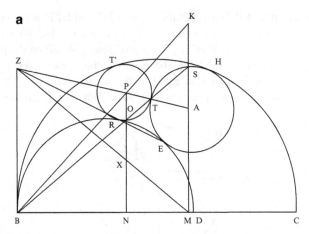

Draw BZ at right angles to BD. Then it is tangent to the circle BHC. And produce AP, after it has been joined, to Z.[1] Since, according to what has been shown before, in the first configuration BM <is> to the radius of the circle EHT, as the sum CBD[2] <is> to their difference CD, whereas in the second and third configuration MB <is> to the radius of the circle EHT as their difference <is> to their sum, i.e.: as the difference of CB and BD <is> to CD,[3] and <since the same is true for the ratio of> BN to the radius of the circle TRT',[4] therefore – alternando – the radius AT of the circle EHT will also be to the radius TP of the circle TRT' as MB <is> to BN.[5] But AZ <is> to ZP as MB <is> to BN (for when ZM is joined, MZ will be to ZX as MB <is> to BN.)[6] And therefore, the radius AT of the circle EHT <is> to the radius TP of the circle TRT' as AZ <is> to ZP.[7]

---

[1] Z is taken as the point of intersection between BZ and AP.

[2] I.e.: CB + BD.

[3] Prop. 14.

[4] I have taken "καὶ ἡ BM πρὸς…" (220, 1) and "καὶ ἡ BN πρὸς…" (220, 6) to be syntactically parallel. We get two corresponding statements about line segments cut off by perpendiculars in relation to radii of corresponding circles. Prop. 14 can be applied directly for BM:AT and NB:TP.

[5] MB:AT = NB:TP, because they are both equal to either (CB + BD):CD (configuration 1) or to (CB − BD):CD (configuration 2/3). V, 16 yields MB:BN = AT:TP. This proportion will be used again in the course of Prop. 15.

[6] X is the point of intersection between MZ and PN. For configurations 1 and 2, consider ΔZBM with intersecting line PX, parallel to BZ. We get: BN:NM = ZX:XM (VI, 2); this transforms to NM:BN = XM:ZX (V, 16), and thus: BM:BN = ZM:ZX (V, 18).

For configuration 3, consider ΔMNX, with intersecting line ZB, parallel to NX. We get: BN:BM = ZX:MZ (VI, 2), and V, 16 yields BM:BN = MZ:ZX.

PN ∥ AM by construction, and therefore: ZM:ZX = ZA:ZP (VI, 4, with V, 16, for ΔZMA, ΔZXP). This argument is applicable in all three possible configurations, and we get: BM:BN = AZ:ZP.

[7] Having shown BM:BN = AT:PT and BM:BN = AZ:ZP (cf. preceding footnotes), we get: AT:PT = AZ:ZP.

And a certain circle BRED touches the circles EHT and RTT' at the points R and E <respectively>. Therefore, on account of the theorem Prop. 13,[1] shown above, the straight line joining the points R and E, when produced, will fall on the point Z, and the rectangle comprised by EZR will be equal to the square over TZ as well.[2] However, the rectangle EZR is also equal to the square over ZB.[3] Therefore, the square over ZB is equal to the square over ZT, also. Therefore, BZ is equal to ZT.

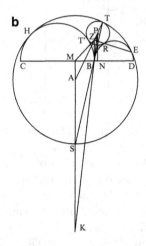

Furthermore, however, since MA, when it is produced, intersects the circumference of the circle EHT in S, whereas PN intersects the circumference of the circle TRT' in the point O, AT is therefore equal to AS, whereas PO <is equal> to PT, and the <straight line> joining the points O and S will pass through T.

For the angle TAS is equal to the alternate angle TPO, and the triangle ATS is equiangular to the triangle PTO, and the <line> AP is a straight line.[4] Therefore, the line drawn through the points S, T, and O is a straight line, also. It will, however, pass through B as well. For the <line> TOB is a straight line, on account of the fact that OP is to PT as BZ <is> to ZT, given that the angles BZT and OPT, adjacent to the parallels BZ and OP, are equal.[5] This, also, has been shown above in Prop. 13.[6]

---

[1] We use the converse of Prop. 13 to establish that E, R, and Z lie on a line. Hultsch (222, 7/8 app.) believes the reference to Prop. 13 is due to an interpolator.

[2] Prop. 13, Addition.

[3] III, 36.

[4] In $\triangle$OPT and $\triangle$SAT, $\angle$A = $\angle$P, because PO $\parallel$ AS (I, 29). Since the triangles are isosceles, they are similar. Therefore, they have equal angles at T. Since PA is a straight line, $\angle$STA +$\angle$ATO = π, and A – T– O is a straight line.

[5] $\angle$BZT = $\angle$OPT, because NZ $\parallel$ PO (I, 29); above, it has been shown that BZ = ZT; obviously, PO = PT, too; therefore, $\triangle$BZT ~ $\triangle$OPT, and TO must pass through B. Otherwise, $\triangle$BZT would not be isosceles.

[6] An argument analogous to the one showing that TO must pass through B was used in the converse to Prop. 13. Hultsch suspects the reference to Prop. 13 to be an interpolation (222, 7/8 app. Hu). It seems also possible that the arbelos treatise as a whole was taken out of a larger treatise, with a more substantial preliminary part, of which only Prop. 13 survives. See the commentary on Prop. 13.

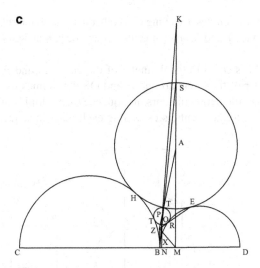

However, produce the <line> BP, after it has been joined, and let it meet MA, after it has been produced, in K. Now, since AZ was to ZP and AT to TP as MB <is> to BN, i.e.: as KB <is> to BP,[1] AS will be to PO, and SK <will be> to PO, as KB <is> to BP. Therefore, AS is equal to SK.[2] Now, since the whole <line> AK is equal to the whole diameter of the circle EHT, and <since> NP is to OP as KM <is> to KS, NP will also be to the diameter of the circle TRT', as MK <is> to KA, i.e.: as MA, together with the diameter of the circle EHT <is> to this diameter[3] – which is what was to be proved.[4]

Prop. 16: Arbelos Theorem

#18 With these things investigated beforehand, assume a semicircle BHC, and on its base choose a point D arbitrarily, and over BD and DC describe the semicircles BED and DYC, and in the space between the three circumferences, the so-called arbelos,

---

[1] ΔBMK ~ ΔBNP by construction ⇒ BK:BM = BP:BN (VI, 4), and BK:BP = BM:BN (V, 16).

[2] We have shown above: AZ:ZP = AT:PT = BM:BN. Now we also have: BM:BN = BK:BP. Obviously, AT = AS and TP = PO (radii). Therefore, AS:PO = AT:PT = BK:BP. Consider that ΔBKS ~ ΔBPO. BP:PO = BK:KS (VI, 4) ⇒ KS:PO = BK:BP (V, 16). It follows that AS:PO = KS:PO, and AS = KS must hold (V, 9).

[3] AK = AS + SK = 2 AS = 2KS. Consider the pairs of similar triangles BKM, BPN and BKS, BPO; We get: PN:KM = BK:BP = KS:OP, and thus (V, 16): PN:OP = KM:KS. Therefore, PN:2OP = KM:2KS = KM:KA. 2OP is the diameter of the circle TRT', and KA is MA + AK = MA + the diameter of the circle EHT.

[4] Some bit of text has been lost at the end of Prop. 15 (cf. 224, 11 app. Hu). As said above, the manuscript A has a figure for the limiting case of Prop. 15 (the case used in Prop. 17 and in Addition 2 to Prop. 16), but no argument. For such an argument, cf. Co p. 78, Lemma in P, and appendix Hu p. 1227 f.

inscribe any number of circles, touching each other and the semicircles,[1] like the ones
with centers A, P, and O, and from their centers draw the perpendiculars AM, PN, and
OS onto BC.

I claim that AM is equal to the diameter of the circle around A, whereas PN is
double the diameter of the circle around P, and OS three times the diameter of the
circle around O, and the perpendiculars in sequence the multiples of their respective
diameters according to the numbers exceeding each other in sequence by a unit.

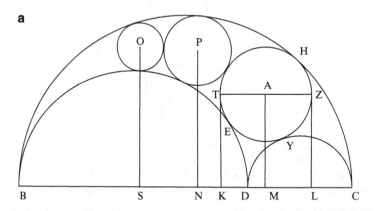

Draw the diameter TZ parallel to BC, and the perpendiculars TK and ZL <onto
BC>. Now, according to what was written down above, the rectangle comprised by
CB and BK is equal to the rectangle comprised by LB and BD, whereas the <rect-
angle comprised> by BC and CL <is equal> to the rectangle KCD.[2] And for this
reason KL is to LC as BK <is> to KL. For both these ratios are the same as the one
of BD to DC. For since the <area comprised> by CB and BK is equal to the <area
comprised> by LB and BD, DB therefore is to BK as CB <is> to BL.[3] Alternando:
as CB <is> to BD, so <is> LB to BK; separando: as CD <is> to DB, <so is> LK to
KB. Conversely, as BD <is> to DC, <so is> BK to KL. Again, since the <area
comprised> by BC and CL is equal to the <area comprised> by KC and CD, CD
therefore is to CL as BC <is> to CK. Alternando: KC <is> to CL as BC <is> to CD.
Separando, therefore, KL is to LC as BD <is> to DC.

However, BK was to KL as BD <is> to CD as well. Therefore, KL is to LC as
BK <is> to KL, also. Therefore, the <area comprised> by BK and LC is equal to
the square over KL. However, it has been shown above that the <area comprised>

---

[1] In the arbelos proper, each circle in the sequence touches the semicircles over BD and BC, and
its own predecessor in the sequence. For the first circle, this is the semicircle over DC.

[2] Prop. 14, intermediate step *.

[3] VI, 16.

by BK and LC is equal to the square over AM as well.[1] Therefore, AM is equal to KL, i.e.: to the diameter ZT of the circle with center A.

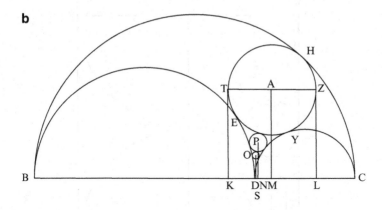

However, since the following has been shown above, also: that PN is to the diameter of the circle around P as AM, taken together with ZT, <is> to ZT,[2] and <since> AM, taken together with ZT is double ZT,[3] PN will be double the diameter of the circle around P.[4] Therefore, PN, taken together with the diameter of the circle around P <is> three times the diameter.[5] And OS stands in the same ratio to the diameter of the circle around O.[6] Therefore, OS is three times the diameter of the circle around O as well.

And similarly <one will see that> the perpendicular belonging to the next circle in sequence is four times the diameter <of that circle>, and the perpendiculars in sequence will be found to be the multiples of the diameters in them according to the sequence of numbers exceeding each other by a unit, and it will be shown that this occurs indefinitely.[7]

<Addition 1 (configuration with two straight lines):>

However, when instead of the circumferences BHC and DYC there are <given> straight lines, at right angles with BD, as in the third configuration, the same will occur concerning the inscribed circles; for right away the perpendicular from center A onto BD turns out to be equal to the diameter of the circle around A.[8]

<Addition 2 (configuration with one straight line):>

---

[1] Prop. 14, Addition 1.

[2] Prop. 15.

[3] It was shown in the first part of Prop. 16 that ZT = AM.

[4] (AM + ZT):ZT = 2 ZT:ZT = PN:diameter of circle P.

[5] PN:diameter ~ 2:1 $\Rightarrow$ PN + diameter:diameter ~ 3:1.

[6] Prop. 15.

[7] The argument in Prop. 16 is related to complete induction. See the commentary.

[8] The first step of the induction is then trivial. The argument can proceed from there, on the basis of Prop. 15, as in Prop. 16. One has to assume the limit case of Prop. 15, for which only the figure, but not the actual argument survives (see notes above). Co p. 80 F provides a direct argument without reference to Prop. 15.

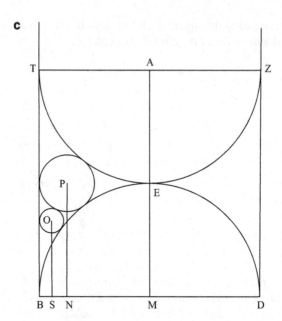

Finally, when the two circumferences BHC and BED remain, and instead of the circumference DYC one assumes a straight line DZ at right angles with BC (as is the case in the fourth configuration), <one gets the following situation:> when BC has to CD a quadratic ratio in numbers,[1] the perpendicular from A will be commensurable with the diameter of the circle around A, whereas when it does not <have such a ratio>, <the perpendicular is> incommensurable <with the diameter>.

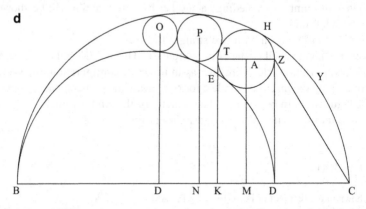

For in general DZ has, in square, the same ratio to the diameter of the circle around A that BC <has> to CD, as is shown <in the next proposition> in sequence.[2]

---

[1] The squares are to have a ratio like two square numbers.

[2] I.e.: Prop. 17 will show hat $DZ^2:d(\text{circle A})^2 = BC:CD$. Because $DZ = AM$, one can see (e.g., using X, 9) that AM will be commensurable in length with $d(A)$ iff BC has to DC a ratio expressible in square numbers.

For example,[1] when BC is four times CD in length, DZ, i.e.: the perpendicular from A, turns out to be double the diameter of the circle around A in length,[2] and the <perpendicular> from P <turns out to be> three times <the corresponding diameter>, whereas the <perpendicular> from O <turns out to be> four times <the corresponding diameter>, and so on in sequence, according to the sequence of numbers.[3]

Prop. 17: Lemma Used in Prop. 16, Addition 2

#19 The lemma[4] that was set aside. <Let there be given> semicircles BHC and BAD,[5] and DE at right angles <to BD>, and a touching circle THZA.

<I claim> that DZ is, in square, to the diameter of the circle THZA as BC is to CD in length.

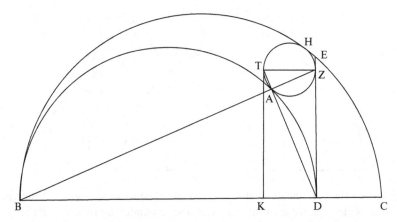

Draw the diameter TZ. Then the <lines> ZAB and TAD are straight lines.[6] Draw the perpendicular TK. Then, on account of what has been shown above, the area comprised by CB and BK is equal to the square over BD.[7] Therefore, as BC <is> to CD,

---

[1] Hultsch (230, 4–8 Hu + app.) believes the example is due to interpolation.

[2] BC = 4CD ⇒ DZ² = 4 diameter²; this entails DZ = 2 diameters.

[3] Prop. 15.

[4] Here, the author uses the word "λῆμμα"; before, in the introduction to the arbelos treatise, the word "λαμβανόμενα" was used.

[5] D is on BC. B – D – C.

[6] Appeal to theorems like *Coll.* VII, 102–106 (tangent circles, parallel chords and lines through the point of touch) seems most likely. ZAB and TAD will be straight lines, because TZ is parallel to BC, and A is the point of touch. As above in Prop. 14, Hultsch ad locum comments that this could be shown via Prop. 13, and refers to his footnote on Prop. 14 to this effect. Co p. 81 B refers to his Lemma p. 74 A.

[7] Prop. 14, passage *. As noted above in the footnotes to the passage, Prop. 17 uses a limiting case for the result in passage *, which was probably included there with a view to Prop. 17.

so <is> BD to DK, i.e.: to TZ.[1] However, as BD <is> to TZ, so <is> DA to TA,[2] and as DA <is> to AT, so <is> the square over ZD to the square over TZ. For TZD is a right-angled triangle, and ZA is a perpendicular onto the hypotenuse ZA.[3] And therefore the square over ZD <is> to the square over the diameter of the circle THZA as BC <is> to CD.[4]

Prop. 18: Addition: Progression Theorem, Odd Numbers

#20 Furthermore, the following, too, has been established through investigation by the lemmata written down above.[5] Let there be <given> the semicircles ABC and ADE,[6] and let the circles with centers Z, H, and T be described, touching their circumferences, and <let> the ones continuing them in the direction of A <be described, also>.

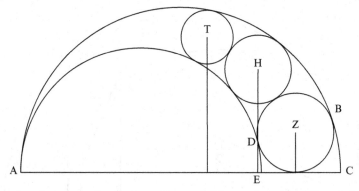

Now, that the perpendicular from Z onto AC is equal to the radius of the circle around Z, is clear. However, I say that, also, the perpendicular from H is three times the radius of the circle around H, whereas the <perpendicular> from T <is> five times <the radius of the circle around T>, and the perpendiculars in sequence <are> the multiples of the radii in accordance with the odd numbers in sequence.

---

[1] BC:BD = BD:BK (VI, 17) ⇒ (BC − BD):(BD − BK) = BC:BD (V, 19), i.e.: CD:DK = BC:BD ⇒ BC:CD = BD:DK (V, 16). TZ = DK by construction, thus: BC:CD = BD:TZ.

[2] ΔBAD ~ ΔTAZ, because TZ ∥ BD and B − A − Z, D − A − T are straight lines (I, 29). Therefore, BD:DA = TZ:TA (VI, 4), and BD:TZ = AD:TA (V, 16).

[3] In the right-angled triangle TZD with height AZ, we have: TZ:TA = TD:TZ, and ZD:TD = AD:ZD (VI, 8); therefore: TZ × TZ = TD × TA, and ZD × ZD = TD × AD (VI, 16). Therefore: (TZ × TZ):(ZD × ZD) = (TD × TA):(TD × AD) = TA:AD (VI, 1). An explicit use of abstract duplicate ratios, interpreted as ratios of squares, as suggested by Hultsch here (p. 233 Hu), can be avoided. Compare also Co p. 81, F.

[4] We have shown above: BC:CD = BD:TZ = DA:TA, and finally: TA:AD = (TZ × TZ):(ZD × ZD). Apply V, 16.

[5] This statement may be an indication that Pappus himself thought that there were at least two layers present in the source he is using.

[6] A − E − C.

For since it has been shown above that as the perpendicular from Z, taken together with the diameter <is> to the diameter, so <is> the perpendicular from H to its respective diameter,[1] and <since> the perpendicular from Z, taken together with the diameter, is 3/2 of the diameter, it <i.e., the perpendicular from H> will therefore be three times the radius.

Again, since the perpendicular from T is to the <corresponding> diameter as the perpendicular from H, taken together with the <corresponding> diameter <is> to the <corresponding> diameter,[2] whereas the perpendicular from H, taken together with the diameter has to the diameter the ratio that 5 has to 2,[3] the perpendicular from T will have that same ratio to the diameter as well. It will therefore be five times the radius. Similarly, it will be shown that the perpendiculars in sequence are multiples of the radii in accordance with the odd numbers in sequence.[4]

Props. 19–22: Archimedean Spiral

Prop. 19: Genesis and Symptoma of the Spiral

#21 The Samian geometer Konon put forth the theorem concerning the spiral described in the plane, whereas Archimedes proved it, employing a certain astonishing plan of attack.[5] The line, however, has a *genesis*[6] of the following sort.

Let there be given a circle with center B and radius BA.[7] Assume that the straight line BA has been set in motion[8] in such a way that, while B remains in its place,

---

[1] Prop. 15.

[2] Prop. 15.

[3] Perpendicular from T: radius ~ 3:1; therefore, (perpendicular from T + 2 radii): 2 radii ~ 5:2. Note that in this phrasing, numbers and magnitudes are again kept apart conceptually.

[4] This is, again, an argument by (complete) induction. Here the odd numbers are viewed as an infinite sequence, in ratios.

[5] This statement is misleading. According to the proem of Archimedes's *Spiral Lines*, it was Archimedes himself who proposed the theorem, challenging Konon to prove it. When the latter died before being able to seriously attempt the task, Archimedes proceeded to publish his own treatment, referencing Konon as the original addressee and intended discussion partner.

[6] The Greek term γένεσις means coming-to-be, creation, growing. It is used in every generation of a motion curve in *Coll.* IV, and I have left it untranslated.

[7] Note that the circle is given from the start, and the spiral inscribed in it. In Archimedes' *Spiral Lines* (*SL*), the spiral is created from two given motions, and the circle is described afterward around it. Only the version in *Coll.* IV will yield the angle section and the squaring of a circle. See the commentary on *SL* versus *Coll.* IV, and on *symptoma*-mathematics of motion curves.

[8] κεκινήσθω. Throughout the descriptions of the motion curves, Pappus will use either κινεῖν or φέρεσθαι. Perhaps the two terms have a slightly different meaning. For lack of examples it is not possible to determine what the difference would amount to. I have chosen to render "κινεῖν" with "move", and φέρεσθαι with "travel". In ordinary usage κινεῖν is the broader term, whereas φέρεσθαι is restricted to locomotion.

A travels uniformly along the circumference of the circle,[1] and together with it
<i.e., together with the rotating BA> a certain point, starting from B, is assumed to
travel uniformly along it, in the direction of A, and assume that within the same
time the point from B passes through BA and A passes through the circumference
of the circle.[2] Now, the point moving along BA will describe a line such as BEZA
during the rotation, and its starting point will be the point B, while the starting point
of the rotation will be BA.

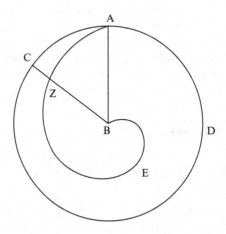

This line,[3] however, is called spiral. And its principal *symptoma*[4] is of the
following sort. Whichever <straight line> is drawn through the interior toward it,
such as BZ, and produced <to C>, the straight line AB is to the <straight line>
BZ as the whole circumference of the circle is to the arc ADC.[5] This, however, is

---

[1] Rotation is in all likelihood clockwise, though counterclockwise rotation is possible, too. The
synchronous linear motion is "inside out."

[2] The two motions have to be synchronized, using the ratio of radius to circumference of a circle, i.e., π.
The implicit inclusion of π is the reason why this version of the Archimedean spiral can be used
to divide an angle in any given ratio (cf. Prop. 35), and also to square the circle (invoking *SL* 18).
It also creates problems for the conceptualization of this version of the spiral. See the commentary.

[3] Reading A's "αυτη" as "αὕτη"; both Hultsch and Treweek prefer "αὐτή" (234, 18 Hu; 101, 7 Tr).

[4] ἀρχικόν σύμπτωμα. The word "archikon" implies the idea of "original" as well as "principal."
In fact, the main, property of the curve, the one on which the mathematical argumentation draws,
stems directly from the curve's origin. A similar use of ἀρχικόν can be found at 252, 21 Hu for
the quadratrix. The word *symptoma* originally denoted a chance happening or casualty. Within
Hippocratic medicine, it was used to label the signs (symptoms) of a disease, the observable characterizing property of a subject of study, the one the expert will look for and work with. Drawing on
this scientific usage, it is then used in geometry for characterizing higher curves, and sometimes
even conic sections. It obviously plays the role of a technical term; and I have left it untranslated.
The *symptoma* of the spiral here, expressible in strictly mathematical terms, derives directly
from the *genesis*, from the origin of the curve. The subsequent mathematical arguments, however,
use the *symptoma* as a principle in the mathematical argumentation, as a quasi-definition of the
curve, avoiding any reference to the *genesis*. For the significance of this move see the commentary
on *symptoma*-mathematics.

[5] Compare *SL* 14 (together with *SL* 2), for a spiral with circumscribed circle.

rather easy to understand from the *genesis* <of the spiral>. For in the time in which the point A passes through the whole circumference of the circle, in that time the <point starting> from B <passes through> BA, also, whereas in the time in which A <passes through> the arc ADC, in that time the <point starting> from B <passes through> the straight line BZ, also. And these motions are of uniform speed,[1] so that the <above mentioned> proportion holds, also.

Prop. 20: Progression of Spiral Radii[2]: Proportional to Rotation Angles

And the following is also obvious: that all straight lines drawn through in the interior from B to the line and containing equal angles exceed each other by the same <line in length.>[3]

Prop. 21: Spiral Area[4] in Relation to the Circle

#22 It is shown, however, that the figure contained by the spiral and the straight line at the starting point of the rotation is the third part of the circle comprising it.[5]

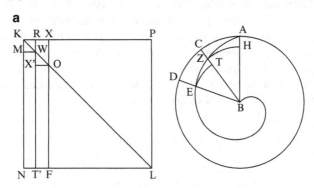

---

[1] Pappus' explanation here is not very felicitous. It unduly fuses concepts of motion and speed. Nevertheless, Knorr (1978a, p. 50 f.) goes too far in concluding that Pappus misunderstood the whole mathematical context.

[2] Line segments between the center of the original circle and the spiral will be called "spiral radii" here.

[3] Spiral radii corresponding to equal angle increments form an arithmetical sequence. Compare *SL* 12 (with *SL* 1). The property follows directly from the *genesis*. For an elementary argument that the spiral radii for equal increments of angles form an arithmetical series see Co p. 83 commentary on Prop. 20.

[4] Areas contained by the spiral line and the spiral radius at some point of the rotation will be called "spiral areas" here. Prop. 21 addresses the spiral area for the first complete rotation.

[5] Compare *SL* 24 (for a spiral with circumscribed circle); note the difference in argumentation. Prop. 21 uses quasi-indivisibles, whereas *SL* 24 has a classical proof via double reductio, and uses a progression of spiral radii (*SL* 12 (with *SL* 10), cf. Prop. 20). See the commentary.

For:

Let there be <given> both the circle and the above-mentioned line, and set out a rectangle KNLP, and cut off, on the one hand, the arc AC as a certain part of the circumference of the circle, and, on the other hand, the straight line KR as the same part of KP,[1] and join both BC and KL,[2] and <draw> the parallel RT' to KN, and the parallel WM to KP, and finally, <describe> the arc ZH around center B.

Now, since as the straight line AB is to AH, i.e.: <as> BC <is> to CZ, so is the whole circumference of the circle to the <arc> CA (for this is the principal *symptoma* of the spiral),[3] whereas as the circumference of the circle <is> to <the arc> CA, <so is> PK to KR, and as PK <is> to KR, <so is> LK to KW, i.e.: RT' to RW, therefore T'R <is> to RW as BC is to CZ, also.[4] And convertendo, therefore, as the square over BC <is> to the square over BZ, so <is> the square over RT' to the square over T'W, also.[5] But, on the one hand, as the square over BC <is> to the square over BZ, so <is> the sector ABC to the sector ZBH.[6] On the other hand, as the square over RT' <is> to the square over T'W, so <is> the cylinder over the rectangle KT' around the axis NT' to the cylinder over the rectangle MT' around the same axis.[7] And therefore, as the sector CBA <is> to the sector ZBH, so <is> the cylinder over the rectangle KT' around the axis NT' to the cylinder over the rectangle MT' around the same axis.

Similarly, however, when we set down, on the one hand, an <arc> CD equal to the <arc> AC, and on the other hand, RX equal to KR, and go through the same constructions, the cylinder over the rectangle RF around the axis T'F will be to the cylinder over the rectangle X'F around the same axis as the sector DBC is to the <sector> EBT. Proceeding in the same manner, however, we will show that as the whole circle <is> to all the figures <constituted> out of sectors that are inscribed in the spiral <taken together>, so <is> the cylinder over the rectangle NP around the axis NL to all the figures <constituted> out of cylinders that are inscribed in the cone over the triangle KNL around the axis LN <taken together>.

---

[1] The ratio for the division is not specified. Most likely, it is $1:2^n$.

[2] Adopting Tr's emendation KL for A's KA (Tr 101, 26).

[3] *Symptoma*: AB:BZ = AB:BH = circumference:arc AC. Thus, AB:(AB − BH) = circumference: (circumference-arc AC) (V, 19).

[4] The path of reasoning in this somewhat lengthy sentence is rather straightforward. Because (AB:AH =) BC:CZ = circumference:arc CA (due to the spiral), while we also have circumference:arc CA = PK:KR (by construction) = LK:KW (VI, 2 and V, 18) = T'R:RW (VI, 4; V, 16; V, 18), we can infer that T'R:RW = BC:CZ.

[5] BC:CZ = T'R:RW implies BC:BZ = T'R:T'W (V, 19, addition). Then the stated proportion holds for the squares (VI, 22).

[6] XII, 2 (circles have the ratio of the squares over the diameters). The sectors in Prop. 21 are the same parts of their respective full circles (use VI, 33 and V, 15).

[7] XII, 11 (cylinders of equal height have the ratio of the circles at their base), and XII, 2 (circles have the ratio of the squares over their diameters).

And again: as the circle <is> to all the figures <constituted> out of sectors circumscribed around the spiral <taken together>, so <is> the cylinder to all the figures
constituted out of cylinders circumscribed around the same cone <taken together>.

From this <result> it is obvious that, as the circle <is> to the figure[1] between the
spiral and the straight line AB, so <is> the cylinder to the cone.[2] However, the cylinder
is three times the cone. Therefore, the circle is three times the said figure, also.[3]

Addition: Spiral Areas and Circumscribed Circles

#23 In the same manner we will show that, when a certain straight line is drawn
through in the interior to the spiral, such as BZ, and the circle through Z around
center B is described, the figure contained by both the spiral ZEB and the straight
line ZB is the third part of the figure contained by both the arc ZHT of the circle
and the straight lines ZB, BT, also.[4]

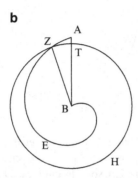

Now, the proof[5] <of Konon's theorem mentioned in the beginning, before Prop. 19>
is of such a sort. In what follows in sequence, however, I will write down a theorem
that holds for the same line and deserves investigation.[6]

---

[1] The manuscript A has a plural (σχήματα, 238, 17 app. Hu). Perhaps we do have a scribal error
here, but it is also possible that Archimedes viewed the spiral area in this argument as actually
composed of spiral sectors with quasi-indivisible arcs.

[2] Archimedes uses an argument that could be called "exhaustion" in the literal sense. It closely
resembles arguments via indivisibles. See the commentary.

[3] XII, 10; Knorr (1978a) p. 55) notes that the reference to XII, 10 leaves a gap. II does not cover
the implicit convergence argument for the spiral-figures, which is, however, crucial here.

[4] The addition targets a spiral segment with circumscribed circle. Thus, it is the true parallel to *SL
24*. The labeling of the spiral is "outside –in", in contrast to the description in Prop. 19.

[5] The Greek term for the argument in Prop. 21 is indeed *apodeixis* (cf. 238, 26 Hu), suggesting that
Pappus may very well have considered it as more than just heuristic exploration.

[6] ἱστορίας ἄξιον; the word ἱστορία does mean "history," among other things. Its original meaning
is "investigation", or "research."

Prop. 22: Ratio of Spiral Areas as Ratio of Cubes over Maximal Spiral Radii

#24 Let there be given both the circle mentioned above in the *genesis* and the spiral AZEB itself.[1]

I claim that, whichever <straight line> is drawn through in the interior <to the spiral>, such as BZ, the cube over AB is to the cube over ZB as the figure contained by the whole spiral and the straight line AB <is> to the <figure> contained by the spiral ZEB and the straight line BZ.

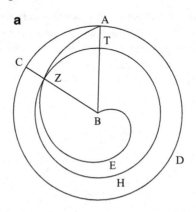

For:

Describe the circle ZHT through Z around center B. Now, since as the figure contained by the line AZEB and the straight line AB <is> to the figure contained by the line ZEB and the straight line ZB, so is the circle ACD to the figure contained by the arc ZHT and the straight lines ZB, BT (for both were shown to be the third part of both <circles>, respectively), whereas the circle ACD has to the area cut off by the straight lines ZB, BT and the arc ZHT the ratio composite of <the ratio> that the circle ACD has to the circle ZHT and <the ratio> that the circle ZHT has to the area cut off by the straight lines ZB, BT and the arc ZHT, but as the circle ACD <is> to the circle ZHT, so <is> the square over AB to the square over BZ,[2] whereas as the circle ZHT <is> to the said area, <so is> its whole circumference to the <arc> ZHT,[3] i.e., <so is> the

---

[1] Prop. 22 does not specify whether the circle is circumscribed or the spiral inscribed in a given circle. In both cases, we have a contribution to the *symptoma*-mathematics of the spiral. In the former case, the theorem would in addition be on a par, conceptually, with the theorems in *SL*. The fact that the circle is mentioned but not used in the theorem may be an indication that we still deal with the inscribed spiral. As in Prop. 22, and in difference from the description in Prop. 19, the spiral is labeled "outside-in."

[2] XII, 2.

[3] Theon's addition to VI, 33 (circles to sectors as circumference to arcs).

circumference of the circle ACD to the <arc> CDA, i.e.: <so is> the straight line AB
to the <straight line> BZ – on account of the *symptoma* of the line.[1]

The figure between the spiral and the straight line AB therefore also has to the figure
between the spiral and the straight line BZ the ratio composite of the ratio that the
square over AB has to the square over ZB and the ratio that AB has to BZ. This ratio,
however, is the same as[2] the one of the cube over AB to the cube over BZ.[3]

Addition: Measurement of Spiral Quadrants

#25 Now, from this <argument> it is obvious that, when, with the spiral and the circle
around it posited, AB is produced to D and CZEK is drawn through the interior at right
angles to it, the area between the line NME and the straight lines NB, BE amounts
to 7 <area units> of the <area units> of which the area between the line BLE and
the straight line BE amounts to 1, whereas the <area> between the line ZTN and the
straight lines ZB, BN amounts to 19 <of these units>, and finally the <area> between
the line AXZ and the straight lines AB, BZ amounts to 37 (for these claims are clear
from the theorem proved above), and that of <the length units of> which AB is 4, of
these ZB is 3, whereas BN is 2, and BE is 1. For this, also, is clear from the *symptoma*
of the line and the fact that the arcs AC, CD, DK, and KA are equal.[4]

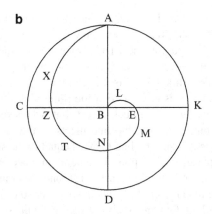

---

[1] Prop. 19.

[2] Prop. 22 uses Prop. 21 and the *symptoma* to express the desired ratio of spiral areas as a composite
ratio. It is the composite of a ratio of squares over radii and one of radii, and this is declared to be
equivalent to a ratio of cubes. The interpretation of composite ratios as quasi-products was not
without its difficulties, though Archimedes seems to have used composite ratios that way without
qualms cf. Saito (1986). Co p. 85, A refers to an Archimedean theorem on centers of gravity for
solids.

[3] Spiral area BA:spiral area BZ = (circle BA:circle BZ) × (BA:BZ) = (BA^2:BZ^2) × (BA:BZ) =
BA^3:BZ^3.

[4] Prop. 19 yields AB:BZ:BN:BE = 4:3:2:1. From Prop. 22, we see that corresponding full spiral
areas are as 64:27:8:1. Subtracting the preceding spiral sectors at each stage, we get 37:19:7:1 for
the spiral quadrants.

Props. 23–25: Conchoid of Nicomedes/Duplication of the Cube

Genesis and Symptoma of the Conchoid

#26 For the duplication of the cube a certain line is introduced by Nicomedes[1] and it has a *genesis* of the following sort.

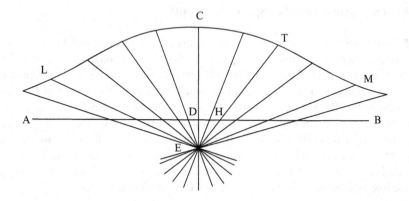

Set out a straight line AB, and a <straight line> CDZ at right angles to it, and take a certain point E on CDZ as *given*.[2] And assume that, while the point E remains in the place where it is, the straight line CDEZ travels along the straight line ADB, dragged via the point E in such a way that D travels on the straight line AB throughout and does not fall outside[3] while CDEZ is dragged via E. Now, when such a motion takes place on both sides, it is obvious that the point C will describe a line such as LCM is, and its *symptoma* is of such a sort that, whenever some straight line <starting> from the point E toward the line meets it, the <straight line> cut off between the straight line AB and the line LCM is equal to the straight line CD. For, while AB remains in place, and the point E remains in place, when D comes to be upon <a point> H, the straight line CD reaches HT, and the point C will fall onto[4] T. Therefore, CD is equal to HT.[5] Similarly, also, whenever some other line <starting> from the point E toward the line meets <it>, it will make the segment cut off by the line and the straight line AB equal to CD.

---

[1] For information on Nicomedes see the commentary. He is associated with the conchoid and the quadratrix (i.e., two of the prominent motion curves used for *symptoma*-mathematics in *Coll.* IV).

[2] δοθέν this is the same term used in geometrical analysis. See the commentary.

[3] The distance CD is kept equal throughout the "dragging process" (*neusis*-property of the curve); DE (corresponding to the pulling rope for a ship) is variable.

[4] Accepting Hu's addition of πεσεῖται (244, 11 + app. Hu), although Tr may be right in preserving the manuscript reading (Tr 105, 11).

[5] The *symptoma* seems to be read off a curve already drawn, not abstracted from the generating motion (as was the case for the spiral). See the commentary.

And, he says, let the straight line AB be called *"canon,"* and the point <E> *"pole,"* and CD *"distance,"* since the <straight> lines drawn toward the line LCM and meeting it are equal to this one, and finally the line LCM itself *"first conchoid"*[1] – since an exposition of a second and third and forth, put to use for other theorems, is also given.[2]

<Further information on the conchoid:>

#27 That, however, the line can be described with an instrument[3] and that it proceeds along the canon at an ever-decreasing distance, i.e.: that of all the perpendiculars <drawn> from any point of the line LCT to the straight line AB the perpendicular CD is the largest, and that a perpendicular drawn closer to CD is always larger than a <perpendicular drawn> further away, and also that when some straight line lies in the space between the canon and the conchoid, that line will, when produced, be intersected by the conchoid, Nicomedes himself has proved, and I myself have used the line mentioned above in the <treatise> on the analemma of Diodorus[4] when I wished to trisect the angle.[5]

Prop. 23: Neusis Construction[6]

Now, on account of what has been said it is obvious that it is possible, when an angle is *given,*[7] like the angle HAB, and a point C outside of it, to draw a <straight line> CH through the interior and to make the <intercept> KH between the line and AB equal to a *given* <straight line>.

---

[1] The spelling of the Greek name for the curve appears as κοχλοειδής in A and in Hu's text through the end of Prop. 25. In almost all occurrences in A, however, the λ was expunged later, and a γ superscripted, changing the name to κογχοειδής. It will be rendered as "conchoid" here.

[2] No documents about Nicomedes' theorems on the other conchoids survive.

[3] Greek for "with an instrument": "ὀργανικῶς". This term should be differentiated from the standard Greek term for "mechanical": μηχανικῶς. Co translates "instrumentaliter" (cf. at Co p. 89). For the significance of this difference see the commentary. What Pappus gives here is not a comment on the conchoid itself, as "mechanical," i.e., generated by motions, but a reference to the use of a concrete instrument, a "conchoid-compass", to draw the curve. Such a compass can be easily constructed from the description of the generation of the curve via motions (cf. Eut., *Comm. in Sph. et. Cyl.* II, pp. 98, 1–100, 14 Heiberg).

[4] We do not have a treatise by Pappus with this title. Information on Diodorus and a work on the *Analemma* is also very scarce (cf. Heath Vol. II, p. 286 f.). Hultsch p. 246 ad locum suspects a corruption of the text, and offers "lemma 1" or "lemma 21" as possible readings.

[5] This side remark documents that Pappus must have been aware of the connection between the angle trisection, the duplication of the cube, the *neusis* construction for which the conchoid operates like a compass, and typical solid problems in general. See the commentary, and Props. 31–33, 42–44.

[6] For a discussion of *neuses* and their role in Greek mathematics see the commentary. In *Coll.* IV, *neuses* are also put to use in Props. 31–34, and in Props. 42–44 (picking up a reference in the meta-theoretical passage between Prop. 30 and Prop. 31).

[7] The Greek text has δοθείσῃ, the term used in geometrical analysis. This suggests an analytic-synthetic background for the *neusis* and the conchoid (as a locus curve). Prop. 23 corresponds to Eut. *In Arch. Sph. et Cyl. II*, pp. 102–104 Heiberg. Compare also the apparatus in Hu ad locum, for parallels and doublets in *Coll.* III, pp. 58–60.

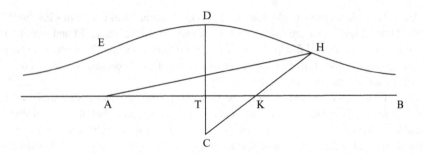

<For:>

Draw the perpendicular CT from the point C onto AB and produce it, and let DT be equal to the *given* <straight line,>[1] and describe the line "*first conchoid*" EDH with pole C, the *given* <line segment>, i.e.: DT, as distance, and AB as canon. On account of what has been said above it will then meet AH. Let it meet <AH> in H, and join CH. Then KH is equal to the *given* <straight line>.

<Neusis via mechanical manipulation, avoiding the use of the conchoid:>

#28 Some people, however, for the sake of usefulness, place a ruler to C and move it until, by trial, the <straight line> cut off between the straight line AB and the line EDH turns out to be equal to the *given* <straight line>.[2] For when this is the case, the <problem> set forth in the beginning is proved (I mean, however, a cube is found that is two times a <*given*> cube). Before that <i.e., before the exposition of the cube duplication itself>, however, two means in continuous proportion for two *given* straight lines are taken. Nicomedes has set out the construction for them only,[3] whereas I have also attached the proof to the construction, in the following manner.[4]

---

[1] D on CT, TD = given line.

[2] Probably Co p. 87 is right in suggesting "(straight) line AH" for "line EDH." Then the procedure by trial and error makes sense, and one avoids having to draw out the conchoid. For once the conchoid is drawn, trial and error is no longer needed, and the sense of Pappus' remark becomes unclear. The use of the term *given* may suggest an analytical context for Nicomedes' original considerations. Co p. 87 nevertheless justifies the success of the ruler manipulation construction with the conchoid.

[3] The Greek text has "μόνην," Hu 246, 22 emends to "μόνον," following Co, and Tr emends as well. The only conceivable sense one might make of the manuscript reading is for Pappus to indicate that Nicomedes furnished a single *neusis* construction, covering both the angle trisection and the cube duplication, whereas Pappus quotes the *apodeixis* of it, adapted to the case of two mean proportionals. Then Prop. 24 would still be essentially by Nicomedes, and Pappus would not claim more than his adaptation of it for the cube duplication here. This would diminish an apparent inconsistency entailed by the emended text: that Eutocius reports much the same *neusis* construction as Nicomedean, whereas in the emended text version Pappus seems to claim it for himself. See also the following footnote, and the commentary. Perhaps the manuscript reading could have been defended, then. Since this is a question of a single letter only, though, I follow the authority of the editors. In any case, the mathematical sense is not affected, and the majority of scholars ascribe the content of Prop. 24 to Nicomedes, even in face of the phrase in the emended text.

[4] Eutocius reports the very same argument in *In Arch. de Sph. et Cyl.* 104–106. Perhaps Eutocius is quoting from Pappus; cf. Ver Eecke (1933b, p. 188, # 3). Jones (1986a) considers the possibility that Eutocius draws on a report by Pappus in *Coll. VII.*

Prop. 24: Two Mean Proportionals via Neusis

Assume that two straight lines CL and LA, at right angles to each other, are *given*, of which to find two means in continuous proportion is the task, and complete the rectangle ABCL, and bisect both AB and BC in the points D and E, and, on the one hand, produce DL after it has been joined, and let it meet CB, after it has been produced, in H, on the other hand, <draw> EZ at right angles to BC and draw CZ toward it, equal to AD, and join ZH, and <draw> CT parallel to it, and since KCT is an angle, draw, from the *given* <point> Z, the <straight> line ZTK through the interior making TK equal to AD or CZ (for that this is possible on account of the conchoid line has been shown), and produce KL, after it has been joined, and let it meet AB, when it is produced, in M.

I say that as LC <is> to KC, <so is> KC to MA, and <so> is MA to AL.

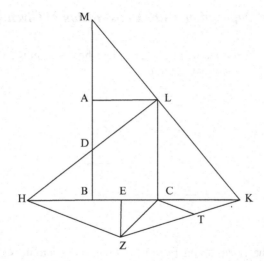

Since BC has been bisected in E, and KC has been added to it, the rectangle BKC, taken together with the square over CE, is therefore equal to the square over EK.[1] Add the common square over EZ. Then the rectangle BKC, taken together with the squares over CE and EZ, i.e.: <with> the square over CZ,[2] is equal to the squares over KE and EZ, i.e.: <to> the square over KZ.[3] And since as MA <is> to AB, <so is> ML to LK, whereas as ML <is> to LK, <so is> BC to CK,[4] therefore as MA <is> to AB, <so is> BC to CK, also. And AD is half of AB, whereas CH is twice BC.[5] Therefore, HC will be to KC as MA <is> to AD.[6] But as HC <is> to CK,

---

[1] II, 6.
[2] I, 47.
[3] I, 47.
[4] VI, 2 with V, 16 (ΔMBL ~ ΔMAL, ΔMBK ~ ΔLCK, on parallel lines).
[5] ΔADL ≅ ΔBDH (I, 26), therefore HB = AL (=BC).
[6] V, 15: MA:AB = BC:CK implies MA:1/2AB = 2BC:CK.

so <is> ZT to TK, on account of the parallels HZ and CT.[1] And therefore, componendo: as MD <is> to DA, <so is> ZK to KT.[2] However, AD has also been posited as equal to TK.[3] Therefore, MD is equal to ZK[4] as well. Therefore, the square over MD is equal to the square over ZK, also. And the rectangle BMA, taken together with the square over DA, is equal to the square over MD,[5] whereas the rectangle BKC taken together with the square over ZC has been shown to be equal to the square over ZK. Of these, the square over AD is equal to the square over CZ (for AD has been posited as equal to CZ). Therefore, the rectangle BMA is equal to the rectangle BKC, also. Therefore, as MB <is> to BK, <so is> CK to MA.[6] But as BM <is> to BK, <so is> LC to CK.[7] Therefore, as LC <is> to CK, <so is> CK to AM. However, MA is to AL as MB <is> to BK, also.[8] And therefore, as LC <is> to CK, <so is> CK to AM, and <so is> AM to AL.

Prop. 25: Cube Duplication, Cube Construction in Given Ratio

#29 After this has been shown, it is very clear how one must, when a cube is given,[9] find another cube in a *given* ratio.

$$\text{—————————}$$

a

$$\text{—————}$$

c

$$\text{———————}$$

d

$$\text{————————}$$

b

For:

Assume that the *given* ratio is that of the straight line a to the <straight line> b, and take c and d as two means in continuous proportion for a and b. Then the cube over a will be to the cube over c as a is to b. For this is clear from the *Elements*.[10]

---

[1] VI, 2 (△HZK ~ △CTK on parallel lines).

[2] V, 18.

[3] By construction (*neusis*).

[4] V, 9.

[5] II, 6.

[6] VI, 16.

[7] VI, 4 (△MBK ~ △LCK).

[8] VI, 4 (△MAL ~ △MBK).

[9] Once again, note the occurrence of derivatives of the technical term δοθέν (250, 26, 27, and 28 Hu).

[10] a and c stand in the triple ratio of a:b (V, def. 11); cube numbers have two mean proportionals, and cube:cube = (side:side)³ (VIII, 12); the cubes with sides a and c stand in that same ratio (XI, 33).

Props. 26–29: Quadratrix[1]

Genesis and Symptoma of the Quadratrix

<u>#30</u> For the squaring of the circle a certain line has been taken up by Dinostratus and Nicomedes[2] and some other more recent (mathematicians). It takes its name from the *symptoma* concerning it. For it is called "quadratrix" by them, and it has a *genesis* of the following sort.

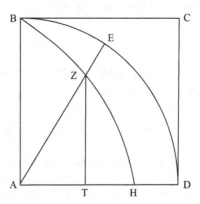

Set out a square ABCD and describe the arc BED of a circle with center A, and assume that AB moves in such a way that while the point A remains in place, <the point> B travels along the arc BED, whereas BC follows along with the traveling point B[3] down the <straight line> BA, remaining parallel to AD throughout, and that in the same time both AB, moving uniformly, completes the angle BAD, i.e.: the point B <completes> the arc BED, and BC passes through the straight line BA, i.e.: the point B travels down BA.[4] Clearly it will come to pass that both AB and

---

[1] The Latin word "quadratrix" (i.e., squaring line) translates the Greek name (τετραγωνίζουσα) for the transcendent curve that will be the subject of Props. 26–29. The Latin version is commonly used as the standard name for this particular curve, though the term can have other meanings, too.

[2] The common author Nicomedes connects the passages on the conchoid and quadratrix curves. Dinostratus was a late fourth century BC mathematician, the brother of Menaechmus, who invented the conics as locus curves. On the authorship concerning the curve quadratrix and its *symptoma*-mathematics see the commentary.

[3] συνακολουθείτω; the basic verb is, once again "ἀκολουθέω" = follow along in order. As in the other instances in *Coll.* IV, it does not have the connotation of strict logical derivation – on the contrary (see below). On the use of "ἀκολουθεῖν" compare the remarks on analysis-synthesis in the introduction to Props. 4–12.

[4] This generation via synchronized motions is reminiscent of the *genesis* of the spiral in Prop. 19; the connection between these two curves has been emphasized by Knorr (e.g., Knorr 1978a, 1986).

BC reach the straight line AD at the same time. Now, while a motion of this kind is taking place, the straight lines BC and BA will intersect each other during their traveling in some point that is always changing its position together with them. By this point a certain line such as BZH is described in the space between the straight lines BA and AD and the arc BED, concave in the same direction <as BED>, which appears to be useful, among other things, for finding a square equal to a given circle.[1]

And its principal *symptoma* is of the following sort. Whichever arbitrary <straight line> is drawn through in the interior toward the arc, such as AZE, the straight line BA will be to the <straight line> ZT as the whole arc <BED is> to the arc ED. For this is obvious from the *genesis* of the line.

Criticism of the Quadratrix Under the Description via Motions (Sporus)

#31 Sporus, however, is with good reason displeased with it, on account of the following <observations.>[2]

For, first of all, he[3] takes into the assumption the very thing for which it <i.e., the quadratrix> seems to be useful. For how is it possible when two points start from B, that they move, the one along the straight line to A, the other along the arc to D, and come to a halt <at their respective end points> at the same time, unless the ratio of the straight line AB to the arc BED is known beforehand? For the velocities of the motions must be in this ratio, also.[4] Also, how do they think that they[5] come to a halt simultaneously, when they use indeterminate velocities, except that it might happen sometime by chance; and how is that not absurd?

---

[1] The quadratrix can be used also for the division of an angle in any given ratio (probably its original use), and for problems related to this construction. Cf. Props. 35–38.

[2] The passage taken from Sporus differs significantly from the mathematical expositions in *Coll.* IV. Note, e.g., the rhetorical questions and the polemical style. Co p. 88 replaces the name "Sporos" with the Latin word "spero." His paraphrase means: "I expect, however, that this line justifiedly and deservedly does not satisfy, for the following reasons." The replacement changes the meaning of the introductory sentence, and indeed of the whole passage criticizing the quadratrix considerably.

[3] The Greek text uses the third person singular. It is unclear whom Sporus' argument targeted.

[4] The use of the notion "velocity" is not quite precise here. However, it is clear what Sporos means, and his argument is valid. In order to synchronize the two motions as required, one must know π – or else use an approximation to stand in for it. However, π is exactly what the curve is supposed to exhibit in construction. Co p. 88 paraphrases "motuum velocitates." Hu 254, 7 emends A's elliptical "ἀναγκαῖον." For a parallel construction, without emendation, see, however 270, 11/12 Hu.

[5] The reading πῶς οἴονται (how do they think) as given in A, was kept. Both Hultsch and Treweek reject it in favor of the reading πῶς οἰόν τε (254, 8 Hu + app/ Tr. 109, 11), attested in the minor manuscripts. Co p. 88 paraphrases "quo pacto arbitrantur."

Furthermore, however, its endpoint, which they use for the squaring of the circle, i.e.: the point in which it intersects the straight line AD, is not found <by the above generation of the line>. Consider what is being said, however, with reference to the diagram set forth. For when the <straight lines> CB and BA, traveling, come to a halt simultaneously, they will <both> reach AD, and they will no longer produce an intersection in each other. For the intersecting stops when AD is reached,[1] and this <last> intersection would have taken place as the endpoint of the line,[2] the <point> where it meets the straight line AD. Except if someone were to say that he considers the line to be produced, as we assume straight lines <to be produced>, up to AD. This, however, does not follow from the underlying principles, but <one proceeds> just as if the point H were taken after the ratio of the arc to the straight line had been taken beforehand.[3] Without this ratio being given,[4] however, one must not,[5] trusting in the opinion of the men who invented the line, accept it,[6] since it is rather mechanical.[7] Much rather, however, one should accept the problem that is shown by means of it.[8]

---

[1] Restoring A's reading πρὸς (when) instead of Hultsch's πρὸ (= before; cf. 254, 16 Hu app).

[2] Restoring, with Tr 109, 20, the reading of A.

[3] For an extension of the quadratrix to the base line one needs to know the direction. As the quadratrix does not have a constant direction, or even curvature, one needs, in the end, to know the position of H, and it would have to be determined beforehand, using the ratio of radius and circumference (π). My translation differs from Hultsch's Latin interpretation. Co has the following Latin paraphrase, rejected by Hultsch (p. 89 Co): Sed ut cumque sumatur punctum …, praecedere debet proportio circumferentiae ad rectam lineam.

[4] The Greek word (δοθῆναι) is the technical term from geometrical analysis. It is not certain (in fact perhaps unlikely) that Sporus, whom Pappus paraphrases here, intended it that way. What is certain, however, is that Pappus is going to interpret it in this strict technical sense for Props. 28 and 29. See below, and see the commentary on Props. 26–29.

[5] Accepting Hultsch's emendation οὐ for the difficult manuscript reading ἤ, kept in Tr 109, 26. Co p. 89 keeps the manuscript reading, and paraphrases as a question: Or should we… ? The disadvantage is that in that case one would have expected the question particle at the beginning of the sentence.

[6] I.e.: accept it as fully geometrical. The quadratrix itself (in the motion description) is not fully accepted; but note the upcoming remark on the mathematics *about* it. It is quite possible that Sporus and Pappus have different opinions on this matter. The issue cannot be pursued here.

[7] Greek: μηχανικώτεραν. This word, used for the curve itself here, and not just for the way in which it is generated, is different from the label "ὀργανικῶς", i.e., "describable with an instrument". The latter was used in connection with Nicomedes' conchoid (cf. footnotes above). Hultsch deletes the phrase "and it is put to use by the students of mechanics for many problems" as an interpolation (254, 24–256, 1+app. Hu). There is indeed no evidence that the quadratrix played a major role in mathematical treatises on mechanics. A similar phrase occurs at 244, 20 Hu. See the introduction to Props. 19–30 in the commentary on "mechanical."

[8] Hultsch has changed the transmitted text considerably. His Latin paraphrase means: "but before I must report (assuming παραδοτέον) the problem that is solved on account of it." With Tr 110, 1–2, I keep the transmitted text. Co's paraphrase on p. 89 is compatible with this reading. See the commentary.

Prop. 26: Rectification of the Arc of a Quadrant

When a square ABCD is \<given\>, and the arc BED with center C,[1] and when the quadratrix BHT has come to be[2] in the above said way, it is shown that as the arc DEB \<is\> to the straight line BC, so \<is\> BC to the straight line CT.[3]

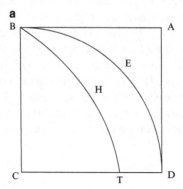

For:

If it is not \<in that ratio to CT\>, it will be \<in that ratio\> either to a \<straight line\> larger than CT or to one smaller.[4]

Assume first that, if this is possible, it is so to a larger \<straight line\> CK, and describe the arc ZHK with center C, intersecting the line in H, and \<draw\> HL as a perpendicular \<onto CD\>, and produce CH, after it has been joined, to E.

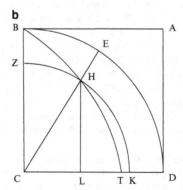

---

[1] Note the change of lettering in the diagram. Perhaps Prop. 26 was taken from a different source (Nicomedes, as opposed to Dinostratus, or else Sporus, for the curve's *genesis*?).

[2] Note that the quadratrix is posited at the outset. The upcoming argument will keep the problematical *genesis* of the curve out of sight, and use its *symptoma* only.

[3] This proportion will yield the construction of a straight line equal to arc DEB (Prop. 27).

[4] We get a classical proof via double reductio (so-called method of exhaustion). Apart from the (short and straightforward) alternative argument for the inverse of Prop. 13, this is the first, and the only, example for this argumentative technique in *Coll.* IV. On Prop. 26 see also Heath (1921, I, pp. 226–229).

Now, since as the arc DEB <is> to the straight line BC, so is BC, i.e.: CD, to CK,[1] whereas as CD <is> to CK, <so is> the arc BED to the arc ZHK (for as the diameter of a circle <is> to the diameter <of a second circle>, <so is> the circumference of the circle to the circumference <of the second circle[2]>), it is obvious that the arc ZHK is equal to the straight line BC.[3] And since, on account of the *symptoma* of the line, BC is to HL as the arc BED <is> to the arc ED, therefore, as the arc ZHK <is> to the arc HK, so <is> the straight line BC to HL,[4] also. And it has been shown that the arc ZHK is equal to the straight line BC. Therefore, the arc HK is equal to the straight line HL as well, which is absurd.[5] Therefore, it is not the case that as the arc BED <is> to the straight line BC, so is BC to a <straight line> larger than CT.

#32 I say, however, that it <i.e., BC> is not <in that ratio> to a <straight line> that is smaller, either.

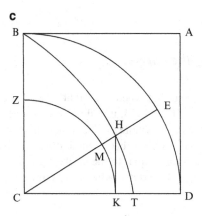

---

[1] By assumption.

[2] This theorem is also used in Props. 36, 39, and 40, and a similar one in Prop. 30 (cf. notes ad locum). An explicit proof is given in *Coll.* V, 11 and *Coll.* VIII, 22. A possible justification might proceed as follows. XII, 2: circles have the ratio of the squares over their diameters; *Circ. mens.* I: circles have the ratio of the rectangles with radius and circumference as sides; V, 16 and VI, 1: circumferences have the ratio of diameters. V, 15: similar arcs have the ratio of diameters. The frequent occurrence of this motif may indicate that it is part of the special "jargon," a kind of basic tool within the "analytic track" of *symptoma*-mathematics of the third kind. Specifically, it might be a typical tool of Nicomedes. Nicomedes apparently systematically exploited properties of spiral lines, taking Archimedean arguments as a starting-point. Compare Pappus' remarks on the study of spiral lines and quadratrices as a central branch of geometry of the linear kind in the upcoming meta-theoretical passage.

[3] BC:CK = CD:CK = arc BED:BC (assumption); CD:CK = arc BED:arc ZHK ⟹ BC = arc ZHK (V, 9).

[4] arc BED:arc ED = BC:HL (*symptoma*). arc BED:arc ED = arc ZHK:arc HK (equal parts).

[5] arc ZHK:arc HK = BC:HL; arc ZHK = BC ⟹ arc HK = HL (V, 9). This is not possible, because 2HL is a chord under two times arc HK.

For if this is possible, assume that it is <in that ratio> to KC, and describe the arc ZMK with center C, and <draw> KH at right angles to CD intersecting the quadratrix in H, and produce CH, after it has been joined, to E. Similarly to what has been written above, then, we will show both that the arc ZMK is equal to the straight line BC, and that as the arc BED <is> to the <arc> ED, i.e.: <as> the <arc> ZMK <is> to the <arc> MK, so <is> the straight line BC to the <straight line> HK.[1] From these <observations> it is obvious that the arc MK will be equal to the straight line KH, which is absurd.[2] Therefore, it will not be the case that as the arc BED <is> to the straight line BC, so is BC to a <straight line> smaller than CT.

It has been shown, however, that it is not <in that ratio> to a larger one, either. Therefore, it <is in that ratio> to CT itself.

Prop. 27: Squaring the Circle

It is obvious, also, however, that when a straight line is taken as the third proportional to the straight lines TC and CB, it will be equal to the arc BED, and its fourfold to the circumference of the whole circle.[3] When, however, a straight line equal to the circumference of the circle has been found, it is very clear that it is rather easy indeed to put together a square equal to the circle itself. For the rectangle between the circumference of the circle and the radius is two times the circle, as Archimedes has shown.[4]

---

[1] Just as in the first part of the "exhaustion," one gets: CD:CK = arc BED:BC (assumption); arc BED:arc ZMK = CD:CK $\Rightarrow$ arc ZMK = BC (V, 9). arc BED:arc ED (= arc ZMK:arc MK) = BC:HK (*symptoma*).

[2] HK must be larger than arc MK. I am not aware of an elementary geometrical argument in ancient geometry for this (correct) statement. Hultsch and Ver Eecke (1933b) ad locum refer to an argument that can be reconstructed from (Ps.-) Euclid, *Catoptrics* 8.

[3] Construct the third proportional s for TC and CB (VI, 11): TC:BC = BC:s; TC:BC = BC:arc BD (Prop. 26 with V, 16) $\Rightarrow$ s = arc BD. Then 4 s is equal in length to the circumference of the circle.

[4] *Circ mens*. I. This rectangle can be transformed into a square via II, 14.

Prop. 28: Analytical Determination of the Quadratrix from an Apollonian Helix

#33 Now, this *genesis* of the curve is, as has been said, rather mechanical[1]; it can, however, be made the subject of a geometrical analysis[2] by means of loci on surfaces in the following way.

<Let> the quadrant ABC of a circle <be *given*> in position, and assume that BD has been drawn through the interior arbitrarily, and also a perpendicular EZ onto BC, which has a *given* ratio to the arc DC.

<I claim> that E lies on a <uniquely determined> line.[3]

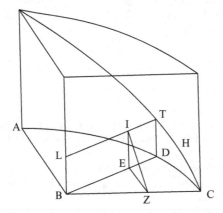

---

[1] Here Pappus picks up the discussion before Prop. 26, on the generation of the quadratrix via motions and the mathematical status of the quadratrix.

[2] ἀναλύεσθαι; since this is a technical term, clearly referring back to the technique of analysis (cf. Props. 4–12, and 31 ff.), Hultsch's Latin paraphrase "problema solvitur" does not capture the meaning and is in fact misleading. What is "analyzed" here is not the problem of squaring the circle, but the *genesis* of the quadratrix. Co paraphrases "lineae ortus … resolvi potest (p. 90). Both Prop. 28 and Prop. 29 provide a *resolutio* in the sense that they show that the quadratrix is *given*, if an Apollonian helix or an Archimedean spiral is posited (i.e., taken as *given*). See the commentary.

[3] With EZ: arc DC *given*, E will be shown to lie on a line that is determined relative to a certain helix, which is assumed as *given*. This characterization is independent from the *genesis* of the line via motions, which has been disqualified as conceptually inconsistent. It is not constructive, however, but rather a characterization via implicit relations. Note that the analysis is quite general in the sense that the ratio which is taken as *given* is not assumed to be the ratio of arc and radius, as in the quadratrix. Co p. 90, B is, in my view, mistaken when he assumes that. For each *given* ratio, the analysis shows that a unique line is determined by it via the intersection of the surface related to a *given* cylindrical helix and a *given* plane. For the special case of a ratio equal to arc ABC:AB, this line will be the quadratrix. Compare the end of Prop. 28, and Hultsch, * on p. 259 and #2 on p. 261.

For:

Consider the surface of a right cylinder over the arc ADC,[1] and described in it a helix CHT, *given* in position,[2] and <let> TD be the side of the cylinder,[3] and draw EI and BL at right angles to the plane of the circle, and finally, draw TL through T as a parallel to BD.[4] Since the ratio of the straight line EI to the arc DC is *given* on account of the helix,[5] whereas the ratio of EZ to the <arc> DC is *given*, also, the ratio of EZ to EI will be *given*, also.[6] And ZE and EI are (*given*) as parallels in position.[7] Therefore, the joining <straight line> ZI is <*given*> as parallel in position, also.[8] And it is a perpendicular onto BC. Therefore, ZI lies in a plane intersecting <the cylinder,>[9] so that I <lies there>, also. It <lies>, however, on a surface belonging to the cylinder as well[10] (for TL travels through both the helix THC and the straight line LB, which is also itself *given* in position, while it remains parallel throughout to the underlying plane). Therefore, I lies on a <uniquely determined> line,[11] so that E does so as well.[12]

---

[1] Co p. 90 D assumes, mistakenly in my view, that the height of the quarter cylinder constructed has to be equal to AB, and that the defining ratio of linear upward motion to rotation is that of arc ADC:BC.

[2] For a definition of the helix cf. Heron, def. 8, 1 and 8, 2.

[3] TD is perpendicular to the plane of the circular quadrant; it is now considered as the height of the cylinder under discussion.

[4] We create a rectangle BDTL, with E on BD and I on HL, and EI parallel to DT.

[5] This ratio is implicit in the helix as the relation of rotation and upward motion in its *genesis*.

[6] *Data* 8. The sentence is truncated in A. Above, I have translated the text as emended by Hultsch (260, 8–10 + app. Hu, see also #3 on p. 261 Hu). Tr 111, 27–112, 3 prints an alternative reconstruction, closer to the actual manuscript reading, and therefore perhaps preferable (see the apparatus in the Greek text).

[7] *Data*, def. 15. When a line is *given* in position, the parallel to it through a *given* point is said to be *given as a parallel in position* (para thesei).

[8] *Data* 41 and 29.

[9] The manuscript is severely damaged by water in this place, and the text is not legible (cf. the apparatus in the Greek text). Hultsch's emendation "ἐν τέμνοντι ἄρα" leaves open the possibility that the intersecting plane is determined by EZ and ZI or else by BC and ZI, whereas Treweek's emendation identifies the plane in question as the one determined by EZ and ZI. The version with BC/ZI as intersecting plane has the drawback that the endpoint Z of the intersection line is not uniquely determined. The plane has BC in common with the garland – shaped surface created by the helix. The version with EZ /ZI has the drawback that one would have to know the exact position of either EZ or ZI, and it is unclear how that could be accomplished at this stage of the analysis. I therefore prefer the former version. See the commentary.

[10] Severe damage to the manuscript text here; cf. apparatus to the Greek text for different emendations suggested.

[11] The point I lies on the line of intersection between the abovementioned plane and the surface created by the ascending line BC in the cylinder.

[12] E lies on the projection of the line created on the cylindrical surface onto the plane.

Now, this has been subjected to analysis in a general way.[1] When, however, the ratio of the straight line EZ to the arc DC is the same as that of BA to the <arc> ADC, the above-mentioned line quadratrix comes to be.

Prop. 29: Analytical Determination of the Quadratrix from the Archimedean Spiral

#34 It can, however, also be made the subject of analysis[2] by means of a spiral described in the plane, in a similar way.

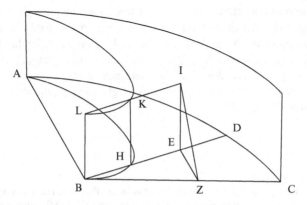

For:

Assume that the ratio of EZ to the arc DC is the same as the <ratio> of AB to the arc ADC,[3] and that in the time in which the straight line AB, moving around B passes through the arc ADC, a point on it, starting from A, arrives at B when AB

---

[1] ἀνελύθη. Compare the introductory phrase of Prop. 28 with note.

[2] The Greek text, again, has ἀναλύεσθαι. Compare the introductory phrase of Prop. 28.

[3] We are starting from a configuration with a section of a circle ABC and a part of it DBC. The arc ADC is not necessarily the arc of a quadrant (Co p. 91 is probably mistaken in assuming so). An Archimedean spiral will be assumed in it, and the analysis will show that any such configuration with spiral will determine a unique quadratrix-type line, though not necessarily the quadratrix itself. When a spiral is chosen with an inbuilt ratio equal to the ratio of the circumference of a quadrant to the radius, we get the quadratrix.

takes the position of CB,[1] and that it creates the spiral BHA. Then the arc ADC is to the <arc> CD as AB is to BH,[2] and alternate <this equation.>[3]

But EZ <is in that ratio> to <arc> DC, also.[4] Therefore, BH is equal to ZE.[5] Draw KH at right angles to the plane, equal to BH. Then K lies in a cylindroid surface over the spiral.[6]

It <lies>, however, also on the surface of a <uniquely determined> cone (for BK, when it is joined, turns out to lie on the surface of a cone inclined at an angle of 45° toward the underlying <plane>, and drawn through the *given* <point> B <as vertex>[7]). Therefore, K <lies> on a <uniquely determined> line.[8]

Draw LKI through K as a parallel to EB, and BL and EI at right angles to the <underlying> plane.[9] Then LKI (lies) on a plectoid[10] surface (for it travels both through the straight line BL, which is *given* in position and through the line, *given* in position, on which K <lies>). Therefore, I lies on a <uniquely determined> surface, also. But it also lies on a <uniquely determined> plane (for ZE is equal to EI, since it is also equal to BH, and ZI turns out to be *given* as a parallel in position, since it is a perpendicular onto BC). Therefore, I <lies> on a <uniquely determined> line,[11] so that E, also, <lies on a uniquely determined line>.

And it is clear that, when the angle ABC is a right <angle>, the above-mentioned line "quadratrix" comes to be.

---

[1] Compare the description of the *genesis* of the spiral before Prop. 19. The direction of the travel through AB and through the circumference is reversed in comparison to the former version. Also, the spiral is inscribed not in a full circle, but in a sector. The above translation accepts Hultsch's emendations in 262, 7–9. Tr 112, 17/18 prints Hultsch' s version, but notes that one might have emended ΓΔΑ in 262, 7 Hu and kept the manuscript reading for the rest of the sentence. Then the spiral is generated exactly like the one in Prop. 19. In the Greek text, Tr's suggestion was implemented (cf. apparatus).

[2] *Symptoma* of the spiral, following directly from the *genesis*.

[3] AB:arc ADC = BH:arc DC.

[4] By assumption.

[5] V, 9 or V, 15.

[6] This surface is built up over the spiral as limiting line of the base. Co p. 91/92 assumes a different situation, with a full cylinder quadrant and an inscribed Apollonian helix, in addition to the cylindroid. For yet another reconstruction cf. Knorr (1986, p. 166 f).

[7] By construction, HK = BH, and $\angle BHK = \pi/2$. Therefore, $\angle HBK = \pi/4$.

[8] K lies on the line created by the intersection of the two surfaces mentioned, cylindroid over the spiral, and surface of the cone with vertex B.

[9] Without loss of generality, L and I can be chosen as the points of intersection between the parallel to BD through T and the straight lines EI, BL.

[10] The Greek word πλεκτοειδῆς (πληκτοειδῆς in A, Tr 112, 27, and Ver Eecke ad locum) is used here as a technical term the context for which is now lost. Following Hultsch, I have left it untranslated. What a plectoid surface looks like can be derived from the description given here by Pappus. There is no other, independent source.

[11] I lies on the line created by the intersection of the surfaces mentioned.

Prop. 30: *Symptoma*-Theorem on the Archimedean Spherical Spiral

Prop. 30: Surfaces Cut Off by a Spiral on a Hemisphere

#35 Just as a certain spiral is contemplated in the plane when a point travels along a straight line that describes a circle, and in solids when a point travels along one of its sides,[1] while it describes a certain surface, so it is in fact a natural next step[2] to contemplate a spiral described on a sphere, in the following way.[3]

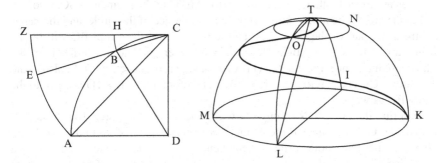

Let KLM be a maximum circle in a sphere with point T as pole, and assume that starting from T the quadrant TNK of a maximum circle is described, and that the arc TNK, traveling around T, which remains in its position, along the surface <of the sphere>, in the direction of the parts <containing> L and M, comes to a halt again in the same position, whereas a certain point traveling on it, starting from T, arrives at K. Now, it describes a certain spiral, such as TOIK on the surface,[4] and

---

[1] Severe damage to the manuscript text; see the apparatus for different conjectures.

[2] The Greek text has ἀκολουθόν; once again, we have a context in which the word cannot signify a logical derivation, and must mean a next step in a somewhat orderly fashion. See the commentary on analysis-synthesis in the introduction to Props. 4–12.

[3] Although this introductory paragraph draws an explicit connection to Props. 19, 28, and 29, the path of reasoning about the spiral line is very different from Props. 28 and 29. It shows affinities to Prop. 21 ("meta-mechanical" path of reasoning about the motion curves, quasi-infinitesimals, limit process, no analysis).

[4] Compare the *genesis* of the plane spiral in Prop. 19. The ratio of the velocities for the two synchronized motions involved in Prop. 30 is simply 4:1. Cf. equations in polar coordinates: spherical spiral $\rho = 1/4\ \omega$, plane spiral in Prop. 19 $\rho = (1/2\pi)\omega$, plane spiral in $SL\ \rho = a\omega$, where a is a natural number or a ratio of two numbers. The spherical spiral by motions can be constructed in thought exactly.

whichever arc of a maximum circle is described starting from T,[1] it will have to the arc KL the ratio that the <arc> LT has to the <arc> TO.[2]

Now, I claim that, when the arc ABC of a quadrant of the maximum circle in the sphere with center D is set out, and CA is joined, the sector ABCD turns out to be to the segment ABC as the surface of the hemisphere <is> to the surface cut off <from it> between the spiral TOIK and the arc KNT.[3]

For:

Draw CZ as a tangent to the arc <ABC>, and describe the arc AEZ <of the circle> through A with center C. Then the sector ABCD is equal to the <sector> AEZC (for the angle at D is two times the angle ACZ, whereas the square over DA is half the square over AC^4). Therefore, <we need to show> that, as the said surfaces are to each other, so <is> the sector AEZC to the segment ABC, also.[5]

Let the arc KL be a part of the whole circumference of the circle, and the <arc> ZE the same part of the <arc> ZA, and join EC. Now, the <arc> BC will be the same part of the <arc> ABC.[6] However, whichever part the <arc> KL is of the whole circumference, the <arc> TO is that same part of the <arc> TOL, also.[7] And the <arc> TOL is equal to the <arc> ABC. Therefore, the <arc> TO is equal to the <arc> BC as well.

Describe the circle ON through O with pole T, and the <arc> BH through B with center C. Now, since as the surface LKT on the sphere <is> to the <surface> OTN, so <is> the whole surface of the hemisphere to the surface of the section the spherical radius of which is TO,[8] whereas as the surface of the hemisphere

---

[1] Cf. the full circle going through LOTI, intersecting the spiral in O. Co p. 93, C is mistaken in assuming that arc KL is fixed as a quarter circle now. A division $1:2^n$ is likely (cf. Prop. 21).

[2] The *symptoma* of the spherical spiral is read off directly from the *genesis* via motions; cf. the plane spiral (Prop. 19) and the quadratrix (before Prop. 26), but contrast the conchoid (before Prop. 23). I have based the translation on Hultsch's emendations in 264, 16/17 Hu. Tr 113, 20–22 prints an emendation that is closer to the manuscript reading and is perhaps preferable (cf. apparatus).

[3] The formulation of the protasis is analogous to Prop. 21. An area theorem is expressed in terms of numerical ratios. Cf. Prop. 16: a theorem on a sequence of ratios of lines is expressed in numerical ratios.

[4] $\angle ADC = \angle ZCD = \pi/2$ (III, 18); $\angle ACZ = \angle ACD = \pi/4$ ($\triangle ADC$ isosceles). $AC^2 = 2AD^2$ (I, 47). 2(sector AZC):sector ACD = $AC^2:AD^2 = 2AD^2:AD^2$ (XII, 2) = 2:1.

[5] The configuration investigated has been transformed to a situation of analogy between surface with surface "inside" and sector with segment "inside"; cf. Prop. 21's use of a parallel auxiliary configuration with rotation cylinders, and investigation via parallel processes of continuous inscription.

[6] arc ZE:arc ZA = $\angle ZCE:\angle ZCA$ (VI, 33); $\angle CDA = 2\angle ZCA$; $\angle CDB = 2\angle ZCE$ (III, 32 and III, 20) ⇒ arc ZC:arc ZA = $\angle ZCE:\angle ZCA = \angle CDB:\angle CDA$ = arc CB:arc CA (VI, 33).

[7] *Symptoma* of the spiral.

[8] V, 15 (surface LKT:surface OTN = surface hemisphere:full surface ONT).

<is> to the surface of the section, so is the square over the straight line joining T and L to the square over the <straight line joining> T and O,[1] or the square over EC[2] to the square over BC, therefore as the sector KLT in the surface <is> to the <sector> OTN, so will the sector EZC be to the <sector> BHC.[3] Similarly we will show that, also, as all the sections in the hemisphere that are equal to KLT, taken together (they are <when put together> the whole of the surface of the hemisphere), <are> to the sections described around the spiral that are of the same order as OTN, taken together, so <are> all the sectors in AZC that are equal to EZC, taken together, i.e.: <so is> the whole sector AZC, to the <sectors> circumscribed around the segment ABC that are of the same order as <the sector> CBH, taken together.

In the same way it will also be shown, however, that as the surface of the hemisphere <is> to the sections inscribed in the spiral, so <is> the sector AZC to the sectors inscribed in the segment ABC, so that as the surface of the hemisphere <is> to the surface cut off by the spiral, so <is> the sector AZC, i.e.: the sector ABCD, to the segment ABC.[4]

Addition:

On account of this result one gathers, however, that the surface cut off between the spiral and the arc TNK is eight times the segment ABC (since the surface of the hemisphere <is eight times> the sector ABCD, also[5]), whereas the surface (cut off) between the spiral and the base of the hemisphere is eight times the triangle ACD, i.e.: <it is> equal to the square over the diameter of the sphere.[6]

---

[1] Surface hemisphere = 2 maximum circle (*Sph. et Cyl.* I, 33); circle with radius TL:maximum circle = TL[2]:(radius hemisphere)[2] (XII, 2) = 2:1 (I, 47); ⇒ surface of hemisphere = circle with radius TL; Surface of sphere through O, N with pole T = circle with radius TO (*Sph. et Cyl.* I, 42:); ⇒ Surface hemisphere:surface ONT = circle TL:circle TO = TL[2]:TO[2]; cf. Co p. 94, K for a slightly different path of reasoning.

[2] By construction, TL = AC = ZC, and BC = TO as chords under equal arcs (III, 29).

[3] XII, 2; V, 15 (circles have ratio of squares over radii); the same proportion holds for equal parts. Ver Eecke (1933b, p. 204, #4) refers to *Sph. et Cyl.* I, 42/43 here.

[4] An implicit limit process is used (cf. Prop. 21). The sought areas are analogously enclosed between all circumscribed and all inscribed composite circular areas/spherical sections. By choosing the arcs involved in the construction ever smaller, the desired lines and areas are approximated.

[5] *Sph. et Cyl.* I, 33 (surface of the complete sphere = 4 area of maximum circle). *Sph. et Cyl.* I, 35: surface hemisphere = 8 quadrants of maximum circle. Thus, surface above spiral = 8 segments.

[6] We compare the remainders after subtraction. Since surface above spiral = 8 segments, we get that surface hemisphere – surface above spiral = surface below spiral = 8$\triangle$ACD. 8$\triangle$ACD = 8(1/2 AD^2) = 4AD^2 = (2AD)^2.

Three Kinds of Mathematical Questions, and Their Appropriate Means of Argumentation[1]

#36 When the ancient geometers wished to trisect a given rectilinear angle, they got into difficulties for a reason such as the following. We say[2] that there are three kinds[3] of problems in geometry, and that some <of the problems> are called "plane," others "solid," and yet others "linear." Now, those that can be solved[4] by means of straight line and circle,[5] one might fittingly call "plane." For the lines by means of which problems of this sort are found have their *genesis* in the plane as well. All those problems, however, that are solved when one employs for their invention either a single one or even several of the conic sections, have been called "solid." For it is necessary to use the surfaces of solid figures – I mean, however, (surfaces) of cones – in their construction.[6] Finally, as a certain third kind of problems the so-called "linear" kind is left over.[7] For different lines, besides the ones mentioned, are taken for their construction, which have a more varied and forced *genesis*, because they are generated out of less structured surfaces, and out of twisted[8] motions. Of such a sort, however, are both the lines found on the so-called loci on surfaces[9] and also others, more varied than those and many in number, which were found by Demetrius of Alexandria in the "linear constitutions,"[10] and

---

[1] Essentially the same statement about the three kinds of geometrical problems is found in *Coll.* III. This passage is somewhat of a locus classicus on methodology. In fact, it is only found in Pappus in this degree of generality. See the commentary.

[2] φαμέν (270, 3 Hu), usually interpreted as equivalent here to "one says." It cannot be excluded, however, that we have another authorial plural here (as in many other places in *Coll.* IV), equivalent to "I say."

[3] γένη (270, 3 Hu); since Aristotle's theory of scientific argumentation (*Analytica Posteriora*), the word had been a standard technical term in Greek theory of science. It has a classificatory connotation (kinds versus species), but it is also used to denote the subject matter of a scientific discipline as a closed field of essential connections. A possible translation for genos is "class", but this obscures the connotation of the word with concepts of kinship, and the connections with an established discourse on scientific methodology.

[4] λύεσθαι (270, 6 Hu); unlike "ana-luein" (Props. 28 and 29), this word means "solve."

[5] Note that the classification of the kinds is derived from the objects needed for a constructive solution, i.e., mathematical lines, not from tools of construction and performance (e.g., ruler and compass).

[6] κατασκευή (270, 11 Hu), the technical term for the construction in a classical apodeixis.

[7] ὑπολείπεται (270,13 Hu), a hapax legomenon in *Coll.* IV. Perhaps it was Pappus himself who lumped all the rest of mathematical problems into one "kind." See the commentary.

[8] ἐπιπεπληγμένων (270, 17 Hu), perhaps related to the term πληκτοειδής/πλεκτοειδής in Prop. 29.

[9] Cf. Props. 28 and 29. The space curves created in the intermediate steps there belong to this group.

[10] γραμμικαὶ ἐπιστάσεις (270, 20/21 Hu); probably a book title. There is no information outside *Coll.* III/IV available on Demetrius. Ver Eecke (1933b, p. 207, # 3) dates Demetrius roughly in the first century BC, because Menelaus (see below) lived in the first century AD; cf. also Tannery (1912, Vol. II, pp. 1–47).

by Philo of Tyana, from the twisting[1] of both plectoids and all sorts of other surfaces on solids[2] and which have many astonishing *symptomata* about them. And some of them were deemed, by the more recent geometers, worthy of rather extensive discussion,[3] and a certain one of them is the line that was also called "the paradox" by Menelaus[4] And of this same kind <i.e., the linear kind> are also the other spiral lines, the quadratrices and the conchoids and the cissoids.[5]

Somehow, however, an error of the following sort seems to be not a small one for geometers, <namely> when a plane problem is found by means of conics or of linear devices[6] by someone, and summarily, whenever it is solved from a non-kindred kind, such as is the problem on the parabola in the fifth book of Apollonius' *Conics*[7] and the *neusis* of a solid on a circle,[8] which was taken by Archimedes in the <book> about the spiral. For it is possible to find the theorem written down by him without using a solid, I mean in fact <it is possible> to show that the circumference of the circle in the first rotation <of the spiral> is equal to the straight line drawn at

---

[1] ἐπιπλοκή (270,21 Hu), perhaps related to the participle "ἐπιπεπλεγμένος" used above. Philo of Tyana is otherwise unknown; Ver Eecke, 1993b p. 207, #4 dates him to the second century BC.

[2] ἐκ ἐπιπλοκῆς πλεκτοειδῶν (or: πληκτοειδῶν) τε καὶ στερεῶν παντοίων; the reference here is certainly to a another book, though probably not directly to a book title. The surfaces used in Prop. 29 probably are examples for such "twisted plectoids".

[3] A considerable corpus of contributions to the geometry of such "higher" curves must have existed.

[4] Menelaus of Alexandria, an astronomer of the first/second century AD, was a predecessor of Ptolemy. His attested works include three books on spherics (preserved in Arabic), a work containing tables of chords in circles, a work on hydrostatics, a treatise on the settings of the signs of the zodiac, Elements of geometry, and a work on higher curves, with connection, *inter alia*, to the duplication of the cube, and to positions of the fixed stars. Hultsch refers to Chasles, *Aperçu historique* for a possible reconstruction of the line called "the paradox" (cf. 271, #4 Hu).

[5] Note the plurals. Pappus has described general quadratrices in Props. 28 and 29. Examples for spirals are mentioned in Props. 19 and 28–30. He has mentioned the existence of several conchoids in Prop. 23. The cissoid was originally invented by Diocles in the third century BC and apparently generalized later; cf. Knorr (1986, pp. 246–263). In Pappus' text, the other curves are indeed labeled as types of spirals, perhaps because all such "higher" curves involve a rotation along with a linear motion.

[6] Apollonius classified plane and solid *neusis* problems, and differentiated them into two classes according to the lines needed for their constructive solution. He may have attempted to develop a complete operational toolbox to solve problems that would fall under those types, determining limiting conditions and providing a scale of increasing complexity via analysis. There is no clear evidence, however, that the demand of "keeping within the kind" ever reached the status of a fundamental claim with universality for all geometry, and all geometers. See the commentary.

[7] Hultsch (273, #1 Hu) believes this must be I, 52; Zeuthen (1886, pp. 280–288), Tannery (1912, vol. I, pp. 302–311) and others show, however, that it could have been the problem of finding the normal to a parabola (*Con.* V, 62 in Toomer's 1990 edition). Apollonius treats it analogously to the (solid) case of the hyperbola and the ellipse. In the case of the parabola, however, a plane construction would have sufficed, if one takes the parabola in question as *given*; cf. Zeuthen (1886, pp. 286–288).

[8] A has a genitive (location) here (and in the parallel phrase in Prop. 44). Hultsch emends to an accusative (direction). I have translated the transmitted text.

right angles to the generator <of the spiral> up to <the point of intersection with> the tangent of the spiral.[1]

Now, since a difference of such a sort belongs to problems, the earlier geometers were not able to find the above mentioned problem on the angle, given that it is by nature solid,[2] and they sought it by means of plane devices. For the conic sections were not yet common knowledge for them, and on account of this they got into difficulties. Later, however, they trisected the angle by means of conic sections, using for the invention the *neusis* described in what follows.

Props. 31–34: Angle Trisection

Prop. 31: Neusis for Angle Trisection

When a rectangle ABCD is *given*, and BC is produced, let it be the task to draw AE through <the interior> and make the straight line EZ equal to a *given* <straight line.>[3] <Analysis[4]>

---

[1] Pappus' objection here, and even more so his upcoming arguments about the *neusis* in question (cf. Props. 42–44 with notes and commentary) have often been misconstrued in secondary literature. The remarks refer to *SL* 18 (subtangent to a spiral of first rotation is equal to the circumference), which invokes *neuses* from *SL* 7/8. These *neuses*, in fact all *neuses* in *SL* 5–9, are indeed solid in Pappus' sense (see the commentary on Props. 42–44 on how far he is able to show this). Pappus also claims that Archimedes could have done with a plane construction for the theorem in *SL* 18. Whether he means that Archimedes could have used a plane *neusis* or that Archimedes could have used some other plane argument, instead of the *neusis*, in *SL*. 18, is unclear. Co p. 95, C, refers to Witelo, *Perspectiva* I, 128 for a plane construction. Since Witelo may very well have had indirect access to the *Collectio* in the thirteenth century (cf. Unguru 1974), this may be significant, and certainly Witelo's suggestion deserves scholarly attention.

[2] τῇ φύσει στερεὸν ὑπάρχον; Pappus ascribes an essential, internal character to mathematical problems. This is in line with the Aristotelian meta-theoretical framework and vocabulary he has been using in this passage, as testified *inter alia* by his use of the term "kind." See the commentary.

[3] The *neusis* can obviously be constructed with the conchoid (cf. Props. 23–25), when one chooses A as *pole*, CD as *canon*, and EZ as *distance*. Perhaps this was what Nicomedes did. Note the relation to the *neusis* that figures in Prop. 24. It seems quite plausible that Nicomedes indeed proposed essentially a single (μόνην) construction for both problems. Cf. above, introductory remarks on Prop. 24.

[4] The analysis was probably added by Pappus to an older argument for the angle trisection that constructed the *neusis* without using conics. He may have excerpted the analysis from a source. In A, the figure for the analysis differs from the one for the synthesis. The manuscript text also shows signs of confusion and incoherent partial corrections (cf. apparatus to the Greek text). Treweek 117a documents the differences in a list. The existence of these differences supports the thesis about the subsistence of an older layer of argument in Pappus' text. They might be used for further investigations. An independent Arabic version, purely synthetic, exists. See the bibliographical references for Props. 31–34 in the commentary. Hultsch has adjusted the lettering of the diagram and of the items used in the argument for the analysis to the features of the synthesis, thus making Prop. 31 conform to regular practice in analysis-synthesis (272 Hu + app; similarly: Co p. 96/97). I have followed him. Treweek 117, 6–118, 2 opts for a more cautious and conservative emendation.

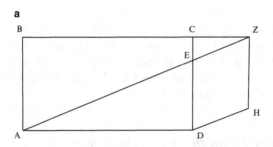

Assume that it has turned out that way,[1] and draw DH and HZ parallel to EZ and ED <respectively.>[2] Now, since ZE is *given* <in length> and it is equal to DH,[3] DH is therefore *given* <in length>, also. And D is *given*. Therefore, H lies on the circumference of a circle *given* in position.[4] And since the rectangle BCD is *given*, and it is equal to the rectangle BZ/ED <in size,>[5] the rectangle BZ/ED is *given*, also, i.e.: the rectangle BZH <is *given*>. Therefore, H lies on a hyperbola. But it also lies on the circumference of a circle *given* in position. Therefore, H is *given*.
<Synthesis>

#37 Now, the problem will be put together[6] in the following way. Let ABCD be the *given* rectangle, and m the straight line *given* in length, and let DK be equal to it, and describe, on the one hand, the hyperbola DHT through D with asymptotes AB/BC (I will provide the proof for this in what follows in order[7]), and, on the other hand, the circular arc KH through K with center D, intersecting the hyperbola in H. And when the parallel HZ to DC is drawn, join ZA.

I claim that EZ is equal to *m*.

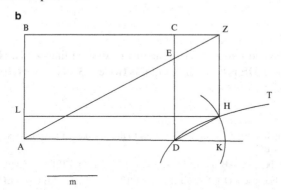

[1] Analysis-assumption. For the structure of analysis-synthesis in general see the introduction to the commentary on Props. 4–12.

[2] This part of the analysis is non-deductive.

[3] The *resolutio* begins here.

[4] *Data*, def. 6.

[5] Complete the rectangle ABZ and apply I, 43.

[6] συντεθήσεται. The synthesis begins here.

[7] Prop. 33.

For:

Join HD and draw the parallel HL to KA. Then the rectangle ZHL, i.e.: the rectangle BZH, is equal to the rectangle CDA, i.e.: to the rectangle BC/CD.[1] Therefore, CD is to ZH as ZB <is> to BC, i.e.: as CD <is> to ED.[2] Therefore, ED is equal to ZH.[3] Therefore, DEZH is a parallelogram.[4] Therefore, EZ is equal to DH, i.e.: to DK, i.e.: to *m*.

Prop. 32: Trisection of the Angle via Neusis

#38 Now, when this has been shown, a *given* rectilinear angle is trisected in the following way.

Let the <angle> ABC, first, be acute,[5] and from a certain point A <draw> the perpendicular AC, and when the rectangle CZ is completed, produce ZA toward E,[6] and since CZ is a rectangle, place the straight line ED between the <straight lines> EA/AC, verging toward B and equal to two times AB (for that this can come about has been written down above).

I claim in fact that the angle EBC is the third part of the *given* angle ABC.

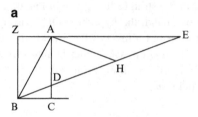

For:

Bisect ED in H, and join AH. Then the three <straight lines> DH, HA, and HE are equal.[7] Therefore, DE is twice AH. But it is twice AB, also.[8] Therefore, BA is equal

---

[1] *Con.* II, 12, paraphrased above in the footnotes to the last part of the analysis. BZ × ZH = ZH × HL = CD × DA = BC × CD.

[2] ΔBZA ~ ΔCZE, because AB ∥ CE; BC:CZ = AE:EZ (VI, 2); ZB:BC = ZA:AE (V, 16/18). ΔAED ~ ΔZEC, because AD ∥ CZ; ZE:EC = AE:ED (VI, 4); ZA:AE = CD:ED (V, 16/18). ⇒ CD:ZH = ZB:BC = CD:ED.

[3] V, 9.

[4] ZH ∥ ED by construction, and we have just seen that ZH = ED.

[5] Prop. 32 is the only example in *Coll.* IV with a *diorismos* fully carried through, in the sense that all possible cases for a problem are covered. But see the commentary on plane sub-cases for this generally solid problem.

[6] The position of E is yet to be determined.

[7] A on the semicircle over DE with center H (III, 31).

[8] By construction.

to AH, and the angle ABD <is equal> to the angle AHD.[1] However, the angle AHD is two times the angle AED,[2] i.e.: <two times> the angle DBC.[3] Therefore, the angle ABD is two times the angle DBC, also. And when we bisect the angle ABD,[4] the angle ABC will be trisected.

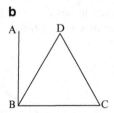

#39 When, however, the *given* angle happens to be a right angle, we will cut off a certain BC and describe over it the equilateral <triangle> BDC.[5] And when we bisect the angle DBC,[6] we will have trisected the angle ABC.[7]

#40 Finally, let the angle be obtuse, and draw BD at right angles to CB, and, on the one hand, cut off the angle DBZ as a third part of the angle DBC, and on the other hand, the angle EBD as the third part of the angle ABD (for I have shown these <two constructions> above). Then the angle EBZ is the third part of the whole angle ABC as well. When, however, we erect an <angle> equal to the angle EBZ along both AB and BC, we will trisect the *given* angle.

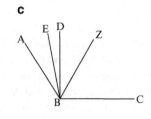

---

[1] I, 5.

[2] III, 20.

[3] I, 29 (parallels ZE and BC with transversal BE).

[4] I, 9.

[5] I, 1.

[6] I, 9.

[7] For an alternative, using a plane *neusis*, cf. Heraclius' construction, which is contained in *Coll.* VII, and also reported in Descartes (1637, pp. 387–389) (188–193 Smith/Latham). It is noteworthy that Pappus did not opt for this route here. See the commentary.

Prop. 33: Analysis-Synthesis for the Hyperbola-Construction in the Trisection Neusis

#41 Now I will provide an analysis for the problem that was postponed.[1] When two straight lines AB and BC are *given* in position, and a point D is *given*, to describe the hyperbola through D with asymptotes AB/BC.

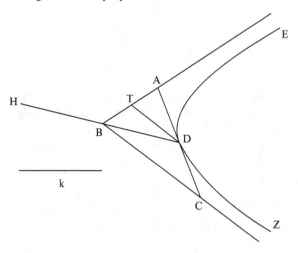

<Analysis>

Assume that it has turned out that way,[2] and that the <hyperbola> EDZ has been described, and from D draw its tangent ADC, and the diameter HBD, and the parallel DT to BC.[3] Then HD and DT are <*given*> in position,[4] and T is *given*.[5] And since AB and BC are the asymptotes of the hyperbola, and AC is a tangent, AD is, therefore, equal to DC, and the square over each of the two of them is equal to one fourth of the figure on HD. For that has been shown in the second <book> of the *Konika*.[6]

---

[1] This phrase introduces the analysis. Co p. 97 translates "resolvemus." This is more accurate than Hu's "solvemus" (277 Hu). Hultsch comments (277, #1 Hu) that a shorter constructive proof would have been possible via *Con* II, 4, though Pappus' resolutio (!) has its merits, too. Such a construction would have been purely synthetic. Pappus' argument here contains analysis and synthesis and serves as exemplary for the methods of argument in "solid" problem solving. Prop. 33 shows strong indications for a close connection to Apollonius' lost analytical-synthetical solution. It is also very close to *Coll.* VII, #204 Hu, by Pappus (commentary on an analytical argument in Apollonius' original *Konika*, Book V). See the commentary.

[2] Analysis-assumption.

[3] This part of the analysis contains an extension of the configuration and is non-deductive.

[4] *Data* 28 (for DT) and 26 (for HD); this sentence marks the beginning of the *resolutio*.

[5] *Data* 25 (for T).

[6] *Con.* II, 3: AC^2 is equal to the figure on HD, and AC = 2AD = 2DC holds. The "figure on HD" is the rectangle constituted of the diameter HD and the latus rectum (parameter) k. Note, however, that Pappus is in all likelihood not referring to the now extant version of the *Konika*. Compare the footnote on the end of the analysis.

Now, since CD is equal to DA, BT is equal to TA, also.[1] And BT is *given*.[2] Therefore, TA is *given*, also.[3] And T is *given*. Therefore, A is *given* as well.[4] Therefore, the <straight line> ADC is <*given*> in position. And AC is *given* in length,[5] so that the square over AC is *given* as well.[6] And it is equal to the figure on HD.[7] Therefore, the figure on HD is *given* <in area>. And HD is *given* – for it is twice BD, which is *given* in length, on account of the fact that B and D are both *given*.[8] Therefore, the latus rectum of the figure is *given*, also.[9]

In fact, the problem has turned out to be of the following sort: when two straight lines, <namely> both HD and the latus rectum, are *given* in position and length, to describe the hyperbola with diameter HD, for which the <straight line> to which the squares are applied is the remaining straight line, and for which the <straight lines> drawn ordinatim to HD will be parallel to a certain straight line AC, *given* in position. This, however, has been subjected to analysis in the first <book> of the *Konika*.[10]

<Synthesis>

#42 Now, it will be put together[11] in the following way. Let AB and BC, on the one hand, be the straight lines *given* in position, and D, on the other hand, the *given* point, and draw DT, on the one hand, parallel to BC, and <draw> TA, on the other hand, equal to BT, and when AD has been joined, produce it to C, and produce BD, after it has been joined, also, and position BH equal to BD, and let the rectangle between HD and a certain other <straight line> k be equal to the square over AC,[12] and describe the hyperbola EDZ with diameter HD and latus rectum k, so that the <straight lines> drawn ordinatim to HD are parallel to AC.[13] Then AC will touch the conic section.[14]

---

[1] CD = AD has just been shown. ΔBAC ~ ΔTAD on parallels CB, DT; BT:TA = CD:DA (VI, 2). Apply V, 9.

[2] *Data* 26.

[3] *Data* 2.

[4] *Data* 27.

[5] *Data* 26: AD is given in length and position, AC is *given* in position. Since AC = 2 AD, AC is *given* in length as well (*Data* 2).

[6] *Data* 52.

[7] *Con.* II, 3, cf. above.

[8] *Data* 26 and *Data* 2.

[9] *Data* 57.

[10] ἀναλύεται. The extant *Konika*, a revision of Apollonius' work on conics by Eutocius (sixth century AD), are purely synthetic and do not contain analyses for the constructions provided. Pappus consistently speaks of Apollonius' treatise on conics as an analytic work in *Coll.* VII, and in *Coll.* IV he handles all problems that are solved by means of conics via analysis-synthesis. Probably the Apollonian work Pappus worked with was analytic-synthetic. For a *synthetic* solution of the construction problem mentioned here by Pappus c.f. *Con.* I, 54/55. Note that Pappus does not mention Apollonius by name. This could mean that in his time, Apollonius' (analytical) *Konika* were the standard reference work, parallel to the *Elements*.

[11] συντεθήσεται. This is the beginning of the synthesis.

[12] *Elem.* I, 45.

[13] *Con.* I, 54/55.

[14] *Con.* I, 32.

And AD is equal to DC (since BT is equal to TA, also[1]), and it is obvious that both the squares over AD and DC are the fourth part of the figure on HD.[2] Therefore, AB and BC are the asymptotes of the hyperbola EDZ.[3] Therefore, the hyperbola through D with the *given* straight lines as asymptotes has been described.

Prop. 34: Alternative Constructions of the Angle Trisection via Solid Loci[4]

<Alternative a>
#43 The third part of a *given* arc is cut off in a different way, also, without the *neusis*, by means of a solid locus of the following sort.

<Assume that> the <straight line> through A and C is <*given*> in position, and that the angle ABC has been bent[5] over the <points> A and C *given* on it, making an angle ACB that is two times the angle CAB.[6]

<I claim> that B lies on a <uniquely determined> hyperbola[7]

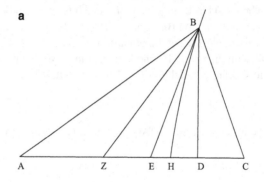

---

[1] △BAC ~ △TAD on parallels BC, TD. Since BT = TA, i.e., BA:TA = 2:1, CA:DA = 2:1 (VI, 2), and DC = DA.

[2] $AC^2 = HD \times k$ by construction. We have just seen that ½AC = AD = DC.

[3] *Con.* II, 1/2.

[4] Prop. 34 gives the essential part of an analysis for the angle trisection via solid loci in two versions. No detailed constructive apodeixes are offered. Pappus is probably drawing on pre-Apollonian treatments of solid loci, perhaps by Aristaeus, and may have an interest in portraying the Apollonian solution, which he presented in detail in Prop. 33, as the classic one in terms of methodology, which nevertheless did not render older contributions utterly superfluous. Prop. 34a is a simplified version of Prop. 34b, using the Apollonian technical apparatus, and may very well be by Pappus himself (cf. Jones 1986a, p. 584). It is the simplest of the three solutions in *Coll.* IV (cf. Heath 1921, I, pp. 241–242; Zeuthen 1886, pp. 210–212). 34b shows clear traces of an older treatise on solid loci (see below for Prop. 34b, and see the commentary).

[5] κεκλάσθω. This word has also been used in Props. 11/12. No construction for the "bending" is offered. Obviously, it is equivalent to the angle trisection. In Prop. 34, the task of trisecting an angle AMC is assumed to have been reduced to the task of trisecting the arc over a chord AC.

[6] Prop. 34a only considers the case where ∠BCA is acute. For the other two possible configurations, one can argue analogously, cf. Co p. 100 f. and appendix Hu p. 1230.

[7] Any point B that meets the condition about the base angles lies on this hyperbola.

Draw the perpendicular BD, and cut off DE, equal to CD. Then BE, when it has been joined, will be equal to AE.[1] Position EZ as equal to DE, also. Then CZ is three times CD. Let AC be three times CH, also.[2]

Now, H will be *given*,[3] and the remaining AZ will be three times HD.[4] And since the square over BD is the difference between the squares over BE and EZ,[5] whereas the rectangle DA/AZ is the difference of these, also,[6] the rectangle DAZ, i.e.: three times the rectangle ADH,[7] will therefore be equal to the square over BD. Therefore, B lies on a hyperbola, the latus transversum of which is AH, and the latus rectum three times AH.[8] And it is obvious that the point C cuts off half the latus transversum AH on the <straight line> CH <drawn> to the vertex H of the conic section.[9] And the synthesis is obvious. For one will have to divide AC so that AH is two times HC,[10] and describe the hyperbola through H with axis AH, the latus transversum of which is three times AH,[11] and to show that it creates the above mentioned twofold ratio of the angles.[12] And that the hyperbola described in this way cuts off the third part of the *given* circular arc is rather easy to understand when the points A and C are posited as the endpoints of the arc.[13]

---

[1] $\triangle$EBD $\cong$ $\triangle$CBD (I, 4) $\Rightarrow$ $\angle$BEC = $\angle$BCA (= 2$\angle$BAE by hypothesis). $\angle$BEC = $\angle$BAE + $\angle$ABE (I, 32) $\Rightarrow$ $\angle$BAE = $\angle$ABE, and $\triangle$ABE is isosceles (I, 6).

[2] Choose H on AC, with HC:AC = 1:3 (VI, 9).

[3] *Data* 2, *Data* 27.

[4] CZ = 3CD and AC = 3CH; 3HD = 3(CH – CD) = 3CH – 3CD = AC – CZ = AZ.

[5] I, 47; ED = EZ by construction.

[6] II, 6: DH $\times$ AZ + EZ^2 = AE^2, i.e., DH $\times$ AZ = AE^2 – EZ^2; AE = BE was shown above.

[7] AZ = 3DH was shown above; VI, 1.

[8] Consider the converse of *Con.* I, 21 (not established as a theorem in itself). *Con.* I, 21 states that for all points B on the hyperbola through H with latus transversum AH, parameter 3AH and ordinate angle $\pi/2$, the above equality holds. In the analysis, we can therefore "conclude" from the equality that B lies on this hyperbola. Note, however, that this justification via Apollonius' *Konika* may be anachronistic in the sense that the alternatives 34a and 34b may very well draw on a pre-Apollonian treatment of the angle trisection via "solid loci" perhaps by Aristaeus. See the commentary.

[9] By construction of H, AH = 2HC.

[10] VI, 9.

[11] *Con.* I, 54/55.

[12] Retrace the steps of the above analysis. All points B on the hyperbola have the property that 2$\angle$BAC = $\angle$BCA. See the commentary for a sketch of the apodeixis suggested here. Co p. 101 gives an extended apodeixis, considering all three possible cases for $\angle$BCA.

[13] For a reconstruction of an angle trisection on the basis of the considerations given here see the commentary. Hultsch p. 285 # 3 refers to a discussion of the synthesis by Commandino (cf. Co pp. 101–102, O). The construction can, of course, be used for a number of "solid" problems, and that may be the reason why Co integrates a longer exposition of a constructive proof. See also the comments on Props. 42–44.

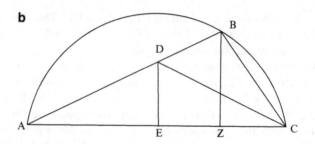

<Alternative b¹>

#44 Some have set out the analysis of trisecting an angle or an arc in yet another way without a *neusis*. Let the argument[2] be about an arc, however. For it makes no difference whether one divides an angle or an arc.

Assume that it has turned out that way in fact, and that BC has been cut off as the third part of the arc ABC,[3] and join AB, BC, and CA. Then the angle ACB is two times the angle BAC.[4] Bisect the angle ACB by CD, and <draw> the perpendiculars DE and ZB. Then AD is equal to DC,[5] so that AE is equal to EC, also.[6] Therefore, E is *given*.[7] Now, since AD is to DB, i.e.: AE to EZ,[8] as AC <is> to CB, alternando, BC is therefore to EZ as CA <is> to AE, also. CA is twice AE, however. Therefore, BC is twice EZ as well. Therefore, the square over BC, i.e.: the <sum of> the squares over BZ and ZC,[9] is four times the square over EZ.

Now, since the two <points> E and C are *given*, and BZ is at right angles <to AC>, and the ratio of the square over EZ to the <sum of)> the squares over BZ and

---

[1] This version is closely related to a (lost) argument from Euclid's *Solid loci*, ultimately resting on a prior argument by Aristaeus. For it is closely connected to Pappus' commentary on such an argument in *Coll.* VII (#237 Hu, Jones (1986a, # 316–318, pp. 365–369, with 583 f); cf. Zeuthen (1886, p. 215) for the connection to Aristaeus). See the commentary, and cf. Heath (1921, I, pp. 243–244, II, pp. 119–121), Zeuthen (1886, pp. 212–215), and Knorr (1986, pp. 128–137 and 327). Knorr expands on Zeuthen's arguments.

[2] λόγος. Hultsch translates "proportio". i.e., "ratio", probably the ratio 3:1. "Logos" can, however, also mean "account", "argument". This translation seemed preferable.

[3] Analysis-assumption.

[4] VI, 33.

[5] ∠ACB = 2∠BAC by assumption; thus, ∠DCA = ∠DAC, and ΔADC is isosceles (I, 6).

[6] DE is the height in the isosceles triangle ADC (I, 26).

[7] *Data* 7, *Data* 27.

[8] ΔABZ ~ ΔADE, on parallels DE and BZ; AD:DB = AE:EZ (VI, 2/ V, 16). In ΔACB, ∠ACB is bisected by DC, with D on AB; AC:BC = AD:DB (VI, 3).

[9] I, 47.

ZC <is *given*>, B lies therefore on a hyperbola.[1] But <it also lies> on an arc that is <*given*> in position. Therefore, B is *given*.

And the synthesis is obvious.[2]

Props. 35–38: Generalization of Solid Problems: Angle Division

Prop. 35: General Angle Division

#45 Now, trisecting a *given* angle or arc is a solid <problem>, as has been shown above,[3] whereas dividing a *given* angle or arc in a *given* ratio is a linear <problem>, and while it has been shown by the more recent <mathematicians>, it will be shown as well in a twofold way by me.[4]

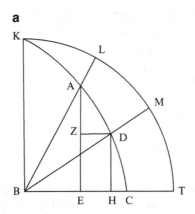

---

[1] In *Coll.* VII, Prop. 237, (cf. Jones 1986a, pp. 365–369, # 316/317), a hyperbola is established via analysis-synthesis, the points of which satisfy the conditions derived in the above analysis. It is the hyperbola with focus C, directrix ED, and eccentricity 2. In the analysis here we can "conclude": B lies on this uniquely determined hyperbola. See the commentary. Co pp. 102–103, E provides an alternative argument.

[2] Bisect AC in E, draw ED as a perpendicular onto AC, and describe the hyperbola with directrix ED, focus C, and eccentricity 2, using *Coll.* VII, 237. The hyperbola intersects the given arc AC in B, in which the arc is divided in the ratio 2:1. For the proof, retrace the steps of the analysis. See the commentary for a list of the decisive steps. For an alternative synthesis for the trisection discussed in Prop. 34b, including a separate treatment of all three possible configurations, see also Co pp. 103–104, E (starting at "et compositio manifesta est").

[3] Apparently, Pappus believes that if an analysis leads to conics, one has shown that the problem in question is (in general) solid. But see the discussion of analysis as a criterion for the determination of problem levels in the commentary on Props. 42–44. Pappus is correct in his assertion that the angle trisection is solid, and his analysis does show that it is not linear (analysis demarcates sharply "upward").

[4] The first of the arguments in Prop. 35 (via the quadratrix) targets acute angles.

For:

Let LT be the arc of a circle KLT, and let the task be to divide it in a *given* ratio.

<Draw> the radii LB and BT, and BK at right angles to BT, and describe the line "quadratrix" KADC through K, and divide the perpendicular AE, after it has been drawn, in Z in such a way that as AZ <is> to ZE, so is the *given* ratio into which one wants to divide the angle up, and <draw> ZD parallel to BC. BD should be joined, however, and the perpendicular DH <from D onto BT> <be drawn>.

Now, since, on account of the *symptoma* of the line, the angle ABC is to the angle DBC as AE <is> to DH, i.e.: to ZE,[1] subtrahendo, the angle ABD is, therefore, to the angle DBC, i.e.: the arc LM <is> to the arc MT, as AZ <is> to ZE, i.e.: as the *given* ratio.[2]

#46 The arc AC of a circle AHC <can be> divided in yet a different way.

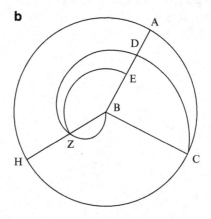

<Draw> the radii AB and BC similarly, and describe the spiral BZDC through B,[3] the generator of which is CB, and let the ratio of DE to EB be the same as the *given* ratio,[4] and through E <draw> the arc EZ of a circle with center B, intersecting the spiral in Z. And produce BZ, after it has been joined, to H. Then, on account of the spiral, the arc AHC is to the arc CH as DB <is> to BZ, i.e.: to BE.[5] And subtrahendo, as DE <is> to EB, so <is> the arc AH to the <arc> HC.[6] The ratio of DE to EB, however, is the same as the *given* ratio. Therefore, the ratio of the arc AH to the arc HC is the same as the *given* ratio, also. Therefore, <the arc AC> has been divided <in the *given* ratio>.

---

[1] Arc KT:arc LT = KB:AE; arc KT:arc MT = KB:DH (*symptoma*) ⇒ arc LT:arc MT = AE:DH (V, 16 and V, 22); DH = ZE by construction.

[2] Apply V, 17 to arc LT:arc MT = AE:ZE; AZ:ZE equals the *given* ratio by construction.

[3] The labeling BZDC suggests motion of the generating point from B to C, as in the *genesis* in Prop. 19. Rotation could be clockwise or counterclockwise. The labeling CB for the generator suggests a motion from C to B, in deviance from the description in Prop. 19.

[4] Divide DB in E in the *given* ratio (VI, 9).

[5] BC:BD = circle:arc AC; BC:BZ = circle:arc HC (*symptoma*) ⇒ BD:BE = arc AC:arc HC (V, 22). Co p. 105 G refers to *SL* 14 instead.

[6] V, 17.

Prop. 36: Equal Arcs of Different Circles

#47 From this <result> it is in fact obvious that it is possible to cut off equal arcs from unequal circles.

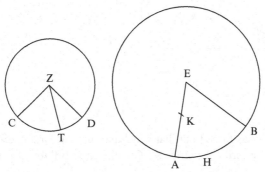

For:

Assume that it has turned out that way,[1] and that the equal arcs AHB and CTD have been cut off. Let the <circle> with center E be the larger one, however. Then the arc that is similar to <the arc> CTD is larger than the <arc> AHB.[2] Now, let the <arc> CT be similar to the <arc> AHB. Then the ratio of the <arc> AHB to the <arc> CT is *given*, for it is the same as the <ratio> of the whole circumferences of the circles, or of the diameters.[3] The <arc> AHB is, however, equal to the <arc> CTD. Therefore, the ratio of the <arc> CTD to the <arc> CT is *given*. And, subtrahendo, <the task> has now become to divide a *given* arc CTD in a *given* ratio in T. This, however, has been written down above.[4]

Prop. 37: Isosceles Triangle with Angles in Given Ratio

#48 <Let the task be> to put together an isosceles triangle with both angles at the base possessing a *given* ratio to the remaining one.[5]

---

[1] Analysis-assumption. Prop. 36 gives only an analysis, reducing the problem to the division of an angle in a *given* ratio. Then Prop. 35 is invoked.

[2] In the smaller circle, the arc over the same angle as AEB (arc CT in the figure) will be smaller than the arc CTD, which was assumed to be equal to arc AHD.

[3] XII, 2 with *Circ. mens* 1 and VI, 1 /V, 15 (similar arcs are in the ratio of the radii (or the circumferences)). Cf. the proof protocol of Prop. 26, section *. The same argument about similar circular arcs and radii was also used in Prop. 26 and will be used in Prop. 39 and 40. A similar argument was used in Prop. 30.

[4] Prop. 35. Co p. 106/107, F provides a constructive proof. See also the commentary.

[5] The problem in Props. 37 and 38 constitutes a generalization of the inscription of a regular pentagon in IV, 10/11 (of the *Elements*). In analogy to the Euclidean construction, a triangle with the required ratio of angles is sought first, and then the polygon is put together from isosceles triangles. See the commentary.

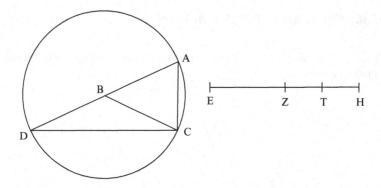

<Analysis>
Assume that it has turned out that way,[1] and that the <triangle> ABC has been put together, and describe the circle ADC with center B through A and C, and produce AB to D, and join DC.[2]

Now, since the ratio of the angle between CA and AB to the <angle> between AB and BC is *given*,[3] and <since> the angle at D is half the angle ABC,[4] the ratio of the angle CAD to the angle ADC is therefore *given*, also,[5] so that the ratio of the arc DC to the <arc> AC <is *given*>, also.[6] Now, since the arc ACD of the semicircle has been divided in a *given* ratio, C is *given*,[7] and the triangle ABC is *given* in kind.[8]

<Synthesis>
<The problem> will, however, be put together[9] in the following way.
For:
Let the *given* ratio, the one which both angles at the base had to have to the remaining one, be the ratio of <a straight line> EZ to <a straight line> ZH, and bisect ZH in T, and set out the circle ADC with center B and diameter AD, and divide the arc ACD in C, so that EZ is to ZT as the arc DC <is> to the <arc> CA (for this <construction> has been written down above,[10] and even generally, somehow, a *given* arc is divided in a *given* ratio), and join BC, CA, and CD. Now, since EZ is to ZT as

---

[1] Analysis-assumption.

[2] Extension of the configuration for the analysis, non-deductive.

[3] In the problem; we are now in the *resolutio*.

[4] III, 20.

[5] *Data* 9.

[6] VI, 33.

[7] C is *given*, because it can be constructed using Prop. 35. From "C is *given*" one might conclude that the triangle is *given* in kind via *Data* 30 and *Data* 40. This is how I would prefer to read the *resolutio*. For an alternative explanation see Hultsch p. 291, * and Co p. 107 E. The phrase "C is *given*" appeared suspicious to him, and he suggests "the straight line BC is *given* in position" in its place.

[8] Cf. *Data*, def. 3 for *given* in kind. A triangle is *given* in kind when its angles are *given*.

[9] συντεθήσεται.

[10] Prop. 35; the proposition is directly applicable only for angles that are at most right angles. Otherwise, divide in half, and put together again after completion of the construction.

the arc DC <is> to the <arc> CA, i.e.: as the angle DAC <is> to the angle ADC,[1] and <equality holds likewise> with respect to the double of the second terms <in the proportion>, therefore as the angle CAB <is> to the angle ABC, so <is> EZ to ZH.[2] Therefore, an isosceles triangle ABC, both of the angles at the base of which possess a *given* ratio to the remaining one, has been constructed.

Prop. 38: Regular Polygon with any Given Number of Sides Inscribed in the Circle

#49 Indeed, when this has been shown, it is obvious that it is possible to inscribe an equilateral and equiangular polygon that has as many sides as anyone might prescribe into a circle.[3]

Props. 39–41: Constructions Based on the Rectification Property of the Quadratrix[4]

Prop. 39: Converse of Circle Rectification

How one finds a circle the circumference of which is equal to a *given* straight line, however, is easy to understand.

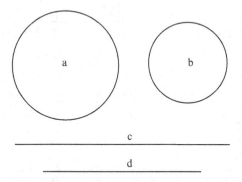

---

[1] VI, 33.

[2] $\angle ABC = 2\angle ADC$ (III, 20); ZH = 2ZT by construction.

[3] The polygon sought for is built up from congruent isosceles triangles in which the angles at the center of the circle stand in a *given* ratio to the full angle. For a polygon with n sides, we get $2\pi/n$ for the vertex angle, and $(\pi - 2\pi/n)/2$ for the angles at the base. The ratio will be $4:(n - 2)$ in modern notation.

[4] In contrast to Props. 35–38, the propositions of this group do not arise from a generalization of plane or solid problems. They are in principle beyond the reach of plane and solid geometry, because they involve the determination of a ratio between a circular arc and a straight line (π).

For:

Assume that the circumference of circle a, equal to the straight line c has
<already> been found,[1] and set out an arbitrary circle b, and find, by means of the
quadratrix, the straight line d, equal to its circumference.[2] Then the radius of circle
a is to the radius of circle b as c <is> to d.[3] The ratio, however, of d to c <is *given*.>[4]
Therefore, the ratio of the radii to each other <is *given*>, also. And the radius of b is
given. Therefore, the radius of a is *given*, also,[5] so that a itself <is *given*>, also.[6]

And the synthesis is obvious.[7]

Prop. 40: Arc over Chord in Given Ratio[8]

<Problem>
#50 When a straight line AB is *given* in position and length, to describe through
A and B the arc of a circle that has to the straight line AB a *given* ratio.

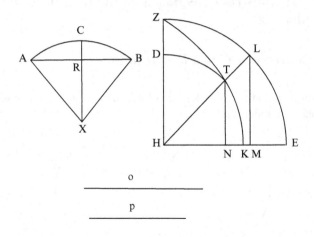

---

[1] Analysis-assumption.

[2] Prop. 26, Addition.

[3] Circumferences have the ratio of diameters, or of radii (XII, 2, *Circ. mens.* 1, VI, 1). A similar
proposition was already used in Props. 26, 30, and 36 and will be used again in Prop. 40. For
details see the section ˙ in the proof protocol of Prop. 26.

[4] *Data* 1 with Prop. 26, Addition.

[5] *Data* 2 with V, 16.

[6] *Data*, def. 5.

[7] Choose a circle b, with radius r, rectify it by means of the quadratrix into a straight line d.
Determine r' with r:r' = c:d (VI, 9), and describe the circle a with radius r'. Then circumference
a:circumference b = r:r' = c:d, and since circumference b = d, we get: circumference a = c. Cf. Co
p. 108/109, F.

[8] Only the situation where the arc is at most a semicircle is envisaged. Thus, the *given* ratio in
Prop. 40 is not arbitrary. It is also, necessarily, larger than 1:1 (in modern terms).

<Analysis>

Assume that the <arc> ACB has been described,[1] and that the quadrant ZHE of a circle, *given* in position, has been set out, and the quadratrix ZTK described, and put together the angle EHL on the arc ZE, equal to the angle that goes through[2] the arc AC,[3] and draw the perpendiculars LM and TN <onto HE>. Now, on account of the property[4] of the line <i.e., the quadratrix>, the arc LE will be to the straight line TN as the arc ELZ <is> to the straight line ZH, i.e.: as LH <is> to HK.[5] But as TH <is> to HL, so <is> TN to LM, also. And therefore, as TH <is> to HK, so <is> the arc EL to the straight line LM.[6] Now, take X as the center of the <circle with> arc ACB, and draw the perpendicular XRC onto AB. Then the angle CXA is equal to the angle EHL. And X and H are the centers <of the circles through C/A and E/L>. Therefore, as the arc AC <is> to the straight line AR, i.e.: <as> TH <is)> to HK,[7] so <is> the arc ACB to the straight line AB.[8]

And the ratio of the <arc> ACB[9] to <the straight line> AB <is *given*>. Therefore, the ratio of TH to HK <is *given*>, also. And HK is *given* <in length.>[10] Therefore, HT is *given* <in length.>[11] Therefore, T lies on the circumference <of a *given* circle.>[12] But it also lies on the line ZTK. Therefore, T is *given*. HTL is *given* in position.[13] Therefore, the angle EHL is *given*.[14] And it is equal to the angle CXA,

---

[1] Analysis-assumption; without loss of generality, C is chosen as the midpoint of arc ACB.

[2] βεβηκυῖα, a hapax legomenon in *Coll*.IV.

[3] ∠EHL = ∠AXC, where X is the midpoint of the sought circle. LM corresponds to ½ AB, i.e., to AR.

[4] ἰδίωμα (where one might have expected σύμπτωμα).

[5] Arc ZE:arc LE = ZH:TN (*symptoma*); arc ZE:ZH = arc LE:TN (V, 16); but arc ZE:ZH = ZH:HK (Prop. 26) = LH:HK; ⇒ arc LE:TN = LH:HK.

[6] ΔHTN ~ ΔHLM on parallels TN and LM; TH:HL = TN:LM (VI, 4); we have just seen: arc LE:TN = LH:HK; thus: TH:HK = arc LE:LM (V, 23).

[7] First, we show that arc AC:AR = arc EL:LM. AR and LM are half-chords under equal angles. Similar arcs are in the ratio of the corresponding radii (this proposition was used in Props. 26, 36, and 39, and a similar one in Prop. 30; see section * in the proof protocol of Prop. 26). arc AC:arc EL = r1:r2. Consider ΔARX ~ ΔLHM ⇒ r1:r2 = AR:LM; thus, arc AC:AR = arc EL:LM (V, 16). Above, it was shown that arc EL:LM = TH:HK. We now get arc AC:AR = TH:HK.

[8] arc AB = 2arc AC, AB = 2AR by construction.

[9] Hultsch prints ABΓ, Tr 125, 16 prints the mathematically correct AΓB.

[10] HK is *given* in the sense that one can posit a freely chosen, but fixed quadrant ZHE with an inscribed quadratrix, and in it, K, and therefore HK are uniquely determined. The quadratrix has to be assumed. The quadrant with quadratrix was assumed to be *given* in position only, not in size, at the outset of the analysis; the actual size of the quadrant with quadratrix is irrelevant for the analysis and synthesis here.

[11] *Data* 2.

[12] *Data*, def. 6.

[13] *Data* 26.

[14] HT (= HL) and HE are *given* in position. Therefore, the angle between them is clearly *given*. There is no directly relevant entry in the *Data*. Hultsch (295, #2/3 Hu) and Ver Eecke (1933b, p. 229, #2) offer a different interpretation for the conclusion of Prop. 40. See the commentary.

and CX is <*given*> in position,[1] and A is *given*. Therefore, AX is <*given*> in position,[2] so that the arc ACB is <*given*>, also.[3]
<Synthesis>

And the synthesis is obvious. For one must <provide the auxiliary construction of quadrant ZHE with quadratrix ZTK,> make the ratio of DH to HK the same as the *given* ratio,[4] and describe the <circular> arc through D with center H, and take T, in which it intersects the quadratrix, and join TH, and <draw> RX, which bisects AB and is erected at right angles to it,[5] and <draw> AX, which comprises with XR the same angle as <the angle> KHT,[6] and describe the arc ACB of a circle with center X through A, which has to the base AB the same ratio as the *given* one.[7]

Prop. 41: Incommensurable Angles

#51 And it is not even incredible <that it is possible> to find incommensurable angles.[8] For with the following <argument> one will even take incommensurable arcs of the same circle, and when we posit one of the angles or arcs as *rational*, the remaining one will turn out to be *irrational*.

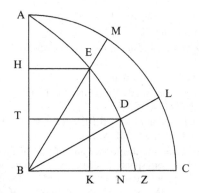

---

[1] CX is the perpendicular bisector of AB, and AB is *given*; *Data* 29.

[2] *Data* 29 (AR is *given* in position, and the angle RAX is *given* in magnitude).

[3] X is *given* (*Data* 25). With X and A *given*, so is the circle with center X and radius XA (*Data* 26); by construction, B lies on it as well.

[4] D on HZ so that HD:HK equals the *given* ratio (VI, 9).

[5] R is the midpoint of AB. The right angle determines the position of RX, whereas the point X is as yet not determined in position.

[6] The easiest way to do this is by constructing a triangle congruent to ΔNHT, or ΔMHL with one side on AB, ∠A = ∠NTH, and producing (if necessary) the sides around A to meet XC in R, and X.

[7] The apodeixis is not given by Pappus. It is easily reconstructed from the analysis. See the commentary.

[8] Incommensurable *lines* were treated in Props. 2 and 3.

Set out the quadrant ABC of a circle, and in it the quadratrix AEDZ, and draw BE through the interior, and EH parallel to BC, and cut off a <straight line> BT from BH, incommensurable with it in length,[1] and draw the parallel DT <to BC>, and join DB. I claim that the angle EBZ is incommensurable with the angle DBZ.

Draw the perpendicular DN <and the perpendicular EK onto BC>. Then, on account of the line <i.e., the quadratrix>, the angle EBZ is to the angle DBZ as EK <is> to DN.[2] EK, however, is incommensurable with DN (since HB <is incommensurable> with BT,[3] also). Therefore, the angle <EBZ> is incommensurable with the angle <DBZ> as well, and when we posit the angle EBZ as *rational*, the angle DBZ will be *irrational*.

Props. 42–44: Analysis of an Archimedean *Neusis*

I have inserted the analysis of the *neusis* that was taken by Archimedes in the book on the Spiral Lines for you, so that you will not get into difficulties when you go through the book.[4] For it, however, the loci described below are taken. They are useful for many other solid problems as well.[5]

Prop. 42: Hyperbola for the Archimedean Neusis

#52 <Let> a straight line AB <be *given*> in position, and from a *given* point C let a certain line CD be drawn forward <so as to fall onto it in D>, and let DE be <drawn> at right angles to AB, and let the ratio of CD to DE be <given.>[6]

---

[1] For the construction of lines incommensurable in length cf. X, 10 ff., e.g., X, 11.

[2] Arc AC:arc MC = AB:EK and arc AC:arc LC = AB:DN (*symptoma*); EK:DN = arc MC:arc LC (V, 16/22). arc MC:arc LC = ∠EBZ:∠DBZ (VI, 33).

[3] EK:DN = HB:BT by construction, and these lines are incommensurable by construction.

[4] Cf. the above meta-theoretical passage before Prop. 31, where it is reported that Archimedes was criticized by some for using a solid *neusis* when a plane argument would have sufficed for *SL* 18. Pappus is going to provide an analysis to show that Archimedes' *neusis* can be determined as the intersection of a parabola and a hyperbola. The *neusis* in Props. 42–44 is closest to *SL* 9, but an analogous argument could be given for *SL* 7 and 8. The hyperbola (Prop. 42) and the parabola (Prop. 43) are considered as solid loci. See the commentary on the use, the power, and the limits of geometrical analysis for the determination of the "degree" of a problem.

[5] This is an indication that there may very well have been some move, on the part of ancient geometers, toward a standardization of "solid" problems via reduction to typical *neusis* with standard constructions.

[6] As in the case of Prop. 40, this ratio is not completely arbitrary. For the upcoming analysis to work, we need CD ≥ DE. In Prop. 44, we will need the equivalent to CD = DE. Note the analogy to Prop. 34a for the starting point of the argument in Prop. 42.

<I claim> that E lies on a <uniquely determined> hyperbola.

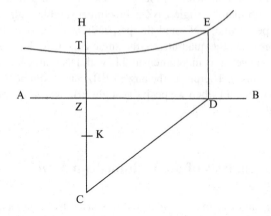

Through C, draw the parallel CZ to the line drawn at right angles <to AB>. Then Z is *given*.[1] <Draw> the parallel EH to AB as well, and let the ratio of CZ to both ZT and ZK be the same as the ratio of CD to DE.[2] Then both T and K are *given*.[3] Now, since the square over CZ is to the square over ZT as the square over CD <is> to the square over DE,[4] the ratio of the remaining square over ZD, i.e.: of the square over EH, to the remaining rectangle between KH/HT is therefore *given*,[5] also. And K and T are *given*. Therefore, E lies on the hyperbola passing through T and E.[6]

---

[1] Z is the point of intersection with AB. It is *given* (*Data* 25, *Data* 28).

[2] Because C – K – Z is assumed, we must have CD ≥ DE, as noted above.

[3] CZ is *given* in position and length (*Data* 26); CD:DE is *given* by hypothesis. ZH and ZK are *given* in length (*Data* 2). They are also *given* in position ⇒ T and K are *given* (*Data* 27).

[4] CZ:ZT = CD:DE by construction. CD:DE is *given* ⇒ $CZ^2:ZT^2 = CD^2:DE^2$, and this ratio is *given* as well (*Data* 50).

[5] The above proportion implies $(CD^2 - CZ^2):(ED^2 - ZT^2) = CZ^2:ZT^2$, so both ratios are *given*. We now show that $CD^2 - CZ^2 = ZD^2$ (I, 47) $= EH^2$, and that $(ED^2 - ZT^2) = KH \times HT$. Then EH^2: KH × HT *is given*.

$ED^2 = ZH^2 = ZT^2 + TH^2 + 2ZT \times TH$ (II, 4). $2ZT \times TH = KT \times TH$ (construction, VI, 1). $TH^2 + KT \times TH = KH \times TH$ (II, 3). So $ED^2 = ZT^2 + KH \times HT$, and $KH \times HT = ED^2 - ZT^2$.

[6] The converse of *Con.* I, 21 is used in analysis. According to *Con.* I, 21, all points on the hyperbola through T with diameter TK, a latus rectum t with t:HK = EH^2:TH × KH, and ordinates parallel to AB fulfill the above equation. For the analysis, we can "conclude": E lies on this hyperbola; cf. Prop. 34a.

Prop. 43: Parabola for the Archimedean Neusis

#53 Let AB be *given* in position and length, and let DC be at right angles <to it>, and assume that the rectangle between AC/CB is equal to the rectangle between a *given* <straight line> and CD.

<I claim> that the point D comes to lie on a parabola that is <*given*> in position.

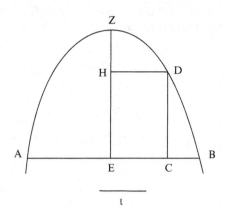

For:

Bisect AB in E, and <draw> EZ at right angles <to AB>, and let the rectangle between the *given* <straight line> and EZ be equal to the square over EB.[1] Then Z is *given*.[2] And <draw> the parallel DH to AB. Then the remaining square over EC, i.e., the square over DH, is equal to the rectangle between the *given* <straight line> and ZH.[3] And Z is *given*; therefore, the point D comes to lie on the parabola passing through A, Z, and B, the axis of which is EZ.[4]

---

[1] Such a rectangle can be constructed using II, 14.

[2] EZ is *given* in length (*Data* 57). Because E is *given* (*Data* 7 and *Data* 27), EZ is also *given* in position (*Data* 29), and so Z is *given* (*Data* 27).

[3] $EC^2 = DH^2$ by construction. Let t be the *given* line. $EB^2 = t \times EZ$ by construction. However, $EB^2 = AC \times CB + EC^2$ (II, 5). By hypothesis, $AC \times CB = t \times DC$, and so $EC^2 = EB^2 -$ [2] $AC \times CB = t \times EZ - t \times CD = t \times ZH$.

[4] Converse of *Con.* I, 20. *Con.* I, 20 shows that all points on a parabola with vertex Z, diameter EZ, parameter t and ordinates parallel to AB fulfill the equation. The analysis can use the converse, even if it is not a theorem.

Prop. 44: Archimedean Neusis (Following Hultsch's Partial Restitution[1])

#54 With these things written down beforehand, the proposed <*neusis* is now subjected to analysis[2]> ... when it has come about beforehand, in the following way.

When a circle ABC is *given* in position, and in it a straight line BC <is *given* in position>, and when A on the circumference is *given*, <the task is> to position between the straight line BC and the circumference BEC a <straight line> that is equal to a posited one and verges toward A.[3]

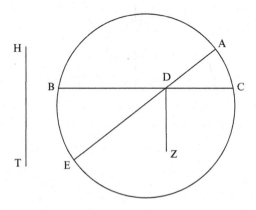

Assume that it has turned out that way, and that it has been positioned and is equal to ED,[4] and draw DZ at right angles to BC,[5] equal to AD.

---

[1] Prop. 44 is not included in Co. Commandino even suggests that *Coll.* IV ends after Prop. 43. The transmitted manuscripts have no figure for Prop. 44. The figure given by Hultsch p. 303 is badly misleading and was not used here. Hultsch himself supplied a correction in the appendix of his edition, pp. 1231–1233. This passage in the appendix also contains a helpful explanation of Props. 42–44 by Baltzer.

[2] Translating the reading in Tr 127, 21/22 for a lacuna in Hu 300, 21/22.

[3] Both the main manuscript A and Hu have "C" here. Hultsch corrected his reading in the appendix to his edition p. 1232/1233. Tr also prints the mathematically correct "A." See the apparatus to the Greek text.

[4] Analysis-assumption.

[5] Z does not necessarily lie on the circumference of the circle.

Now, since AD has been drawn forward from the *given* <point> A <up> to BC, which is <*given*> in position, and since it is equal to the <straight line DZ> erected at right angles starting from <D, the point Z> lies on a <uniquely determined> hyperbola[1] (since the rectangle BDC is equal to the rectangle ADE,[2] i.e.: to the rectangle ZDE[3]). And DE is *given*.[4] Therefore, the rectangle BDC is equal to the rectangle between a *given* <straight line> and DZ. Therefore, Z lies on a <uniquely determined> parabola.[5] Therefore, Z is *given*[6]....[7]

Archimedes used the problem <i.e., the *neusis* for which Pappus has provided an analysis in terms of solid loci here> in order to show a straight line equal to the circumference of a circle.[8] Some, however, reproach him, alleging that he did make use of a solid problem in an inappropriate way ...[9] they show that it is also possible to find a straight line equal to the circumference of a circle by plane means, when one makes use of the theorems pronounced on the spiral.[10]

<This is the end of> Book IV of the *Collection* of Pappus, which is <made up> of splendid theorems, plane, solid, and linear.

---

[1] For the lacuna at 302, 8 Hu, Tr 128, 3 prints ἡ ΑΔ τὸ Ζ σημεῖον ἄρα, the resulting meaning coincides with the paraphrase given above, and with Hultsch's conjecture ad locum.

[2] III, 35.

[3] AD = ZD by construction. Prop. 42 states that Z lies on a (uniquely determined) hyperbola.

[4] DE is *given* in the problem.

[5] Prop. 43.

[6] Z is the point of intersection of the parabola and the hyperbola.

[7] For this lacuna at 302, 12 Hu, Tr 128, 7/8 prints: ἀναλύεται ἄρα. Τούτῳ (τῷ προβλήματι). Therefore, it is subjected to analysis. This problem.

[8] *SL* 18 uses *SL* 7/8.

[9] For the lacuna at 302, 15 Hu, Tr 128, 11 prints: δυνατόν γὰρ ὡς ἀπο(δείκνουσιν) (For it is possible, as they show).This makes Hu's addition of ὡς and ἔστιν in the following line unnecessary.

[10] Unfortunately, no such argument survives. For reconstructions and bibliographical references see the commentary.

Part II Commentary

Introductory Remarks on Part II

The purpose of this commentary is to complement the text, translation, and especially the notes in Part I. It is meant to be read in conjunction with Part I, although some of its prose passages can also be read independently. They typically pick up a topic or keyword as it comes up in the course of Pappus' text, and provide an outline of the historical or mathematical context, or of the historical or methodological significance, or of the main thrust of scholarly discussion with regard to it. No detailed analysis and argumentation will be given, since the present edition is intended primarily as a source text. Instead, I have restricted myself to information that will yield a preliminary orientation as necessary to understand Pappus' text, leading up to the identification of topics that might deserve further investigation, with bibliographical references for such further study. The format of these inserted vignettes is non-uniform, as is their content. For example, the commentary on Props. 2 and 3 contains an excursus on *Elements* X, more or less restricted to a summary of the book's content, with a table that visualizes the content of X. The commentary on Props. 4–12 contains an excursus on the method of analysis-synthesis, in the form of a series of paragraphs targeting different facets of Greek geometrical analysis. And the commentary on Props. 26–29 contains a section on the history of the problem of circle quadrature. The vignettes are primarily intended as a help to understand a particular passage in Pappus, by providing a "horizon" for it, but they can, in many cases, be also read independently from Part I. Apart from the vignettes, the format of the commentary is uniform. Each group of propositions, and most individual propositions, will receive a section with introductory remarks, followed by a schema that visualizes the main characteristics of the group or proposition, and each section also contains proof protocols (or their equivalent in the case of arguments that are not full-fledged proofs) for each individual proposition. The purpose of the protocols is to aid the reader in surveying the mathematical arguments in Part I at a glance, identifying their overall structure and the decisive argumentative moves. The above-mentioned overview schemata go beyond this local level, and aim toward general characterizations and a placement of groups or propositions within *Collectio* IV (*Coll.* IV). They lead up to selective bibliographical references for further study, and they have the following set-up:

H. Sefrin-Weis, *Pappus of Alexandria: Book 4 of the Collection,*
Sources and Studies in the History of Mathematics and Physical Sciences,
DOI 10.1007/978-1-84996-005-2, © Springer-Verlag London Limited 2010

context: e.g., squaring the circle (Props. 26–29).
sources: e.g., Nicomedes' treatise on the quadratrix (Props. 35–41).
means: e.g., *Elements* I, II, III, VI, X (Props. 2 and 3).
method: e.g., apodeixis, analysis-synthesis, analysis.
format: theorem or problem.
reception/historical significance: e.g., reception of the angle trisection in Props. 31/32 in Islamic culture, significance of Props. 31–34 as our only complete surviving sources on the angle trisection by means of conics.
embedding in *Coll.* IV: list of motivic or conceptual connections to other propositions in *Coll.* IV.
purpose: e.g., illustration of the structure of analysis-synthesis (Prop. 4).
literature: reference to treatments in general standard histories, notably (Heath 1921), and to specific scholarly articles, e.g., (Knorr 1978) for Props. 19–22.

These schematized profiles summarize the main ideas on the content, style and purpose of each of the groups or propositions. For a survey of *Coll.* IV, and the groups I have identified, see the general introduction. As said there, the reader should view the way in which I have put propositions together in groups, and my remarks on the purpose of propositions as directly resulting from my general thesis: *Coll.* IV can be read, and was intended to be read, as a unified, coherent survey of the classical geometric tradition from the point of view of methods. The meta-theoretical passage, on the three methodologically defined kinds of geometry, with a homogeneity criterion in place, is to be understood as Pappus' guiding motif in selecting and presenting the material.

As in the translation, Euclid's *Elements* are used as reference work for the justification of intermediate steps. References to individual propositions in the *Elements* will be given in Roman numerals, followed by Arabic numerals (e.g., I, 47 refers to *Elements* Book I, Proposition 47 in Heath 1926). References to books will be given in Roman numerals alone. For all other references, standard techniques and abbreviations will be used.

II, 1 Plane Geometry, Euclidean Style

1 Props. 1–6: Plane Geometry, Euclidean Style

The first six propositions cover material from *Elements* I, (II), III, VI, X. Since IV contains special construction problems in plane geometry of the circle, V contains the general Eudoxean theory of proportions, VII–IX number theory, and XI–XIII stereometry; one can say that Pappus has given an illustration of plane geometry as given in the *Elements* by means of exemplary arguments.

We miss the beginning of the text of *Coll.* IV, certainly the proem. On my thesis about the purpose of *Coll.* IV, it is reasonable to assume (as others have done, also), that we do not miss much of the actual mathematical content at the beginning, for there could hardly be a more suitable starting point for an overall portrait of the methods of plane Greek geometry but *Elements* I, specifically the Pythagorean theorem.

1.1 Prop. 1: Generalization of the Pythagorean Theorem

The so-called theorem of Pythagoras (I, 47) is perhaps the most famous theorem of elementary geometry, the culmination of *Elements* I. It states that in a right-angled triangle, the sum of the squares over the kathetes is equal to the square over the hypotenuse. It is often taken, and not without good reason, as a paradigm of what "Euclidean" argumentation, that is, a classical Greek apodeixis, looks like. And this appears to be the way Pappus viewed it, too. For he gives us, in Prop. 1, a generalization of I, 47 that is very close to I, 47 itself, and is a very good example for the method of classical apodeixis. Within the *Elements*, the Pythagorean theorem is generalized in VI, 31, for similar and similarly positioned parallelograms. VI, 31 uses the Eudoxean theory of proportions (V). But the theorem can easily be generalized further, with or without the use of proportions. One such example, perhaps due to Heron, builds on the proof strategy in VI, 31. Prop. 1 is another example; it avoids the use of proportions and relies solely on means from I (congruence geometry, areas of parallelograms).

> context: Pythagorean theorem.
> sources: /.[1]
> means: I.
> method: classical apodeixis (synthesis).
> format: theorem.

---

[1] Tannery (1912, I, pp. 157–167) considers the possibility that Prop. 1 may be due to Heron, but decides instead in favor of VI, 31, together with a possible further extension of the latter, as due to Heron and his school. Hultsch was probably justified in bracketing a remark, in the conclusion of Prop. 1, that links Prop. 1 to VI, 31 (see the notes to the translation). It may have been this remark that led Proclus to associating Pappus' and Heron's extensions of I, 47, with each other. For an argument by Heron, commenting on I, 47, and showing some connection to Prop. 1, see *Anaritii commentarii in Euclidis Elementa*, ed. Curtze, pp. 78–84; see also Heath (1926, II, pp. 366–368).

H. Sefrin-Weis, *Pappus of Alexandria: Book 4 of the Collection*,
Sources and Studies in the History of Mathematics and Physical Sciences,
DOI 10.1007/978-1-84996-005-2, © Springer-Verlag London Limited 2010

reception/historical significance: transmitted and discussed in connection with the
Pythagorean theorem, e.g., An-Nairizi; Clavius 1574 refers to Pappus explicitly.
embedding in *Coll.* IV: /.
purpose: illustration of classical apodeixis.
literature: Prop. 1 received some attention in secondary literature, mostly in connection
with I, 47; e.g., see Tropfke IV (1923, pp. 135 ff.[1]), Heath (1926, I, pp. 350–368, 1921, II,
pp. 369–370).

The content of Prop. 1 is not used anywhere in *Coll.* IV. As said above, an elementary generalization of the theorem of Pythagoras would appear to be a very fitting
starting point for a survey of ancient Greek plane geometry from a methodological
perspective, and Pappus' proof is a classical apodeixis.

1.1.1   Schema of a Classical Apodeixis, According to Proclus[2]:

1. Protasis (propositio): proposition
2. Ekthesis (enuntiatio): setting-out
3. (if necessary) Diorismos (determinatio): specification and determination
4. Kataskeue (dispositio): construction
5. Epideixis (demonstratio): proof (in reference to the specific configuration)
6. Symperasma (conclusio): conclusion

Sometimes, the term *apodeixis* (proof) is used more specifically for steps 5 and 6
together, or even for steps 4–6. Nevertheless, the picture of a fairly settled standardized
pattern arises.

The following proof protocol of Prop. 1 shows how this pattern, this method of
argumentation, is realized in mathematical discourse.[3] It was therefore given in
some detail.

1.1.2   Proof Protocol Prop. 1

1. Protasis
Let there be given a triangle ABC. Over its sides AB and BC describe parallelograms ABED and CZHB. Produce DE and ZH to their point of intersection T, and
join TB.

Then ABED + CZHB is equal to a parallelogram with sides equal to AC and TB,
and an angle at its base equal to ∠BAC + ∠DTB.

---

[1] This source contains numerous bibliographical references on the Pythagorean theorem and its
history/context.

[2] Proclus, *Commentary on Euclid's Elements*, pp. 203–205 Friedlein, cf. Heath (1921, I, p. 370,
1926, pp. 129–131). The technical terms, and the pattern, are standardized in ancient mathematics,
in fact already in the *Elements*; contrast the situation for analysis-synthesis, described in the introduction to Prop. 4.

[3] Prop. 1 is atypical in that it does not contain a full generalization/abstraction step in the
symperasma.

2. Ekthesis/ 4. Kataskeue
Produce TB to its point of intersection with AC, K. Draw the parallels CM and AL
to KN, obtaining the parallelogram ALMC. The task is to show that ALMC fulfills
the requirement.

5. Epideixis
   5.1. ABTL is a parallelogram $\Rightarrow$ AL = TB
       CMTB is a parallelogram $\Rightarrow$ CM = TB   [I, 34]
       $\Rightarrow$ ALMC is a parallelogram with sides equal to AC, TB
   5.2 $\angle$LAC = $\angle$LAB + $\angle$BAC
       and $\angle$LAB = $\angle$TBH = $\angle$DTB   [I, 29]
   5.3 ABED = ABTL = AKNL, and CZHB = CMTB = CMNK   [I, 35]
       $\Rightarrow$ ACML (= AKNL + CMNK) = ABED + CZHB
6. Symperasma
We have shown that ABED + CZHB is equal to a parallelogram (ACML) with sides
equal to AC and TB, and an angle at the base equal to $\angle$BAC + $\angle$DTB.

Prop. 1 is closely analogous to I, 47.[1] This is why I would disagree with claims that
Heron could be the source for Prop. 1. His generalization took a different route. Pappus
even makes a point of stating that his theorem is "far more general" than a theorem from
the *Elements*. I take him to be referring to I, 47. Though this cannot be affirmed with
certainty, it seems not unlikely to me that Pappus himself is the author of Prop. 1.

1.2   *Props. 2 and 3: Construction of Euclidean Irrationals*

   context: X (geometrical classification of 13 types of irrationals (all first-order
   irrationals)).
   sources: XIII, 11 as model for Prop. 2; both Props. 2 and 3 are by Pappus.
   means: I, II, III, VI, X.
   method: synthesis.
   format: problem.
   historical significance/reception: /.
   embedding in *Coll.* IV: motifs "semicircles, tangents and chords": Props. 4–6, 11, and 12;
   motif "commensurable/incommensurable magnitudes": Props. 17, 41.
   purpose: illustrate operation with the theory of irrationals in X.
   literature: The two propositions have so far been neglected by secondary literature.[2] They
   could, however, be very useful for the reconstruction of the ancient understanding of the
   theory in X, in addition to Pappus' commentary on X.[3] For they show how an ancient author
   operates with that theory. The only surviving ancient actual use of the theory outside Pappus
   is XIII, 11.[4] Prop. 2 appears to be modeled on XIII, 11 (see below). Our understanding of X

---

[1] No construction of similar figure over hypotenuse in I, 47, since it is already given there. But as in
Prop. 1, one constructs two parallelograms, each of which can be shown, via equal areas of parallelograms with equal heights, to be equal to a corresponding parallelogram over one of the kathetes.

[2] Brief reference in Heath (1926, III, pp. 9/10).

[3] With regard to the significance of Pappus' commentary, I am somewhat more optimistic than
Jones (1986a, p. 11), who judges it to be "of only modest historical value."

[4] References to results from X, although no actual work with the concepts, are to be found also in
XIII, 5, 6, 16, and 17.

has for a long time been impeded by a tendency, even in such influential scholars as Heath[1] and Knorr,[2] to view the book as essentially quasi-algebraic. More recent approaches more appropriately emphasize its geometrical character. The best current interpretations are by Taisbak (1982) and Fowler (1992). Knorr (1975a) also contains a very helpful discussion. Taisbak and Fowler are quite compatible with Pappus' commentary (surviving in Arabic). An Arabic text of Pappus' commentary, with English translation and commentary, was published by Thompson and Junge in 1930. Taisbak and Fowler did not consult Pappus on X, and they did not mention Props. 2 and 3 of *Coll*. IV. Therefore, taking these propositions into account could yield independent additional support for the Taisbak/Fowler reading of X.

1.2.1   Excursus: Remarks on *Elements* X (Irrational Lines)

The (geometrical) theory of irrationals in X poses problems for the modern reader. The concepts used overlap with modern notions, but are not synonymous with them. For example, "rational" and "irrational" do not mean what one would assume them to mean, and the concept of "square root" does not exist in Greek geometry (see below: the diagonal of a unit square is rational in the sense of X). X is the longest book of the *Elements*; it yields a classification of rationals (1–20), and a complete classification of all irrationals that can be exhibited by a single operation of the application of areas (21–35 and 36 ff.). Irrationals from X, 21 on are introduced as sums or differences of lines, but characterized geometrically, via application of areas. This dual procedure with its resulting complexity adds to the modern reader's difficulty. And even after one understands the structure, the question remains: what are the *irrationals* good for? What is their mathematical use? Except for the *Medial*, the *Minor* and the *Major*, they seem not to have been used in ancient mathematics outside of X. Is X sheer art pour l'art, for the sake of showing that the known irrationals can be embedded in a complete structural theory? At present, it almost looks like that.[3] The names for the irrationals in X are, for the most part, obviously made up ad hoc. The exceptions: *Medial*, *Minor*, and *Major* are attested for pre-Euclidean geometry (the *Medial* for Theaetetus, the *Minor*, anonymous, in the construction of the golden section; the *Major* is probably owed to Eudoxus[4]).

In what follows, I will give a brief informal explanation of *rational, irrational*, commensurable, incommensurable, and a survey of X in the form of a table. The intention is to give the reader enough information to follow Pappus' proofs in Props. 2 and 3. For more detailed discussions see the above-mentioned secondary literature.

---

[1] Cf. Heath (1926) on X and Heath (1921, I, pp. 401–412). The algebraic notation used there is, in my view, something of a hindrance to the reader's understanding of X.

[2] Knorr's interpretation developed by expanding certain speculative trends in his brilliant reconstruction of the evolution of the Euclidean *Elements* (Knorr 1975a, relying for the reconstruction of the pre-Euclidean theory of irrationals to a significant degree on earlier results by Becker). In my opinion, his later contributions in this area (Knorr 1978c, 1983a, 1985), departed too far from the actual source material. In this regard, I find Fowler/Taisbak preferable.

[3] For positions, differing at least partially, from the one endorsed here on the purpose of X cf. Mueller (1981, chapter XII); Knorr (1975a, 1983a, 1985, 1986).

[4] Cf. Knorr (1975a).

Rational: a line is picked as the magnitude of reference; it is called the *Rational*; other lines are called "rational" or "irrational" in relation to it.

Commensurable

(a) Commensurable in length: two lines are commensurable in length, if they have to one another a ratio expressible in numbers (e.g., 2:3); the ratio of the squares over them is then expressible as a ratio of square numbers (e.g., 4:9)
[X, def. 1; X, 5, 6, and 9].
(b) Commensurable in square: two lines are commensurable in square, if their ratio is not expressible in numbers, but the ratio of the squares over them is (e.g., square on a: square on b is as 2:3)   [X, def. 2; X, 5, 6, 9].

X, 9, serves as the crucial criterion in determining if lines and squares occurring in a geometrical argument are commensurable. Its essence goes back to the pre-Euclidean mathematician Theaetetus.

Incommensurable

(a) Two lines are incommensurable in length, if they do not meet the criterion for commensurability in length (they can still be commensurable in square, e.g., lines with squares in relation 2:3, but also those with an inexpressible ratio for the squares).
(b) Two lines are incommensurable in square, or incommensurable simply speaking, if they meet neither of the above criteria for commensurability (the ratio of their squares is not expressible in numbers)   [X, def. 1; X, 7, 8, 9].

Rational

The basic line of reference is *rational*; also *rational* are all lines that are commensurable with it, either in length or in square (note the difference to the modern concept: the diagonal of the unit square (in our terminology: $\sqrt{2}$) is *rational*)
[X, def. 3 and 4].

Irrational

All lines that are incommensurable with the *Rational* are *irrational*
[X, def. 3 and 4].

1.2.1.1 Survey of Elements X

1.2.1.2   Contents of X, 36–110

	name for a + b	#	name for a − b	#
a, b rat.,[1] comm. squ.	*Binomial*	36	*Apotome*	73
a, b med., comm. squ. a × b rat.	*First Bimedial*	37	*First medial Apotome*	74
a, b med., comm. squ. a × b med.	*Second Bimedial*	38	*Second medial Apotome*	75
a, b incom. $a^2 + b^2$ rat., a × b medial	*Major*	39	*Minor*	76
a, b incom. $a^2 + b^2$ med., 2(a × b) rat.	*Side of a rational plus a medial area*	40	*Line which produces with a rational a medial whole*	77
a, b incom., $a^2 + b^2$ med., 2(a × b) med. $a^2 + b^2$ incom. with 2(a × b)	*Side of the sum of two medial areas a medial whole*	41	*Line which produces with a medial*	78

Following the two groups of propositions that introduce the 12 irrationals, one has:

#42–47 and # 79–84:   Uniqueness of the representations as sums/differences
#47a and # 84a: Six types of the *Binomial* and *Apotome*
#48–53 and # 85–90:   Construction of the types of *Binomial* and *Apotome*
#54–59 and # 91–96:   Construction of the 12 irrationals, using the types
#60–65 and # 97–102:   Uniqueness of the geometrical representations
#67–72 and # 105–110:   The irrationals form complete classes

See also the tables in Fowler(1992, pp. 244–245) and Taisbak (1982, p. 50). Props. 2 and 3 give a surprisingly simple construction for complex *irrationals*. The configurations for Props. 2 and 3 are very similar. Prop 2 constructs a *Minor*[2]. Prop. 3 uses X to go beyond Euclid. The constructed irrationality is not one of those covered in X, but a "higher" *irrational*. It is not named, its status is not defined, and

---

[1] I am using the following, rather obvious abbreviations: rat. = rational, comm. = commensurable, squ. = in square, med. = medial, incomm. = incommensurable.

[2] The *Minor* turns up in the golden section; according to Knorr (1975a), it is one of the three central items that gave rise to the classification theory in X. The *Minor* also appears as the side of a regular pentagon inscribed in a circle with *rational* diameter.

it is not shown to be uniquely determined: a geometrical characterization via application of areas is not given. Perhaps no embedding theory for the higher *irrationals* was available in antiquity.[1] A geometric characterization for the next stage of irrationals, analogous to X, would have been much too voluminous to be covered in a single ancient book.

1.2.2   Prop. 2: Construction of a *Minor*

1.2.2.1   Proof Protocol Prop. 2

Extension of the configuration: Z center of circle, H base of perpendicular from E onto ZC, BT = 1/4 BZ, draw connecting lines. AB (= ZC) is the *Rational*.

1. HT is a *fourth Apotome*.
   1.1 HT is an *Apotome*:
   ZT is *rational*, ZH is *rational*,
   and ZT is commensurable in square only with ZH.
   $[ZC^2:ZT^2 = 64:25^2; ZC^2: ZH^2 = 64:12;$
   $ZT^2:ZH^2 = 25:12; X, 9]$
   $\Rightarrow$ ZT – ZH = HT is an *Apotome*.   [X, 73]
   1.2 HT is a *fourth* Apotome.
   The square for $(ZT^2 – ZH^2)$ has a side that is incommensurable in length with ZT, and ZT is commensurable in length with the *Rational*.
   $[(ZT^2 – ZH^2):ZT^2 = 13:25;$ apply X, 9]
   Thus, ZT – ZH = TH is a *fourth Apotome*   [X def. III, 4].
2. $CE^2 = 2CZ \times TH$, i.e., CE is a *Minor*   [X, 94][3]
   2CZ is rational, TH is a *fourth Apotome*.
   $\Rightarrow$ X with $X^2 = 2CZ \times TH$ is a *Minor*   [X, 94]
   Show that $CE^2 = 2CZ \times TH$ [II, 12/13 generalized]

Compare the following proof protocol for XIII, 11. The close parallel indicates that Prop. 2 may very well be modeled on XIII, 11.

---

[1] According to Pappus, Apollonius studied higher irrationals (cf. Junge and Thompson 1930, p. 64). No traces of his treatment, or a theory around it, survive.

[2] The equation sign is used here for the sake of abbreviation. Pappus himself does not, usually, equate ratios of magnitudes with ratios of numbers directly. In Greek mathematics, numbers and magnitudes are not directly comparable, they are different kinds of entities. Expressions like ZC^2 are abbreviations for "the square with side ZC". They are not to be understood as numbers.

[3] While the auxiliary magnitude TH was established as an irrational via the difference definition, the target magnitude CE is shown to be a *Minor* by reference to the geometrical characterization.

1.2.2.2   Proof Protocol for XIII, 11

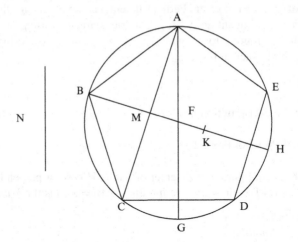

Extension of the configuration: F center of the circle, M base of perpendicular from A onto BH, FK = 1/4BF, draw connecting lines. AG (=BH) is the *Rational*.

1. MB is a *fourth Apotome*
    1. 1. MB is an *Apotome*
    BK is *rational*, MK is *rational*,
    and MK is commensurable with BK in square only.
    [BK^2:FK^2 = 25:1; MK^2:FK^2 = 5:1; BK^2:MK^2 = 5:1; X, 9]
    $\Rightarrow$ BK – MK = MB is an *Apotome*.   [X, 73]
    1.2 MB is a *fourth Apotome*
    The square for (BK^2 – MK^2) has a side that is incommensurable in length with
    BK, and BK is commensurable in length with the *Rational*
    [BK^2 – MK^2):BK^2 = 4:5; X, 9]
    $\Rightarrow$ BK – MK = MB is a *fourth Apotome*.   [X def. III, 4]
2. AB^2 = BH × MB, i.e., AB is a *Minor*
    BH is rational, and MB is a *fourth Apotome*.
    $\Rightarrow$ X with X^2 = BH × MB is a *Minor*   [X, 94].
    Show that AB^2 = BH × MB   [similar triangles ABH, ABM].

1.2.3   Prop. 3: Construction of an Irrational Beyond X, with the Notions from X

1.2.3.1   Proof Protocol Prop. 3

Extension of the configuration: H center of the circle, semicircle HBD, center C, K on semicircle and DZ extended, L, T bases of perpendiculars from B, K onto HD. Draw connecting lines. AC (= HD) is the *Rational*.

1. DK is a *Binomial* [X, 54][1]
     1.1 DL is a *first Binomial*
       1.1.1 DL is a *Binomial*
       DC is *rational*, CL is *rational*,
       and DC and CL are commensurable in square only
       $\Rightarrow$ DC + CL = DL is a *Binomial*.   [X, 36]
       [$DC^2{:}DH^2 = 1{:}4$; $DC^2{:}CL^2 = 4{:}3$;
       $DC^2{:}CL^2 = 1{:}3$; X, 9]
       1.1.2 DL is a *first Binomial*
       The square for ($DC^2 - CL^2$) has a side that is commensurable with DC, and
       DC is *rational*.
       [$(DC^2 - CL^2){:}DC^2 = 1{:}4$; X, 9]
       $\Rightarrow$ DL is a *first Binomial*   [X, def. II, 1]
     1.2 DK is a *Binomial*
       HD is *rational*
       $\Rightarrow$ X with $X^2 = DL \times HD$ is a *Binomial*.   [X, 54]
       Show that DK = DL × HD.
2. KZ is a *Line which produces with a rational a medial whole*   [X, 95]
     2.1 LT is a *fifth Apotome*
       2.1.1 LT is an *Apotome*
       LC and CT are *rational*, and comm. in square only
       [LC: see above, 1.1.1; CT:HD = 1:4;
       $LC^2{:}CT^2 = 1{:}3$; X, 9]
       $\Rightarrow$ LC – CT = LT is an *Apotome*   [X, 36]
       2.1.2 LT is a *fifth Apotome*
       CT, the lesser one of the pair LC, CT, is commensurable with the *Rational*
       HD, and the square for $LC^2 - CT^2$ has a side that is incommensurable with
       LC in length. Thus, LT is a *fifth Apotome*   [X, def. III, 5]
     2.2 KZ is a *Line which produces with a rational a medial whole*
       Show that $KZ^2 = LT \times DH$.
       Since LT is a *fifth Apotome*, and HD is *rational*, KZ is a *Line which produces*
       with a rational a medial whole   [X, 95]
3. Since KZ is a *Line which produces with a rational a medial whole*, and DK is a
   Binomial, DZ = KZ – DK meets the claims made in Prop. 3.

1.3 Props. 4–6: Plane Geometrical Analysis in the Context of Euclidean Geometry

context: Analysis-synthesis in plane geometry.
possible sources: /.
means: I, III, VI.

---

[1] Note the general strategy: construct an auxiliary line, identified via addition definition, then show
via geometrical characterization that the target line is an irrational of the type claimed.

method: analysis–synthesis.
format: theorem.
historical significance/reception: /.
embedding in *Coll*. IV: motif "circle, chords and tangents": Props. 2, 3, 5, 6; features of the
method of plane geometrical analysis: Props. 7–10, 12; motif "analysis" outside of plane
geometrical analysis: Props. 28 and 29, 31–34, 36–40, 42–44.
purpose: illustration of the structural features of plane analysis-synthesis, with analysis
aspects to be spelled out in Props. 5–12.[1]
literature on analysis-synthesis in Prop. 4: (Hintikka and Remes 1974, especially pp.
22–30); their approach is parallel to my interpretation in that they take Prop. 4, just as I do,
to be an argument that exhibits the structure of analysis-synthesis as a method of argumentation. In my opinion, Prop. 4 was designed by Pappus himself for precisely that purpose.
literature on Greek geometric analysis-synthesis in general: (Hintikka and Remes 1974,
1976); survey of analysis-synthesis as an argumentative technique: (Zeuthen 1886, pp.
98–104; Jones 1986a, pp. 66–70); some influential contributions to the discussion of Greek
analysis-synthesis in secondary literature besides the ones mentioned above include:
Cornford (1932), Robinson (1936), Gulley (1958, 1962), Mahoney (1968), Szabo (1974),
Lakatos (1978, pp. 70–103), Mueller (1981), Jones (1986a) passim (especially the essays
on analytical works), Mäenpää (1997), and Netz (2000b).

The result of Prop. 4 is not used in *Coll*. IV. From a methodological point of view,
Prop. 4 sets the stage for the illustration of various aspects of the method of plane
geometrical analysis in Props. 4–12. It should be, and has been, read as programmatic. Props. 5–12 then complete Pappus' portrait of plane analysis (e.g., Prop. 7:
use of Data, Props. 8–10: Apollonian analysis, *resolutio*, determination of data,
Prop.12: analysis, *apagoge/epagoge*). For a schema of ancient analysis-synthesis
see below. As said in the introduction, analysis-synthesis is a two-part method. In
the analysis, one assumes what one wants to construct or show as already established and applies different strategies for identifying features that are crucial and
constitutive of the target situation (either as actual elements in it, or as consequences from auxiliary constructions that one can apply), until one arrives this way
at a situation that can be verified or constructed from elsewhere. Then one shows
that this end stage is independent from the analysis-assumption, i.e., that it can be
reached from what is given, not just from what one wants to establish. The synthesis,
essentially a classical apodeixis (cf. Prop. 1), follows. It corroborates the result.
There has been much scholarly discussion in secondary literature about Greek geometrical analysis, its nature, its goals, and even its practice and status within Greek
mathematics. A large volume of analytical Greek geometry once existed, but the
sources are for the most part no longer directly accessible to us. Scholarly discussion must focus on the slim evidence we have, in addition to "meta-theoretical"
characterizations, notably the proem of *Coll*. VII. *Coll*. IV provides quite a few
examples for different types and usages of analysis in practice (see below), which
have not yet been fully exploited. Perhaps the full documentation of these examples
in their original context can serve as a basis for further investigation. It seems to me

---

[1] Cf. Heath (1921, II, p. 371). He states that the content of Props. 4–6, 7, 11/12 is not of any intrinsic
mathematical interest.

that the fact that Pappus has a methodological perspective on the examples he presents may be an advantage for any inquiry that looks for the significance of ancient geometrical analysis as a method of argumentation.

1.3.1 Excursus: Greek Geometrical Analysis as a Method: Sketch of the Status Quaestionis

Greek geometrical analysis has been the subject of intense, controversial scholarly discussion in recent decades. In my opinion, we are still rather far from reaching a communis opinio, even from sifting through all the relevant material.[1] The following remarks are therefore rather general and sketchy. No attempt will be made to convey a settled view on the details of ancient analysis. In fact, the documentation of various, varying examples of ancient analysis in practice provided here is intended to help complete our picture, when taken into account for further discussion. The purpose of my remarks is merely to convey a rough idea about the status quaestionis on ancient analysis: what sources are available, what can be made out in outline, what issues and problems were raised in recent scholarly discussion, and from what perspectives. I will take a stand and say a little more about the topic of Greek analysis as an analysis of configurations. For more details, and for alternative views, the reader is referred to the literature mentioned above.

1.3.1.1   Sources on Greek Geometrical Analysis

Perhaps the crucial problem in the discussions about ancient geometric analysis is the scarceness of sources and examples. Of the vast corpus of contributions that used the analytical method in a programmatic and thoroughgoing way (Apollonius), only a fraction survives. For the most part, we have to rely on indirect sources (*Coll.* VII), schematic characterizations (proem to *Coll.* VII), and a few isolated examples. The following lists may illustrate this point.

> Ancient descriptions of the method: *Coll.* VII, proem,[2] *Elem.* XIII, 1, Heron on Euclid II (Al-Nairizi), Proclus, *Commentary on Euclid's Elements.*
> Indirect sources: *Data*[3]; *Coll.* VII.[4]

---

[1] A thorough study of Apollonius' *Sectio rationis* from the perspective of how analysis operates there is still a desideratum. Not all of the examples in *Coll.* IV, illustrating a broad spectrum of types and usages of analysis, have been accounted for. Scholarly discussion has so far concentrated on the interpretation of the proem of *Coll.* VII, and a very narrow selection of actual examples (e.g., Hintikka and Remes (1974) use Prop. 4, Mahoney (1968) uses Prop. 12 as exclusive typifiers).

[2] This is the most detailed account. It was the basis for most of the more recent scholarly discussion of the ancient method of analysis.

[3] Euclid's *Data* are not, in themselves, an analytical work. They are closely linked to the *Elements*, and provide a set of (synthetically proved) propositions useful for analytical work in practice.

[4] cf. Jones (1986a). *Coll.* VII is a commentary on the lost analytical works, not a direct source for analytical arguments.

Sources on ancient analysis in practice: Aristotle, *Meteor.* III, 5, Scholia and diorismoi on/
in Euclid's *Elements* (e.g., additions to XIII, 1–5), Archimedes, *Sph. et Cyl.* II, 3–7.[1]
Sectio rationis, fragments and testimonies of other minor analytical works by Apollonius,
Eutocius *in Arch. Sph. et Cyl.* II, 78–85 (Menaechmus) and, so far neglected, for the most
part: *Coll.* IV, Props. 4, 7, 8–10, 12, 28, 29, 31–34, 36–40, 42–44.

1.3.1.2   Pappus' Outline of Analysis-Synthesis in *Coll.* VII

In this situation, it is perhaps understandable that modern discussions have concentrated on Pappus' longer methodological characterization of analysis-synthesis in
Coll. VII as a starting-point. This passage is the proem to his commentary on the
treasury of analysis. Its goal is not to give a historically accurate description of the
actual practice, but an attempt to explain its structure as a method in a general way.
Therefore, statements made in this proem cannot suffice to reconstruct the details
of the actual practice of the method. Examples of the method at work are needed
for a thorough and comprehensive assessment. Even so, reflecting the general trend
in scholarship, the following brief overview also starts with a sketch of the proem,
highlighting in the initial comments the main points that have given rise to scholarly
discussion.

According to Pappus, analysis is a heuristic technique. There are two general
types of analysis: problem-oriented and theorem-oriented, but they are methodologically equivalent. This second claim of Pappus' has met with criticism from several
commentators, as overly schematic and out of touch with the actual mathematical
practice. In fact it looks as though analysis was indeed a heuristic technique, developed and employed primarily for problem solving, as a research tool. Our sources
on theoretical analysis are all relatively late and seem to point to a didactic context.
With this modification, however, it is perhaps fair to say that the theoretical analyses
are nevertheless analyses in the full sense, methodologically equivalent to the
problem-oriented ones, and helpful to learn and understand the method. Pappus
gives examples for both theoretical and problem-oriented analysis, as well as for
further, yet different usages of analysis in the course of *Coll.* IV (see below).[2]

Analysis is the first part of a two-partite method. Its job is essentially to furnish
the grounds for a synthetic proof.[3] It consists of two parts. In a first part (called
apagoge throughout the present study), one assumes what one wants to prove as
true, or what one wants to construct as already established, and proceeds from there
in what Pappus chose to call with the exasperatingly vague term "orderly fashion"

---

[1] These are the only examples for analytical argumentation in Archimedes. Analysis was not
Archimedes' favored method of investigation.

[2] For those who place a heavy emphasis on creative research and problem-solving as the essence
of mathematics, a less favorable assessment will arise. Some would go so far as to discard the
didactic examples as not "real analyses" at all.

[3] Many scholars disagree vehemently here (e.g., Hankel, Robinson, Mahoney, and Szabo). They
view analysis as an independent technique that furnishes validation on its own. See below.

until one reaches something that is already known from elsewhere. Then one has to show that this end stage is independent from the above assumption, and constructible without it (this part will be called *resolutio* throughout the present study). In Pappus' description, two apparently conflicting visions of the overall direction of analysis are combined. He portrays it as an upward move toward principles for a proof, but also uses language that suggests it operates primarily via deductions: a downward move,[1] toward conclusions. Specifically, the Greek word used for "orderly fashion" is ἀκολούθως. Many have thought this means that analysis proceeds by logical deduction.[2] Even though the word does not in fact have such a narrow meaning, the impression remains that analysis might be primarily deductive. This immediately raises the questions of how it can, at the same time, be seen as an upward move, whether it is a method of proof of sorts, and the question of convertibility of analysis-steps. For if analysis is to be a validation technique, the steps have to be reversible. The bulk of scholarship has focused in on the items deduction, calculus-character, validation character, and convertibility. Perhaps one can gain a new perspective by taking the nature of Greek analysis as an analysis of configurations into account. This would shift the focus away from concerns stemming from propositional calculus and its logic (see below).

In Pappus' account, and in the ancient examples we have, the analysis is followed by a synthesis, whenever an argument is presented as complete. The synthesis has the form of a classical *apodeixis* and retraces the steps of the analysis.

---

[1] Gulley (1958) provides a convincing argument for the thesis that the "upward" global view derives from philosophical considerations and contexts, while the "downward" items reflect "local" ingredients of the mathematical practice. He holds that Pappus did not succeed in combining them in a coherent overall picture. Here I disagree with him. In my opinion, the two views constitute two complementary perspectives, both relevant and pertinent, not two conflicting and competing strategies.

[2] Hankel (1874) argued that analysis is indeed a fully valid proof strategy, rendering the synthesis superfluous, and ascribed the persistent presence of syntheses in analytic-synthetic arguments to the "Nationalcharakter" of the Greeks – a kind of genetic meticulousness. Other authors, who do not go quite this far, but still maintain that analysis is a deductive method, aiming at establishing conclusions, include Robinson (1936), Gulley (1958), Mahoney (1968), and Szabo (1974). The word "akolouthos," interpreted as implying logical derivation, is strongly emphasized by Robinson (1936) and Gulley (1958). Robinson and Szabo also place special emphasis on indirect (dis-) proof: an analysis that derives a contradiction from p proves that p is false. A famous example from pre-Euclidean geometry would be the proof for the incommensurability of side and diagonal in a unit square. It seems to me, though, that such indirect arguments were not the primary focus of operation for mathematical analysis. Hintikka and Remes (1974) have argued, in my view convincingly, that analysis is not a method of proof, nor can it stand alone as a (direct) method of validation. Perhaps they overemphasize the non-deductive ingredients and features of analysis. Hintikka (1973), discussing a passage from Aristotle's *De interpretatione*, also showed that "akolouthos" does not necessarily have the force ascribed to it by some scholars. In fact, *Collectio* IV contains at least three distinct instances where the word does not have that specific meaning: Prop. 12: The synthesis is said to follow the analysis in orderly fashion. *Genesis* of the quadratrix: The upper side of the square "follows along in orderly fashion," as the left-hand side describes a quadrant. Prop. 30: Archimedes claims that since spirals in the plane, and on cylinders (or cones) have been considered, it "follows in order" that one should envisage spirals on hemispheres as well.

It seems to me that this quite obviously means that analysis was not viewed as an independent proof strategy by the Greeks. Nevertheless, if the synthesis simply retraces the analysis, it would seem as though it is a mere formality, and one virtually "has it all" after the analysis. The synthesis, furthermore, clearly has to be deductive. So the issue of the convertibility of analysis-steps arises anew, as does the question if the burden of proof lies with the analysis (where all the decisive ideas and moves come from) or with the synthesis. Scholarly opinions differ widely here, and will probably continue to do so. And it helps very little to realize that the synthesis can certainly stand alone, while the analysis cannot.

1.3.1.3   Analysis in Outline

The following tentative general considerations focus on analysis: its setting, its ingredients, its nature. I will leave the question of the exact interdependence of analysis and synthesis aside, while claiming that analysis was not, and could not be, a proof strategy on its own. And I will avoid as much as possible the technical vocabulary of the standard discourse, because I am pleading for a fresh look at the issues.

Recent scholarship, notably Netz (1999), has put the spotlight on an aspect of Greek mathematical discourse that has perhaps not been sufficiently recognized in the more recent past: the importance of figures, and the essential role of configurations in geometrical argumentation. Picking up on considerations in Hintikka and Remes (1974), I should like to bring this aspect to bear for a preliminary global evaluation of Greek geometrical analysis. It is important to note that Greek geometrical analysis is not an analysis of propositions, not a method for manipulating propositional content, but an analysis of configurations.[1] It looks for dependencies and interdependencies of items within geometrical configurations, and in this context it hunts for grounds of argumentation. Perhaps we are better off not to assimilate this strategy too closely to our ways of operating in terms of propositional calculus. If analysis is successful in clarifying crucial interdependencies in a configuration, it may be convincing, even convey certainty. It can be methodical and carry validation. But it need not be a self-sufficient, closed proof strategy. In fact, there are good reasons for wanting it to be more open and not restricted to deductive procedures. Let me explain.

In an analysis-setting, we have a complex configuration, part of which is hypothetical, and we are trying to make sense of it, corroborate it. One strategy that suggests itself immediately is the idea of breaking down the whole into constitutive building blocks, on which we hope to have a handle. This strategy amounts to a downward movement, in the form of reduction and deduction. Obviously, that is a reasonable path to take. In all likelihood, reduction/deduction was the historical

---

[1] When Hintikka and Remes put forth considerations to this effect in 1974, they met with a lot of resistance among historians of ancient Greek geometry and others. It appears to me, though, that many of their points are valid and deserve reconsideration today.

core of the ancient method of analysis (see Prop. 12). The building-blocks (partial configurations) we are isolating would be already actually present in the configuration, they would turn up as "conclusions" of a derivative chain, but function at the same time as starting-points for putting together the whole again.[1] The issue of convertibility is important for the overall success, but it seems reasonable to assume that an investigator would ignore it initially.[2] The attention would be on recognizing patterns, identifying standard partial configurations, and on learning how to reduce complex configurations to standard types. This strategy has the additional advantage that it can be brought into a quasi-algorithmic framework, with predictable standard moves. It is straightforwardly operational, and very powerful. I agree with those scholars who insist that deductive moves were the preferred strategy within analysis.

Yet this is not all, and one would not want to be restricted to it. In addition to making use of patterns that are already present in the target configuration, one will also try to detect incomplete patterns, complete them, and work from there. One must resort, inter alia, to extensions of the configuration, auxiliary constructions. This is a non-programmable, non-deductive (upward) move. It relies on intuition. Nevertheless, it can be methodical, if one keeps in mind that the context of argumentation is the configuration at hand, and doing it is still hunting for grounds of argumentation within the configuration, in service of the overall goal to reduce to something you can handle. In other words: it is a vital part of *apagoge*. When Hintikka and Remes pointed to the pervasive presence of auxiliary constructions in ancient analyses, insisted that they were an important part of the method, and stressed that this feature implies that analysis is not a deduction method (not essentially downward), and does not establish conclusions (is no valid proof), they were right, I think. Given the context of the discussion, their contribution was received as one with philosophical, meta-theoretical focus, and critics insisted that it has nothing to do with the ancient mathematical practice. This is unfortunate. For in my opinion, the importance of operating with extensions has everything to do with the practice of conducting an *apagoge* in full view of a concrete configuration. Both ingredients, the deductions and the extensions, are means to provide grounds for argumentation by reduction to familiar patterns. It is a needless impoverishment of one's analytical toolbox to restrict oneself to strategies that can be conceptualized as propositional deduction moves. This is also why I find the discussion of whether analysis is primarily "upward" or primarily "downward" helpful only to a certain degree, and likewise the attempt to capture the fruitfulness of analysis by the degree to which it is proof-like.

Even after a successful *apagoge*, when one has reached a configuration over which one has control, one still has to secure, in the *resolutio*, that the identified end stage of the *apagoge* is indeed fully controllable even without the initial

---

[1] Some might argue that this is not, in fact, a downward move to begin with, inasmuch as the elements are prior to the composite. I will leave this question aside, and just accept that we are dealing with a deductive move.

[2] This does not imply that he thinks it is not an issue, or thinks he automatically has convertibility. Nor does it imply that he sees himself as primarily hunting for conclusions/valid derivations.

assumption made at the beginning of the *apagoge*, and that all the items needed can be determined, or derived, from what was either given or can be provided independently. In both cases, one establishes the status of the necessary ingredients as "*given*" (for this term, see the introduction to Prop. 7). Also, one must determine conditions for solvability, and discriminate between different cases. This part of the *resolutio* is called *diorismos*. In general, the *resolutio* of the analysis will be mostly deductive as well, but there are instances in *Coll.* IV (Props. 33 and 34, e.g.), where the *resolutio* is not deductive throughout. In general, it appears that the *resolutio*, the stage in which analysis culminates and "hands over" to synthesis, most clearly has the ensuing synthesis in view. Very often, the first part of the synthesis will more or less repeat the final steps of the *resolutio*, in the same order (cf. Props. 4 and 12). Even so, it was evidently seen as necessary that the synthesis repeat those steps, and this means: this part of the proof was not seen as covered by the analysis itself, either.

To sum up this sketch of analysis: Greek geometrical analysis is an analysis of geometrical configurations. One operates with deductions, and with non-deductive moves such as extensions of the configuration, with the overall goal to reduce the target configuration to familiar patterns. After this phase, one has to make sure that the identified items in the end stage of the analysis are constructible. A synthesis follows, which repeats the constructions, and proceeds with a regular apodeixis that retraces the steps of the analysis. The synthesis will often be easy after a successful analysis, but it is the part that carries the formal proof. One will often find that the synthesis is left to the reader as obvious. But then exactly this will be stated. An analysis never ends with "q.e.d."

1.3.1.4   Examples of Analyses in *Coll.* IV

Most of the examples for analyses in *Coll.* IV illustrate the dominant and primary use of the technique of analysis as a heuristic tool that was described above. Specifically, Props. 4, 7, 8–10, and 12 illustrate different aspects of plane geometrical analysis, Props. 31, 33, 34, and 42–44 illustrate the use of the technique for problem-solving in solid geometry, and in Props. 36–40, analysis is employed to reduce "linear" problems to basic *symptomata* (defining properties) of the quadratrix. There are, however, different examples as well. Analysis was, apparently, not restricted to furnishing a heuristic toolbox. In Props. 28 and 29, it is employed to characterize the transcendent motion curve quadratrix, via its *symptoma*, through an analysis of loci on surfaces. This use of analysis is directed toward determining the properties a certain geometric object has in relation to other geometric objects within the same configuration. In Props. 42–44, analysis is used as a device to determine the "level" of a problem, independently of an actual solution. Such analyses were probably followed, in practice, by actual solutions. Pappus may be somewhat idiosyncratic here in picking out the analysis only, to make a point about procedures to establish homogeneity for already existing mathematical contributions. Nevertheless, the fact that analysis can be so used shows that ancient geometrical analysis was not seen only as a toolbox for actual problem solving. It had intrinsic merits, beyond facilitating operations.

Especially because of these other usages of the analysis technique, it is quite apparent that the standard reading of ancient Greek analysis is too narrow and has not yet made full use of the source material (limited as it may be). As Mahoney has suggested, we should look at the usage made of analysis in actual Greek geometry in the transmitted texts to determine what analysis meant for the Greeks. *Coll.* IV provides quite a few examples that can broaden and deepen our understanding of it. We have the additional advantage that the author of *Coll.* IV, presenting these actual examples of ancient geometrical analysis, is also the author of the proem of *Coll.* VII, our most extensive testimony for the characterization of the method as such. We can use the material to substantiate what was left open or vague by Pappus in his methodological description, filter out what was solely due to the fact that the text from *Coll.* VII is a proem to a treatise, and come to a more comprehensive understanding of what at least this one well-informed ancient mathematician thought about analysis as a mathematical method.

1.3.1.5   Structural Schema of Analysis-Synthesis[1]

Protasis (Proposition)
Ekthesis (Setting-out)
1. Analysis
   1.1. Assumption: problem solved/proposition true
   1.2. Analysis proper: *apagoge*, or *epagoge*
   (transform the desired proposition/state of affairs into something known to be constructible/true: use reduction, which can, but need not be, deductive,[2] and use suitable extensions of the configuration, amounting to a strategy of transposition; this second aspect is non-algorithmic, relies on intuition, one needs to bring additional information to bear[3])
   1.3. *Resolutio*: *diorismos* and determination of *data*[4]
   (determine conditions for solvability/provability, often resulting in sub-cases for the desired proposition; show that the endpoint of the analysis is reachable/true, independently of assumption 1.1.)

---

[1] Cf. Prop. 1 for the structural schema of a synthetic proof
[2] Examples for non-deductive steps in an analysis: in Props. 31–34, see below.
[3] See Hintikka and Remes (1974, especially pp. 22–30 and 41–48).
[4] The *resolutio*, with both its components, was central in Apollonius' works (cf. Jones 1986a). *Diorismos* is not adequately represented in *Coll.* IV, cf. the introduction to Prop. 7. Since the analysis proper assumed that the problem is solved (the theorem true), the *resolutio* has to do two things: (i) show that the items in the end stage configuration are determined within the configuration, from the problem, independently of the analysis-assumption (ii) determine conditions for solvability, including, perhaps, a split-up into cases. This way, it establishes necessary, but not sufficient conditions for the existence of the figure at the end stage of the analysis. The synthesis, with kataskeue and apodeixis, will establish the latter; cf. Hintikka and Remes (1974, pp. 49–69).

2. Synthesis (cf. Prop. 1)
   2.1. Kataskeue (Construction)
   2.2. Apodeixis (Proof)
   2.3. Symperasma (Conclusion)

Analysis started out as a heuristic technique; it was never "settled" in the mathematical inventory to the degree that synthesis was. An indication for this is the lack of proper names for some of the features: *diorismos*[1] and *data* (Greek: δοθέντα) were used consistently. There was no name for the crucial first stage, the analysis-assumption, and for stage 1. 3. *Resolutio* is a modern term, coined in the nineteenth century (Hankel 1874). Nevertheless, the stages themselves are detectable as essential parts of a pattern that is present in the ancient sources. As to the labels "*apagoge*" and "*epagoge*," they are both attested, but it is not clear whether they were interchangeable, or whether there was a conscious differentiation in meaning, where *apagoge* replaced *epagoge* as the name for analysis proper, with *epagoge* taking on the meaning of "reduction" strictly speaking, and *apagoge* covering other aspects of stage 1.2 as well. The above schema, going back to Zeuthen (1886, pp. 98–104), is a good description of the actual structure of ancient geometrical analysis as one finds it in the sources. It is not, however, an ancient description.

1.3.2   Prop. 4: The Structure of Plane Analysis-Synthesis

1.3.2.1   Proof Protocol Prop 4

The following protocol, like the protocol for Prop. 1, is somewhat more detailed so as to illustrate how the steps of the synthesis mirror the steps of the analysis.

1. Analysis
   1.1 Assumption: EK = EL
   1.2 Analysis proper:
   Extension of the configuration: TM, EN[2]
   Then 1.1implies

MX = XT	(1) [VI, 4; V, 16]
ZN = NT	(2) [III, 3]
NX ∥ ZM	(3) [VI, 2]
∠TNX = ∠NZM = ∠TAX	(4) [I, 29, III, 21]
T, N, X, and A lie on a circle	(5) [III, 21, conv.]
∠ANT = ∠AXT = ∠AEL = ∠AED	(6) [III, 21, I, 29]
A, N, E, and D lie on a circle	(7) [III, 21, conv.]

   1.3 *Resolutio*

(7) is true independently of assumption 1.1	[III, 18, III, 31]

---

[1] For "diorismos," cf. Procl. in Eucl. 202, 2 ff. Friedlein, and Heath (1921, I, p. 371). It should perhaps be noted that this technical term was interpreted in slightly different ways by different authors.

[2] This step is non-deductive, as Hintikka–Remes have pointed out. The success of an analysis can rest on the insightful choice of such extensions rather than on successful deductions.

2. Synthesis
   [2.1 Kataskeue: construct M, X, N as in the extension in the analysis]
   2.2 *Apodeixis*

A, N, E, and D lie on a circle	$(7')^1$ [III, 18, III, 31]
∠AND = ∠AED = ∠AXT	$(6')$ [III, 21, I, 29]
A, N, X, and T lie on a circle	$(5')$ [III, 21, conv.²]
∠TAX = ∠TNX = ∠TZM	$(4')$ [III, 21]
NX ∥ ZM $(3')$	[I, 27³]
MX:XT = ZN:NT = 1:1	$(2', 1')$ [III, 3, VI, 4]
EK = EL	[I, 29, VI, 4]

1.3.3   Props. 5/6: Reciprocity in Plane Geometry

In the above excursus on analysis-synthesis, I briefly touched upon the discussion about the question of the convertibility of steps in plane analysis. Already in Aristotle's *Posterior Analytics* (78 a 7 ff.), the problem of convertibility in analysis-synthesis was characterized as a non-trivial question, when he said that induction-deduction would be (!) easy, if analysis were always directly convertible. In plane geometry, we have, as a rule, a kind of reciprocity: if feature a entails feature b, then feature b entails feature a. But even in plane geometry, this holds only for the most part, not always. Even in plane geometry, one may have to use additional devices, differing intermediate steps, to convert an argument. Props. 5 and 6 can be used to illustrate this aspect of plane geometry. They are reciprocal. In both cases, we have a circle, a chord, two tangents in the endpoints, and a line segment E–H–Z, intersecting the circle in T and K. If EH = HZ, then TK = HK (Prop. 5), and if TK = HK, then EH = HZ (Prop. 6). However, not all steps in the proof of Prop. 5 can be used in Prop. 6. The content of Props. 5 and 6 is rather trivial. The twin propositions illustrate the following point quite nicely: even in plane geometry, where usually

---

[1] Note that the endpoint of the *resolutio* has to be proved again in the synthesis, as its first step. For the synthesis has to provide a complete proof. In the case of Prop. 4, the first step of the synthesis turns out to be identical with the last step of the analysis. Nevertheless, it is no mere repetition, and certainly it is put down by Pappus intentionally. Whenever in an analysis-synthesis this first step of the synthesis is passed over, this means it is not given explicitly, because trivial, but it does not mean that the *resolutio* is taken to be part of the proof. The situation is comparable to the omission of the *kataskeue* in Prop. 4.

[2] Whereas the use of the converse of a proposition is unproblematic in the analysis, it is not permissible in the synthesis. One has two options here. Either the converse was known to be a theorem (though not attested in the surviving texts), and then that theorem is invoked here. In fact, Pappus uses what amounts to the converse of III, 21 in several places outside of Prop. 4 as well. So this would appear to be the preferable option. Or, one has to assume that Pappus passes over an implicit intermediate step here, showing X to be on the circle through A, N, and T; e.g.: Assume that X does not lie on the circle. Draw AX. It intersects the circle in V, with either A–V–X or A–X–V. In both cases, one has AVT = ANT = AXT, in contradiction to I, 16. Therefore, X must lie on the circle.

[3] I, 27 is the converse of I, 29 within Euclidean geometry.

the converse of a proposition or deductive step is valid, if the proposition or deductive step is so, we cannot always assume that the argument from a to b just has to be retraced, with the very same steps, from b to a, in order to get a deductive derivation of a from b. And this point, once it is granted, obviously has consequences for the nature of plane analysis-synthesis as a technique of argumentation, and specifically for the relation of analysis to synthesis in general.

> context: plane analysis-synthesis, convertibility of argument steps.
> sources: /.
> means: I, III.
> method: synthesis.
> format: theorem.
> historical significance/ reception: /.
> embedding in *Coll.* IV: motif "circles, chords, tangents and relative measures in length":
> Props. 2, 3, 4, 8, 9, 10, 39–41, motif "convertibility in geometry of circle": Prop. 4.
> purpose: illustrate reciprocity in plane geometry, specifically: geometry of the circle (III).

The results of Props. 5 and 6 are not used in *Coll.* IV.

1.3.3.1   Proof Protocol Prop. 5

(Assume EH = HZ, show TH = HK)
1. Ekthesis
With center of the circle L, construct an isosceles triangle ZLM.
2. Apodeixis
   2.1. EL = LM = LZ
   2.2. ΔEHL ≅ ΔZHL, HL ⊥ EZ, TH = HK

1.3.3.2   Proof Protocol Prop. 6

(Assume TH = HK, show EH = HZ)
1. Ekthesis
With center of the circle L, construct triangles EAL, EHL, LZC.
2. Apodeixis:
   2.1. Δ EAL ≅ Δ ZCL, EL = LZ
   2. 2. HL ⊥ EZ, EH = HZ

II, 2 Plane Geometry, Apollonian Style

2 Props. 7–10: Plane Geometry, Apollonian Style

2.1 Overview Props. 7–10

Announcement of Prop. 10
  7: Illustration of how to work with the *Data* in a *resolutio*
  8: Auxiliary lemma for Prop. 9, with connection to Apollonius, *Tangencies* I, 16/17
  9: Auxiliary lemma for Prop. 10
  10: *Resolutio* for a special case of the Apollonian problem, with implicit restrictions
  so as to make Prop. 9 applicable

The purpose of Props. 7–10 is to elucidate the *resolutio* part of the analysis technique in plane geometry (see above, introduction to Prop. 4, structural schema of analysis-synthesis). Among the classical authors, Apollonius was the main representative of analysis-dominated plane geometry. It is therefore no coincidence that Pappus chooses an example visibly associated with Apollonius in order to illustrate the technique. As will become clear below, his portrait suffers some limitations.

The *resolutio* in an analysis fulfills two tasks. First, it determines *data*, i.e., it determines that the entities identified in the analysis proper as crucial for construction and that proofs are *given*, independent of the initial analysis-assumption. This aspect is indeed represented in Pappus' account. He also makes a point of adding in Euclid's *Data*, quite probably because, in his opinion, this work is the basis for the technique in plane geometry, just as the *Elements* are for plane mathematical synthesis.

The other aspect of the *resolutio*, and the one that is clearly dominant in Apollonius' work on plane geometry is the *diorismos*, i.e., the determination of conditions of solvability, not only on the local, but also the macro-level (in Apollonius, it determines the pattern of exposition for complete treatises). It includes the split-up of problems/propositions into cases which then are treated separately. Pappus' account does not represent this aspect adequately (there are two cases only for Prop. 7, and those are not treated exhaustively). The only explicit *diorismos* in *Coll.* IV can be found in Prop. 32 (angle trisection). If Pappus intended a full portrait of plane analytical geometry, Apollonian style, from a methodological point of view, one must say he was not quite successful. Only a fraction of Apollonius' works in this area survives. Even so, it is quite apparent that the split-up into cases and sub-cases, their arrangement in order of complexity, and working them off step by step, and in this way exhausting the original question, is the strategy that dominates the set-up of the Apollonian works, and determines their presentation on the local level as well. The feature is quite idiosyncratic, and quite pronounced. It is not far-fetched to assume that

H. Sefrin-Weis, *Pappus of Alexandria: Book 4 of the Collection*,
Sources and Studies in the History of Mathematics and Physical Sciences,
DOI 10.1007/978-1-84996-005-2, © Springer-Verlag London Limited 2010

it was programmatic in Apollonius[1]; it certainly does have drastic consequences for the presentation of mathematics on the macro-level, and even for the question of what plane geometry essentially is. The program, if it existed, seems to have pointed toward operationalization, focusing on relations as opposed to objects: mathematics as a toolbox, proto-algebraic in methodological emphasis. This is a controversial, much debated issue. It cannot be pursued here; however, since any material on analysis, including Props. 7–10 of *Coll.* IV would have a bearing on a further clarification and discussion on the nature and purpose of Greek geometrical analysis, some brief remarks may be appropriate here, in addition to naming the general trend. According to Jones (1986a, p. 531, # 40), a "katholou pragmateia" (general treatise) by Apollonius once existed. It is attested in *Eucl. Op.* 6. p. 234, and in Menelaus' *Spherics* (pp. 229–240). Jones' supposition that the several general statements from Apollonius in Proclus' *Commentary*,[2] among other things on the status of the cylindrical helix as equivalent to circle and straight line, and also on definitions and postulates, ultimately derived from this lost treatise, is quite convincing. Further investigation, perhaps using the clearly methodologically framed material from *Coll.* IV here (and below, Props. 28, 29, 31, 33, and 42–44) would seem warranted. Perhaps one would get closer to what Jones labels as an "interesting, but unanswerable question": "What in the character of Apollonius, a mathematician of enormous ability and perhaps genius, led him to devote so much effort to tedious programmatic writings of this kind?" (Jones 1986a, p. 530).

Perhaps Pappus was blind to this important programmatic aspect in Apollonius, tied to extensive application of *diorismos*, and to its revolutionary potential. Perhaps, however, he was quite aware of the potential, and chose to ignore it, because he did not favor its operationalist outlook on methodology for geometry. The question can, of course, not be pursued in the present commentary. A detailed investigation of the remains of Apollonius' works and their implicit/explicit methodology, including their reception in antiquity, is still a desideratum.[3] For now, it

---

[1] The *resolutio* becomes central; cf. Jones (1986a, pp. 510–527), specifically, p. 524: "long-winded approach", "desirable in mathematical treatises perhaps not only for beginners". Apollonius achieves thoroughgoing operationalization, generalization, and schematization (disambiguate a configuration, then proceed algorithmically). For a similar global assessment cf. Jones (1986a, p. 400). The result can make for rather tedious reading, even monotony. Compare Jones (1986, p. 524) "taste for exhaustiveness", and p. 530 for the judgment that this kind of set-up was programmatic. See also Hogendijk (1986, pp. 218–223). It is certainly very different from the synthetic treatment in Euclid or Archimedes.
[2] cf. *Procl. in Eucl.* ed. Friedlein passim; compare also Tannery (1912, I, pp. 124–138), concerning Apollonius' attempts to revise the set-up of the *Elements*.
[3] Jones (1986a) provides an excellent basis for the study of the lost works; on the *Tangencies* see pp. 66–88 and 510 ff. In addition, there is the fully preserved *Sectio rationis* (in Arabic), Latin translation by Halley from 1706, English translation by Macierowski/Schmidt from 1987, the latter somewhat flawed. From the *Sectio rationis*, the other two *Sectiones* can be reconstructed, cf. Jones (1986a, pp. 510–527). Further remarks can be found in Pappus, and in such authors as Proclus, who is interested predominantly in the methodological and meta-theoretical aspects.

must suffice to state that Pappus' portrait of the methods of plane analysis is selective. In general, *Coll.* IV only includes such features as can be represented on a micro-level (argumentative devices within a proof), or on an intermediate, local level (set-up of at most a small group of theorems, preferably one). Pappus chose not to include a set of propositions that would illustrate Apollonian *diorismos*-strategy in a miniature format, as he has done for Archimedean monographic mathematics (see Props. 13–18).

2.1.1 Historical Context for Props. 7–10: Apollonius' *Tangencies* and the Apollonian Problem

As said above, Prop. 10 is the *resolutio* to a very special case of the so-called Apollonian problem: given three circles, find a fourth one touching them. Prop. 10 also connects to Props. 13–16, because each of the circles in the arbelos-figure is a solution to the Apollonian problem (where the three given circles are mutually tangent in a specific way; on Props. 13 ff. see below), and this may have been one of the reasons why Pappus chose this particular version of the Apollonian problem in Prop. 10. Props. 7–10 have so far been neglected. There is some discussion in secondary literature, however, for the work out of which the problem is taken: Apollonius, *Tangencies*. This work addressed the following more general questions: given three entities, each of which can be either a point (P), or a straight line (L), or a circle (C), find a circle that touches all of them. Apollonius proceeds case by case, building up from the easiest one (obviously P–P–P), and reducing more complex cases to the ones already solved. The cases dealing with C–C–C took up the whole of Book II. The *Tangencies* are lost. Our information on them goes back, in essence, to Pappus' commentary in *Coll.* VII.[1] My analysis of Props. 7–10 suggests that Prop. 8 could be a further testimony for Apoll. *Tangencies* I, 16/17 (see the translation, and see below). If so, Prop. 8 may be an important source text, so far over-looked. The question whether Prop. 8 deserves the status of a testimony, or even fragment for Apollonius, *Tangencies,* cannot be decided within the present frame-work. Around 1600, Commandino's edition of Pappus' *Collectio* played a major role in the development of a new kind of analytical mathematical techne (art), in the context of the new algebra/analysis, for example, in Vieta and Fermat. It is certainly no coincidence that this reception of Pappus/Apollonius is firmly placed in a context of a programmatic renewal of analytical methods and methodology. Attempts were made, *inter alia*, at a reconstruction of the ancient procedures of investigation and problem solving, in order to develop them further and integrate them within a more

---

[1] *Coll.* VII, 636 Hu and 820–836 Hu; cf. also Heath (1921, II, pp. 181–185), Jones (1986a, pp. 534–539). In addition, see also Hogendijk (1986), fragments in Arabic translation, especially pp. 218–223.

far-reaching scope. Vieta (1600, under the title: *Apollonius Gallus*[1]) uses his newly developed tools to solve the Apollonian problem – geometrically, with ruler and compass only, chiding Adrian Van Roomen, to whom he had put the problem, and who had not been able to come up with a plane solution; Van Roomen used conic sections, which meant he presented a solid solution for a plane problem.[2]

2.2   Prop. 7: Determining Given Features Using the Data

The term *given* does not have a direct equivalent in modern mathematical terminology. There is, also, not a complete agreement on what the ancient term signified (cf. Jones 1986a). A final clarification would have to come from an exhaustive (philological) study of the actual uses made of the term in the ancient sources – at present still a desideratum. The following rough description should be uncontroversial, however, and should suffice as a preliminary orientation to facilitate understanding of the arguments in *Coll.* IV.[3]

A *given* magnitude, area, figure, or ratio is determined by the context of a given configuration, though it need not be uniquely determined. Usually, this means that the entity itself, or a congruent/equivalent entity is constructible, even though the construction could yield other solutions as well. For example: take a line and a point outside of it. Then the point on the line that has a certain distance to the point outside is *given* (*Data* 31), although the obvious construction (circle around the point with distance as radius) yields zero, one, or two solutions. What is captured in the label *given* is the fact that one of them – the one of them one is interested in – is fixed on the line, as it were (if it exists at all).

Points can be *given* in position; then they are constructible from the information implicit in the given configuration/conditions.

---

[1] Vieta (1600, cf. pp. 74–80) in Schooten's 1646 edition. De Fermat (1679, pp. 74–88) solves the related problem for four spheres. The original problem was also treated by Newton in the *Arithmetica Universalis* Problems XLII–XLVII in Horsley's 1779 edition (pp. 132–137), in Whiteside (1972, Vol. V, pp. 252–267) and in *Principia* I, Lemma XVI (pp. 70/71 in the 1726 edition, generalization of Prop. 10, leading to conic sections), by Casey (1882, pp. 121–123) (limit process, coaxial circles, points of similarity), and by Monge (according to (Hilbert and Cohn-Vossen 1932, pp. 120–121), again in connection with projective geometry. Cf. also Hofmann (1990, II, pp. 146–151), with additional references to modern solutions by Bieberbach and Coxeter, Ver Eecke (1933b, I, pp. LXVI–LXXII), Chasles (1875, p. 53) and Notes XXVIII (pp. 372–375). Hofmann (1990, II, pp. 146–164) contains a discussion of Props. 8, 13, 15, and 16 in connection with projective geometry and points of similarity. For a reconstruction of a possible ancient context for something like a theory of points of similarity, connecting Props. 10 and 13, cf. Zeuthen (1886, pp. 381–383); see also Heath (1921, II, pp. 182–185). Heath goes beyond Zeuthen in postulating the nucleus of a projective geometry for the ancients.

[2] cf. Van Roomen (1596). For the connection of geometrical analysis for the determination of the appropriate "level" of a solution see also Props. 42–44.

[3] See also Taisbak (2003).

Lines (i.e., line segments) can be *given* in position or size (or both). In the first case, the straight line on which they lie is constructible, in the second, one can construct a congruent line segment.

Figures can be *given* in size, or kind, or position (or a combination of some of these). In the first case, one can construct a figure with the same area (usually, a square or rectangle), in the second, a similar figure (e.g., a triangle with the same angles), in the third, the whole figure is fixed in the given plane.

Ratios are *given* if one can construct two lines that stand in that ratio.

Euclid's work with the title *Data* closely corresponds to the *Elements*. It was transmitted and known throughout the Middle Ages. It gives a catalog of lemmata, derived from the *Elements*, useful for analyses in practice, specifically for the *resolutio*-stage of actual problem solving. One can refer to lemmata in the *Data* instead of showing, in each case again, that an entity is *given*; though the *Data* belong to the "analytical topos" of argumentation, as a kind of encyclopedia, or toolbox for standardized situations, they are not, in themselves, an analytical work. In Pappus' opinion, they clearly represent the base of reference for plane geometrical analysis. He presents them this way in *Coll.* VII, and that is also how they are used in Props. 7–10.[1]

As said above, in Greek analytical arguments of the "plane" type, there is a noticeable tendency toward maximizing calculatory automatisms in the *resolutio*-strategy. The result is a quasi-algorithmical procedure, avoiding the need to constantly refer in detail to specific geometrical configurations and abstract from them.[2] This can be observed in the extant *Sectio rationis* by Apollonius, and it is reconstructible from Pappus' commentary on Apollonius' lost works in the area of plane geometry.[3] Pappus illustrates this aspect of plane *resolutio* in Prop. 7 in the following way. He creates stereotyped situations, where the same set of propositions from the *Data* can be invoked in an analogous manner (compare the translation). Pappus does not himself explicitly refer to the *Data*, and he goes through the steps in detail only once, while hinting at the repetition of the pattern in the other cases. The reader should be aware that in presentation of Prop. 7 as given here, the *Data* have taken on the same role in the footnotes that the *Elements* play elsewhere. I have assumed that Pappus wants his readers to refer to a background knowledge that is also expounded in the *Data*, and the references are intended as a possible path of justification for those who do not have the whole context present in their mind. The path suggested in the notes and commentary is not the only possible one. Even so, it might be useful for an initial orientation, and for illustrating the point made above, about the schematic, quasi-calculatory operation in Prop. 7. To further stress this point, because it captures the essential feature of

---

[1] On the *Data* and their relevance cf. also Heath (1921, I, pp. 421–425), especially p. 422.

[2] Contrast the role of diagrams in apodeictic procedure, as characterized in Netz (1999).

[3] From Pappus' examples of solid geometry (notably 31, 34, 42–44), a similar picture arises. But the number of examples is too small to arrive, at the present stage of research, at a sufficiently based conclusion as to the tendency toward algorithmization in the field of solid geometrical analysis.

Prop. 7, a list of tools and devices was compiled, and the proof protocols for 7a and 7b then boil down to identifying suitable triangles, and appealing to almost identical sequences of applying the items on this list, in a quasi-automatic way. The other cases for the configuration targeted in Prop. 7, not treated by Pappus, could be handled in exactly the same way.

Toolbox for Props. 7a and 7b.
a I, 47 (Pythagoras): squares over sides of right-angled triangles.
b II, 12/13, generalization: area theorem for triangles that are not right-angled.
c VI, 4.
d VI, 8, Porisma.
e VI, 17: ratio and area theorems for similar triangles.
f V, 16: enallax.
g *Data* 1:   a, b, *given* ⇒ a:b *given.*
h *Data* 2:   a:b *given,* a or b *given* ⇒ the other entity is *given.*
i *Data* 3:   a, b *given* ⇒ a + b *given.*
j *Data* 4:   a, b *given* ⇒ a – b *given.*
k *Data* 52:   the figure of a *given* kind over a *given* side is *given* in size.
l *Data* 55:   area *given* in kind and size, ⇒ sides *given* (used only for squares in Prop. 7).
m *Data* 57:   area, applied to *given* line, in *given* angle *given* ⇒ remaining side *given* (used for rectangles with one side *given*).

The mathematical content of Prop. 7 is rather trivial: in a quadrilateral with all four sides and one angle (a right angle) *given,* both diagonals are *given* in position and size. Such a quadrilateral is obviously constructible, if it exists at all.[1] Also, Pappus discusses only two of quite a few possible sub-cases. Therefore, his point in presenting the argument cannot have been to establish the truth of the claim as such. It probably was to show the reader, in a case where the facts are clear, how one operates in order to establish *givens*: Prop. 7 is of a purely methodological interest.

context: method of analysis, specifically: *resolutio,* specifically: *given* features.
sources: /.
means: *Data,* and some selected corresponding theorems from the *Elements.*
method: analysis.
format: proposition and corroboration (not a full-fledged apodeixis (proof)).
historical significance/reception: /.
embedding in Coll. IV: used in Props. 8, 9, though not essential there; motif "aspects of plane analysis": Props. 4–12.
purpose: illustration of the operation of Euclid's *Data* in the *resolutio* of an analysis.

As indicated above, Prop. 7 is used in Props. 8 and 9. In both cases, however, one could have easily avoided a reference to Prop. 7. This is an indication that Prop. 7 and its usage in 8/9 were inserted by Pappus into source material of an independent provenance (possibly: from Apollonius, *Tangencies,* cf. commentary on Prop. 8), where the *Data* were not instrumentalized in the way illustrated by Pappus.

---

[1] Construct a right-angled triangle ABC, with the *given* AB, BC, and ∠B; this yields AC. Then construct the triangle ADC, with the *given* sides AC, AD, DC. Draw DB.

2.2.1   Proof Protocol Prop. 7a

Sub-case of: quadrilateral has an acute angle at D
   Create triangles and show, successively, that line segments are *given*
Resolutio

1. AE, EC, AC, BE are *given* (ΔABC with height BE).
Apply a, k, i, l (AC); d, e, k, m (EC); j (AE); a, k, j, l (BE).

2. DH, HC, AH are *given* (ΔADC with height AH).
Apply b, k, j, m, j, h (DH); j (HC); a, k, j, l (AH).

3. ZC, EZ, ZB are *given* (ΔAHC ~ ΔCEZ).
Apply c, f, g, h (ZC, EZ); i (ZB).

4. ZT, TC, BT are *given* (ΔZBT ~ ΔZEC).
Apply c, g, h (ZT); j (TC); a, k, j, l (BT).

5. DB is *given*; apply a, k, i, l.

2.2.2   Proof Protocol Prop. 7b

Sub-case of: quadrilateral has an obtuse angle at D
   Create triangles and show, successively, that line segments are *given*
Resolutio

1. AC, AE, EC, DE are *given*.
(ΔABC for AC, ΔADC with height DE)
Apply a, k, j, l (AC); b, k, i, m, j, h; j; a, k, j, l.

2. EZ, DZ, CZ, BZ are *given*.
(ΔABC ~ ΔCEZ); c, g, h; i; c, g, h; j.

3. ZH, HC, BH, HD are *given* (ΔDZC with height DH)
b, l, j, m, j, h; j; j; a, k, j, l.

4. DB is *given*; a, k, i, l.

2.3   Prop. 8: Resolutio for an Intermediate Step in the Apollonian Problem

context: determine *givens* within a *resolutio* in plane analysis.
possible source: Apollonius, *Tangencies* I, 16/17 (extract).
means: *Data* (via Prop. 7, in one place only; but see comments).
method: analysis.
format: proposition and corroboration (not an apodeixis).
reception/historical significance: possible evidence for Apollonius, *Tangencies*, I, 16/17.

embedding in *Coll.* IV: auxiliary theorem for Prop. 10; motif "aspects of plane analysis":
Props. 4–12; motif "touching circles and their diameters": Props. 13–18; motif "Apollonius":
Props. 31–33, perhaps Prop. 28.
purpose: illustration of the technique of determination of *givens*
Prop. 8 is used in Prop. 9.
literature: (Hofmann 1990 II, p. 151).

As noted above, Prop. 8 is an auxiliary lemma. It shows that, when two equal
circles and a point outside are *given*, the diameter for the circle through the *given*
point, touching the two circles outside and inside, respectively, is *given*. The
proposition provides only the *resolutio*, not the *kataskeue* and a*podeixis*.[1] Prop. 8
is the most complex of the propositions in the group 7–10. Its style of argumentation differs from the one used in the other plane arguments. As said above, it shows
some features that indicate that Pappus has used an independent source that has
not been "worked up" fully. The source sidestepped the *Data*, and Pappus re-
introduces references to an argumentation via the *Data* through Prop. 7. The
resulting argument is not completely smoothed out.[2] Also, connections to Pappus'
commentary on Apollonius, *Tangencies* I, 16/17 can be established in several
places. The argument as given in Prop. 8 leaves out steps that are presented in *Coll.*
VII as part of Pappus' commentary on *Tangencies* I, 16/17 (*Coll.* VII, Props.
102–106), and this means that Prop. 8 could very well be taken from the original
argument on which Pappus commented.[3] The crucial passages are indicated in the
footnotes of the translation. For easier reference, keywords are added in the proof
protocol below, with "two layers" indicating the places where the transmitted text
shows signs that a source text was not fully integrated into the argument as presented by Pappus. Taken together, these indications corroborate the thesis that the
source which Pappus used for Prop. 8 could have been a fragment from the lost
work of Apollonius. If so, Prop. 8, so far neglected in secondary literature, would
gain considerable significance. It is not mentioned in Jones (1986a), and may
simply have been overlooked so far. The question cannot be pursued in detail here;
it certainly deserves scholarly attention.

---

[1] They pose no problem after the resolutio. This means that Pappus intentionally restricts his
presentation to the analysis only, puts the emphasis on the methods, not the actual mathematical
result.

[2] We encounter insertion of trivial argumentative steps that point to Prop. 7 and do not completely
fit their immediate context; Hultsch suspects interpolation, but the insertions may very well have
been brought in by Pappus himself: 196, 9–16 Hu, 186, 18/19 Hu, 196, 25 Hu. Double script
occurs at 196, 22–23 Hu, the text is uncertain at 196, 27 Hu, and in one place, the text transmitted
includes a truncated phrase that doesn't fit the context; see translation with footnotes.

[3] Commandino p. 67 plausibly argues that the original author of Prop. 8 intended an argument via
similar arcs, while Pappus offers a more elementary argument. Heath (1921, II, p. 371) remarks
about Prop. 8: "the proof is in many places rather obscure and assumes lemmas of the same kind
as those given later a propos of Apollonius' treatise".

2.3.1   Proof Protocol Prop. 8

*Analysis assumption (implicit): the desired circle CEZ has already been found.[1]
Resolutio

1. *Diorismos* (determine conditions for solvability)
   1.1 Extend the configuration on assumption *
   1.2 HT ∥ CE and TK ∥ PC must hold (cf. *Coll.* VII, 102)
   1.3 DE = ZH must hold (cf. Coll. VII, 106)
   1.4 BM = MA, LM = LS must hold
2. Determination of *givens*
   2.1 AM is *given* in position and size.
   (i.e.: M is *given,* cf. *Data;* 2 layers)
   2.2 CM is *given* in position and size (cf. Prop. 7; 2 layers).
   2.3 CP is *given* in position and size (i.e.: P is *given,* cf. *Data*).
   2.4 CPZ is *given* (i.e. the circle through C, P, Z is *given;*
   cf. *Coll.* VII, 104/105).

2.4   Prop. 9: Auxiliary Lemma for Prop. 10

2.4.1   Proof Protocol Prop. 9

1. Extension of the configuration
With the *given* d = AD − DC = DC − DB, construct circles around A, B with radii
AE = BZ = d.

2. *Resolutio*
2.1 The diameter of circle ZCE is *given* ⟹ DZ (radius) is *given* (Prop. 8).
2.2 AD, DC, DB are *given* (use 1).

2.5   Prop. 10: Resolutio for a Special Case of the Apollonian Problem

Prop. 10, as given by Pappus, is not quite complete. At a certain stage in the argument, an implicit additional condition, namely, HB−HC = HC−HA, is used, so as to make Prop. 9 applicable (see proof protocol). Without an additional restriction in the *protasis* for Prop. 10, a non-trivial gap in the argument results. Hu 201, #3 suggests a path how the gap might have been filled, but refrains from making any explicit

---

[1] Commandino p. 66 A points out that the circles in the starting configuration have to be *given* both in position and size, whereas Pappus only mentions position.

suggestions as to how Pappus would have effected the proof for the missing intermediate step. Hultsch' s solution seems to lead to conics.[1] A proof was also suggested by Ver Eecke (1933b, pp. 147/148, #8). The center of the sought circle is needed for it, however. Jones (1986a, p. 537), Heath (1921), and Hofmann (1990, II, pp. 151/152) point out the connection of Prop. 10 in its present form to Prop. 13. This may be the reason for Pappus' transformation of Prop. 10. For it was probably not one of Apollonius' cases in this form.[2] Pappus' argument as given is flawed, but not false. Pappus himself, at the end of Prop. 10, adds the following remark: "Let this have an end for me now right here; I'll write down the rest later." This remark, quite unusual in the context of *Coll.* IV, suggests that he was somehow aware that something was missing, either from Prop. 10, or from the whole group Props. 7–10. It is unclear what that "rest" was, and also what "later" means. Pappus does not come back to the material treated in Props. 7–10 within *Coll.* IV. Maybe he intended to add a lemma that would fill the gap in 10 (assuming he was aware of it); maybe he intended to revise the whole argument for Props. 8–10, smoothing it out. For the purposes of the present commentary, I have left the argument as given in Pappus, and have made the tacit assumption made in the reference to Prop. 9 explicit. It can be viewed either as an additional condition for Prop. 10, omitted in the protasis, or as a marker for a gap in the argument which Pappus failed to fill in.

2.5.1   Proof Protocol Prop. 10

1.[*] Assume task is accomplished.
Circle with center H, touching in Z, E, D.

2. Extension of the configuration
Draw H-B-Z, H-C-E, H-A-D, BA, BC, AC.

3. *Resolutio*
3.1 BA, BC, AC are *given*.
3.2 HB–HC and HC–HA are *given*.
Assume that these differences are also equal
(additional restriction, implicit in Pappus).
3.3 AH is *given* (Prop. 9).
3.4 DH, and therefore 2 DH, is *given*.

---

[1] As does Newton's in *Principia* I, L. XVI, mentioned above.
[2] Generalizations of Prop. 10 toward C–C–C seem to lead to conics, but the *Tangencies* use only plane methods.

II, 3 Plane Geometry, Archaic Style

3 Analysis-Synthesis Pre-Euclidean Style

3.1 Props. 11 and 12: Chords and Angles in a Semicircle

The configurations and the mathematical content of the propositions, as well as the argumentative means, are reminiscent of pre-Euclidean mathematics, as exemplified in the Hippocrates fragment[1]: circles with inscribed triangles, geometry of the circle (presented in III/IV of the *Elements*), and argumentation via congruent angles in extended configurations. All this gives Props. 11 and 12 an old-fashioned character. On the other hand, no sources are known for Props. 11 and 12. Language and style show no signs of archaism. Props. 11 and 12 were probably constructed by Pappus. What was their purpose? They could be an illustration of what the operation with the technique of analysis-synthesis originally looked like. For Prop. 12, this surely is the case,[2] whereas the situation is less clear for Prop. 11, and only a tentative thesis can be formulated.

Prop. 11 as given, is purely synthetic, with no trace of the heuristic background. The proof's core is a suitable extension of the configuration (the introduction of T). Therefore, a successful heuristics may reasonably be assumed to consist in the determination of the crucial role of T, and this observation suggests the following consideration in the context of the overall structural schema of analysis-synthesis.[3] When analysis comes down to a determination of a suitable extension of the configuration, it cannot be coined out as a propositional device, as a step of reasoning that could be logically inverted in the synthesis. As Hintikka/Remes have pointed out, this aspect of analysis can in general not be schematized (it is also nondeductive). Such an analysis, and all parts of an analysis that consist in determining suitable extensions cannot be instrumentalized as making any explicit contribution to the apodeixis proper. The crucial analysis information would be tacitly integrated in the kataskeue of the synthesis. Otherwise, it would leave no visible traces in the resulting proof. Prop. 11, since it is put within the group of propositions illustrating plane analysis-synthesis, could be an illustration of the effects of an analysis that consisted solely in determining a suitable extension of the configuration. It has to be admitted, however, that such an explanation is somewhat speculative and perhaps not wholly satisfactory. Among other things: why did Pappus not give the analysis, since putting

---

[1] On the Hippocrates fragment see Simpl. in Phys. 61–68 Diels, Heath (1921, Vol. I, p. 183 and pp.195–196), Knorr (1986, pp. 32–34), and Netz (2004).

[2] Prop. 12 has been read this way by Mahoney (1968), e.g.

[3] See the introduction to Props. 4–12.

H. Sefrin-Weis, *Pappus of Alexandria: Book 4 of the Collection*,
Sources and Studies in the History of Mathematics and Physical Sciences,
DOI 10.1007/978-1-84996-005-2, © Springer-Verlag London Limited 2010

it beside the synthesis would have made the point about its not showing up in the synthesis quite transparent? Another possibility for the purpose of Prop. 11, which does, however, not explain why it is put into a group of propositions on plane analysis-synthesis, is that it illustrates the argumentative style of pre-Euclidean synthetic plane geometry. But in this case, Pappus might have chosen a more attractive example, as he has done for most of his vignettes in his portrait of plane geometry (Pythagoras, Apollonian problem, and Arbelos) and, for the later parts of *Coll.* IV (quadrature of the circle, spiral lines, conchoid, quadratrix, and angle trisection). Perhaps further investigation will throw more light onto the question of the purpose of Prop. 11. The reading assumed here must be understood as tentative.

Prop. 12, like Prop. 4, has a full analysis and synthesis. The analysis essentially comes down to reduction, and is predominantly deductive (resting on steps, however, that are convertible). The synthesis simply retraces the steps of the analysis. Prop. 12 lends support to the thesis that analysis as an identifiable technique was originally identical with reduction. In my opinion, one should nevertheless refrain from inferring that Greek geometrical analysis therefore was and remained essentially deductive and reductive. Reduction does not exhaust the conceptual horizon of Greek geometrical analysis as the technique developed over time.[1] Within *Coll.* IV there are several examples of other, non-reductive usages of analysis.

3.2   Prop. 11: Representation of a Chord as Segment of the Diameter

context: analysis–synthesis (tentative, see above).
possible sources: /.
means: I, III, VI (V).
method: synthesis (analysis would have consisted in suitable extension alone).
format: theorem.
reception/historical significance: /.
embedding in *Coll.* IV: motif "overall structural components of analysis–synthesis": Prop. 4; motif "triangles, chords in semicircles": Props. 2–6, 12; motif "perpendiculars on diameter compared with diameter in length": Props. 13–18.
(suggested) purpose: illustration of outcome when analysis consists of finding out how the configuration must be extended to provide a proof.

The content and result of Prop. 11 are not used in *Coll.* IV.

3.2.1   Proof Protocol Prop. 11

We want to show: EB = 1/2AZ.

1. Ekthesis/Kataskeue
T, then K, H, auxiliary lines.
Several right-angled triangles and pairs of similar triangles are created.

---

[1] For a very different assessment, cf. Mahoney (1968).

2. Apodeixis
   2.1 Show that $\triangle$ATH $\sim$ $\triangle$CTZ, and HK = KZ
   (i.e.: EK = 1/2AZ)
   2.2 Show that BE = EK (i.e., EB = 1/2AZ)

3.3 Prop. 12: Angle Over a Segment of the Diameter

context: overall structure of analysis–synthesis in plane geometry.
means: I, III, VI, (V).
method: analysis restricted to *epagoge* (reduction), with synthesis mirroring analysis.
format: theorem.
history and reception: /.
embedding in Coll. IV: analysis–synthesis as a technique: Prop. 4; components of plane
geometrical analysis: Props. 7–10; motif "chords and angles in semicircles": Props. 2–6, 11.
purpose: illustration of argumentative technique of reduction in analysis proper.
literature: Mahoney (1968) contains an extensive discussion of the analysis in Prop. 12.

The mathematical content and the result of Prop. 12 are not used in *Coll.* IV.

3.3.1 Proof Protocol Prop. 12

We want to show: $\angle$BED = $\angle$DET

1. Analysis
   1.1 Assumption[*]: problem solved
   1.2 Analysis proper
   Extension of configuration

Reduce claim to DK $\parallel$ ET	(1)
Reduce claim to LD:HD = DT:TZ	(2)
Reduce claim to LA = DH, LB = BH	(3)
Reduce claim to BL = LD	(4)

   1.3 *Resolutio*

BL = LD holds in fact[1]	(5)

2. Synthesis
   Apodeixis (Kataskeue not given explicitly)

BL = LD	(5') (4')
BL = BH, LA = DH	(3')
LD:HD = DT:TZ	(2')
KD $\parallel$ ET	(1')
$\angle$BED = $\angle$DET	

---

[1] Note that the *resolutio* is minimal here.

A comment on the word "akolouthos," used in 206, 12 Hu, at the beginning of the synthesis of Prop. 12 ("and the synthesis follows the analysis step by step (akolouthos)") seems appropriate. The meaning of akolouthos cannot be restricted to "follows logically, deductively." There are several places in *Coll.* IV where akolouthos, or a related word, need not, even cannot, have that narrow meaning. This is one of them. The others are "sunakoloutheito" for the second moving point in the generation of the quadratrix (252,10 Hu) and "akolouthon" for the consideration of the spherical spiral as a natural next step after plane and conical ones in the introduction of Prop. 30 (264, 7 Hu). Whereas the synthesis of Prop. 12 is clearly derived from the analysis, in that it retraces its steps in order, it is not inferred from it by deduction. The word is not used here in the description of the sequence of the analysis-steps themselves, although they are deductive. On the role of "akolouthos" in the scholarly discussion of the nature of analysis see the introduction to Props. 4–12.

II, 4 Plane Geometry, Archimedean

4 Props. 13–18: Arbelos (Plane Geometry, Archimedean Style)

4.1 Observations on Props. 13–18

context: no traces of ancient sources providing a context of similar problems, but: connection to non-trivial theorems/methodological devices with "potential" for future mathematical theories:

  (i)   Diameters in configuration of tangent circles, theorem of Menelaus
        (→ points of similarity (projective geometry))
 (ii)   Arithmetical progression (→ complete induction)
(iii)   Capture infinity using a quasi-mapping onto natural number progression

possible sources: lost monograph by Archimedes, with intermediate transmission stages (controversial, see below).
means: beyond *Elements*, but for the most part strictly "orthodox"; unusual means: nucleus form of complete induction (Props. 16, 18).
method: synthesis.
format: monograph in miniature form, lemmata, main theorems, corollaries.
history and reception: *Liber assumptorum* (deteriorated form, see below).
embedding in *Coll.* IV: tangent circles: Props. 8–10 (Apollonian problem, connection to Prop. 10 especially close); motif "chords and circles": Props. 2–6, 11 and 12; motif "commensurable versus incommensurable straight lines in circle configuration": Props. 2 and 3, motif "progression towards infinity": Props. 19–21, 30; motif "association with Archimedes": Props. 19–22, 30, 35b, 42–44.
purpose: illustration of plane synthetic geometry, monographic style: Archimedean. Through the connection with Archimedes (in style, if not in person), Props. 13–18 form a bridge to the second part of *Coll.* IV, specifically to Props. 19–22, which are indeed by Archimedes.
literature: Heath (1921, II, pp. 371–377); Buchner (1824), (arbelos via classical geometry, and via analytical geometry), Casey (1882, pp. 95–112) (involutions, limit processes), Hofmann (projective geometry, Zweiecke in: Hofmann (1990, II, pp. 146–164); see also Hofmann (1990, I, pp. 273–281)). The alternative treatments are interesting for a comparison in terms of methodology.

4.1.1 Archimedean Character of Props. 13–18

The group of theorems on the arbelos has been associated by quite a few scholars with Archimedes. There are two reasons for this. First of all, a treatise in Arabic (the *Liber assumptorum* (*Lib. ass.*)), transmitted under the name of Archimedes,[1] contains some theorems that are closely connected to Props. 14–16. This basic indicator, however, turns out to be much weaker than one might hope. A second reason for connecting the little arbelos treatise with Archimedes is that its mathe-

---

[1] Latin translation in: Heiberg, *Archimedes, Opera omnia*, Vol. II, 510–525.

H. Sefrin-Weis, *Pappus of Alexandria: Book 4 of the Collection*,
Sources and Studies in the History of Mathematics and Physical Sciences,
DOI 10.1007/978-1-84996-005-2, © Springer-Verlag London Limited 2010

matical content is quite "worthy" of Archimedes once one takes a closer look at the ideas and devices involved. Heath, for example, thinks the arbelos treatise could very well be by Archimedes, and says it is "extremely interesting and clever, and I wish that I had space to reproduce it completely" (Heath 1921, II, pp. 371). Unfortunately, a direct Archimedean authorship for Props. 13–18 cannot be taken for certain. For the *Lib. ass.* cannot, in the form preserved, be by Archimedes. His name is mentioned in it.[1] Furthermore, the relative triviality of the content of *Lib. ass.* also makes it unlikely that it stems directly from Archimedes. This weakens our evidence for a connection of Props. 13–18 to Archimedes. It does not rule out the possibility for an indirect connection, though some authors have concluded that any connection is unlikely.[2] It is probably not possible to prove that Props.13–18 are essentially by Archimedes. Even Pappus did not know the author, and spoke of an "ancient proposition," transmitted in "some documents."[3] I will therefore not claim that the arbelos treatise is Archimedean in the sense that Archimedes is the direct source for Props.13–18, although, not unlike Heath, I am inclined to believe that he may very well have been the author of some original form of the argument (now lost, and originally probably more extended[4]). Even if Props. 13–18 cannot be shown to be by Archimedes, the arbelos treatise does show a number of features that allow for an association with geometry in the style of Archimedes. These features are quite distinct and differentiate Props. 13–18 from the Euclidean geometry as portrayed in Props.1–6, 11 and 12, and from the analytical (Apollonian) geometry in Props. 7–10. Together with documented connections to the *Lib. ass.*, they justify the label "plane geometry, Archimedean style" for Props. 13–18. In what follows, I give a survey of these characterizing features. They will also be marked in the commentary on the single propositions.

1. Global characteristics: set-up, structure, and order of exposition.
   Props. 13–18 treat a well-defined topic quite exhaustively, in a self-contained quasi-monographic piece of text, divided into preliminary lemma, technical lemmata, then theorems, and additions/corollaries. We find no trace of a heuristic background. The exposition is polished, purely synthetic. Simple, orthodox means are made brilliant and original use of, by an appropriate and ingenious choice of perspective, so as to create a venue for unexpected insights. The global set-up is analogous to Archimedes' monographs, which are also self-contained and structured

---

[1] *Archimedis Opera Omnia* Vol. II p. 514 Heiberg.

[2] Jones (1986a, pp. 538–539), for example, denies any connection with Archimedes for the *Lib. ass.* One of his reasons is the above-mentioned low level of sophistication in the *Lib. ass.* In my opinion, this low level could perhaps be explained as the result of progressive deterioration in transmission, and need not speak against an ultimate provenance of the material from Archimedes. Another point he makes is that there is no connection to the content of other works by Archimedes. In my opinion, this observation, too, can be relativized in its weight by pointing to such treatises as the *Sand reckoner*. I also think there is more coherence to Props. 13–18 than Jones' remarks on p. 539 op. cit. and Hofmann (1990, I, pp. 146ff.) suggest.

[3] Cf. translation, beginning of Props. 13–18.

[4] See the list of indications pointing towards a larger extension of the original treatise below.

similarly (announcement of theorem(s) – preliminary lemmata, set apart and used as quasi-axioms, – technical lemmata – theorem(s)). Archimedes, too, has a very polished, purely synthetic and orthodox style of exposition on the local level, showing no traces of the heuristics and giving far-reaching results with relatively, sometimes astonishingly, simple means, by choosing an unexpected perspective. The style of exposition is different from Apollonius' treatises (which are monographic, but have a very different set-up and are analytical), and also from the *Elements* (which are not monographic).

2. The main proposition is surprising and simple. It draws an unsuspected connection between diameters and circles on the one hand, and natural numbers on the other, formulated in simple ratios.

   In the extant longer works of Archimedes, a preference for similar theorems in terms of numerical ratios can be found. Examples include *SL* 24 (area in spiral:area of circle = 1:3), *QP* 24 (Parabola segment:inscribed triangle = 4:3), *Sph. et Cyl.* I, 34 with corollary (cylinder:inscribed sphere = 3:2, surface cylinder:surface sphere = 3:2), *Sph. et Cyl.* I, 33 (surface of sphere:maximum circle = 4:1). One might also compare the Archimedean Props. 21 and 30. No such theorems are to be found in Apollonius (this much can be said, even though we do no longer possess the complete texts of his original treatises), and they are very rare in the geometrical books of the *Elements*.

3. Infinite progression of inscribed figures, and use of a prototype of complete induction (in main theorem Prop. 16).

   For the use of a progression of inscribed figures compare *Coll.* IV, Props. 21 and 30, by Archimedes. There we also find the use of indivisibles. For an example in Archimedes' monographs, compare *Circ. mens.* I (there are many more examples). *Circ. mens I* gives a proof by exhaustion (no indivisibles). The examples from Archimedes' attested works using inscription processes imply the filling up of a given area. This is not the idea in the arbelos treatise. There are no parallels for either the use of infinite progression, or attempts to deal with infinity in a mathematical way in Apollonius. The parallels in *Elements* XII are proved by exhaustion (and are probably by Eudoxus). Very few examples for what amounts to complete induction are attested in ancient geometry. The ones known to me are by Archimedes: *SL* 10 and 11 and *QP* 23.

4. Role of Prop. 13.

   This proposition's role is analogous to the one preliminary lemmata play in Archimedes' monographs. The results of 13 are labeled as "lambanomena" (assumptions). In *SL*, Archimedes uses and labels *SL* 1–11, which he sets off from the main part of the treatise, as if they were assumptions (lambanomena), quasi-archai, for the following purely geometrical treatise. The separation is emphasized by the fact that the definitions for the treatise appear after *SL* 1–11. In *SL*, the preliminary results are called lemmata, but also lambanomena. For a similar phenomenon, cf. *QP* 1–5.

5. Doublet in the proof of the converse for Prop. 13.

   The converse is proved (i) by exhaustion (ii) directly. The direct proof in (ii) is sufficient, and, from the point of view of Aristotelian theory of science, it would be preferable. Perhaps it is by Pappus.[1] The indirect proof (i) is, however, also presented. Proof by exhaustion was a trademark of Archimedes, who used the method with great virtuosity. There are no examples for proof by exhaustion in Apollonius.

6. Role of Prop. 14.

Prop. 14 is a technical lemma, the most elaborate proposition in the arbelos group. It provides the means for the argumentation in Props. 15 and 16. The proposition itself is referred to in Prop. 15. In Prop. 16, and passim, it is nevertheless mostly an intermediate result within the proof of Prop. 14 (and also an addition to Prop. 14 that uses it) that is actually used later on. The relevant information could have been proved ad locum within the later propositions just as easily. The author, however, prefers a presentation within a theoretical, systematic setting before-hand. After this proposition, Props. 15 and 16 (the main theorem) are transparent and easy. The reader of Props. 14–16 is forced, however, to keep the whole of each argument in mind, because results that will become important later are not emphasized. This set-up is analogous to the set-up of Archimedes' monographs: most of the detailed technical work occurs in preparatory theorems, main theorems draw on these (and often on intermediate results within), so that the main theorems become slim and elegant, giving full sight of the core mathematical idea. For the technical lemmata within the treatises, the emphasis is on systematic rather than linear development, not result-oriented in its local presentation. The reader needs to remain aware of intermediate results in relation to the "telos" of the treatise at all times, even if they are not marked out by the way the material is presented; compare, again, SL 18 and 24, and their "setting," as well as the structure of QP, especially 18–24. Contrast, again, Apollonius' monographs: there we have a linear exposition, exhaustion of all possible cases in a list, step by step.

7. Handling of proportion theory (Prop. 14 especially, but also Prop. 15).

The handling of proportions is an "orthodox" application of V. It is analogous to Archimedes's extensive use of proportions, e.g., in SL and Sph. et Cyl. passim. In this respect, the mathematical style in Props. 13–18 is again decidedly different from Apollonian mathematics. Apollonius does not use abstract proportions, his arguments involving ratios rely on equalities that correspond directly to equations between areas (i.e., II, 14, VI, 27–29, cf. Zeuthen 1886 passim).

8. Addition, not from the arbelos in Coll. IV, but from comparison with Lib. ass.

The Lib. ass., transmitted in Arabic, claims to be a work by Archimedes. Lib. ass. V and VI are simple versions of Coll. IV, Props. 14 and 16. In addition, one might note that in Prop. 14, a lemma that is equivalent to Lib. ass. I is invoked at one step. Heiberg (cf. Archimedes, Opera Omnia Vol. II, pp. 513, #2, 518, #1, 523, #1) claims that Lib. ass, I, IV–VI, VIII, and XIV are probably by Archimedes. Perhaps this cannot be asserted. Heath (1921, II, p. 372) observes, however, that Lib ass. IV and VI are simple versions of Props. 14, and 16. Lib. ass. VIII is indeed closely associated with Archimedes by most authors.[2] It yields an angle

---

[1] The proof is exactly analogous to a group of proofs by Pappus in Coll. VII (64, 118, 128, and 130 Hu; Jones 1986a, # 118, 184, 195, and 198).

[2] Cf., e.g., Heath (1921, I, pp. 240–241) and Knorr (1986, pp. 185–186).

trisection via *neusis* and connects to Archimedes, *SL* 5–9. A similar *neusis* also appears in *Coll.* IV, Props. 42–44, as a discussion of Archimedes' *neusis* in *SL*. Thus, we have evidence for a connection between the *Lib. ass.* and the work attested for Archimedes, and if Props. 14–16 can be associated with *Lib. ass.*, we have an additional evidence that they can be connected (indirectly) with Archimedes, too.

1–8 will be taken as the main elements of a description of the Archimedean style of Props. 13–18, and mentioned ad locum below. In my view, it is once again this mathematical style, and through it the mathematical methods that are the center of gravity for the presentation of the arbelos theorem, despite the fact that it is also very appealing in terms of its content.

4.1.2 Factors that Point Toward an Original Larger Extension of the Arbelos Treatise

There are several passages or instances in Props. 13–18 where the reader gets the distinct impression that the text as we have it is a truncated version of a treatise that was once more extended, even though the arguments as given are not in themselves incomplete. Specifically, the possibility that Props. 14–18 were preceded by a more extensive treatment on points of similarity in configurations of tangent circles has captured the interest of some commentators; cf. Hofmann (1990, II, 153 ff.) for the potential mathematical context. That more extensive investigations of configurations with points of similarity must have existed was argued for by Zeuthen and others. In antiquity, the theoretical background was captured by the theorem of Menelaus (cf. Zeuthen (1886, pp. 381–383), also for the connection to the arbelos configuration with points of similarity). Although this topic cannot be pursued here, I will give a list of the most pronounced indicators for the thesis that the arbelos treatise may have been part of a larger monograph, with a potentially broader scope.

Prop. 13 is labeled as "ta lambanomena." This label would make more sense if there were a larger number of general preliminary lemmas.

The protasis in Prop. 13 appears to have been reformulated so as to assimilate[1] an existing proposition with a longer protasis more closely to what is explicitly claimed in later instances where appeal to Prop. 13 is made.

The first step in Prop. 14 could be verified by reference to a more general version of Prop. 13.

In Prop. 14 a limit case is inserted. Within Prop. 14, it appears as a side thought, and looks as if it had been inserted specifically to prepare for the appendix Prop. 17.

For Prop. 15, the main manuscript A includes a figure for a limit case that is not actually treated in the text, but used in Prop. 17 and in Prop. 16, Addition 2.

---

[1] See below, remarks on Prop. 13.

Prop. 17 appears as a later insert, probably by a different author, to cover a step concerning ratios and proportions in Prop. 16, Addition 2.

In several places (notably in Props. 14 and 15), more far-reaching results than needed further in the treatise are established.

4.1.3   Arbelos Theorem

A configuration with three semicircles is given. They have their base on a common line segment; one semicircle is described over the full segment; it is then divided into two parts arbitrarily, and two smaller semicircles are described over those parts. In the remaining space, which has the shape of a shoemaker's knife (arbelos[1]), a non-finite series of touching circles is inscribed. Each of these touches the outer semicircle, one of the inner ones, and its predecessor (i.e., each of the circles in the progression is a solution to a version of the Apollonian problem). One compares the diameters of these circles with the length of the perpendiculars from their centers onto the base, and one finds that they stand in the ratio $1:n$, with n being the number of the circle in the progression.

4.1.4   Structure of Props. 13–18

Prop. 13: preliminary lemma (general, no explicit connection to arbelos
        configuration).
Prop. 14: technical lemma: proportion involving perpendiculars and radii in compari-
        son to diameter of one of the semicircles in the starting configuration; two
        additions.
Prop. 15: lemma for induction.
Prop. 16: arbelos theorem; two additions for limit cases.
Prop. 17: auxiliary lemma for one of the limit cases in 16.
Prop. 18: appendix: theorem for a progression of inscribed circles when the starting
        configuration has only one inner semicircle.

4.2   Prop. 13: Preliminary Lemma

Prop. 13 plays the role of a preliminary lemma, much like the preliminary lemmata in Archimedes' geometrical treatises (cf. #4). Not Prop. 13, but the converse and the addition will be used in Props. 15 and 17 – perhaps an indication that the arbelos

---

[1] As noted in the translation, the connection between the word "arbelos" and the cobbler's tool is not securely established.

treatise as we have it was originally embedded in a more extensive work.[1] The converse of Prop. 13 is proved twice. The first proof is an (admittedly rather trivial) exhaustion proof (double reductio). This type of argument was favored by Archimedes. The second proof is a direct proof and probably by Pappus himself (cf. #5). The mathematical context for Prop. 13 is the theory of points of similarity.[2]

4.2.1 Proof Protocol Prop. 13

13 a Theorem

1. Protasis
Assume D as the point of intersection between KL and circle(A); then AH ∥ CD, and: AB:BC = AE:EC (E is an outer point of similarity).
2. Kataskeue
Draw CD.
3. Apodeixis.
    3.1. AKDC is a trapezoid: AH ∥ CD.
    3.2. Δ KEA ~ Δ DEC ⇒ AE:EC = AB:BC.   [VI, 7; I, 27; VI, 4]

13 b Converse

Protasis
Assume AB:BC = AE:EC, and take D as the point of intersection between the parallel to AK through C and circle(C). Then K–D–E is a straight line (D on KE).

1. Apodeixis by exhaustion
    Draw CD.

---

[1] Prop. 13 holds also for circles that touch internally, even though it is given only for circles that touch externally. On Hultsch's reading of Prop. 14, the proposition uses the version for internally touching circles. Another explanation for the step in Prop. 14, one that does not imply that Pappus left a gap in the argument in the arbelos treatise is that the reference in Prop. 14 is rather to elementary lemmata for which Pappus gives a proof in *Coll.* VII 102ff. This is the explanation I preferred in the notes to the translation of Prop. 14. Independently from the question of completeness of the argument as given by Pappus, I think it is quite possible that the treatise from which Prop. 13 ultimately stems was more extensive and contained a greater number of preliminary lemmata, dealing with touching circles and points of similarity.

[2] In the configuration of Prop. 13, E is a point of similarity for the three circles concerned. A general theorem for such points was provided by G. Monge, according to Hilbert and Cohn-Vossen (1932, pp. 120–121). Compare also the contributions by Hofmann and Casey in the literature list above. As pointed out in the remarks on the mathematical context for Props. 7–10, Props. 13 and 14 are connected to Prop. 10, and this may be one of the reasons why Prop. 10 was formulated by Pappus the way it is.

If it does not intersect KE in D, then
either C–T–D, with T as the point of intersection,
or C–D–N, with N as the point of intersection.
Both these assumptions lead to a contradiction.
Therefore, CD intersects KE in D.   [VI, 4; V, 16]
2. Apodeixis by direct proof
Complete the parallelogram AKNC.
KN:ND = EC:CD.   [V, 17; V, 16]
Δ EDC ~ Δ NDK, ∠EDC = ∠NDK.
⇒ Since C, D, and N lie on a straight line, so do K, D, and E.

13 c Addition

In the configuration of Prop. 13, we have: $EB^2 = KE \times EL$.
    This follows from AH ∥ CD (i.e., from Prop. 13a, step 1) [VI, 2; III, 36].

4.2.2   Defense of My Reading of Prop. 13

As indicated in the notes to the translation, my reading of Prop. 13 and its converse
differ from the one suggested by Hultsch (endorsed by Ver Eecke, and based on
the text as transmitted). On my reading, the claim that DC ∥ AH, i.e., that AKDC is
a trapezoid, should be included in the protasis. The text as transmitted, however,
clearly suggests that the protasis (claim) concerns the proportion only ("deixai"
is used before the claim about the proportions[1]), and that the trapezoid ACDK is
already given. Therefore, I must defend my interpretation, and I shall do so by
showing that the text as given creates severe problems, and that my reading can
eliminate them. The severest problem of the text as given concerns the logic of Prop.
13 and the converse. If one posits, to begin with, that AKDC is a trapezoid, Prop. 13
would not even need a proof. For we would have two similar triangles EAK, ECD,
and VI, 4 yields: EC:CD = AE:AK, thus: AE:EC = AK:CD (V, 16), and obviously
AK = AB, CD = BC. Inclusion of the fact that AKDC is a trapezoid in the ekthesis
renders Prop. 13 superfluous. Also, in the apodeixis of Prop. 13, we are first
prompted to draw CD – which would already be given in the ekthesis. Hultsch
resorts to an elimination of the phrase concerning CD from the text.[2] But even then,
the argument is still skewed. For we proceed to prove that DC ∥ AH, i.e., that we
really have a trapezoid AKDC. This would not make any sense, if the trapezoid
is posited to begin with. Finally, the converse uses the fact that DC ∥ AH in both its
proofs. Hultsch and Ver Eecke are wrong in assuming that this could be shown

---

[1] 210, 6 Hu.
[2] 210, 8 + app. Hu.

as in the proposition itself[1]; for the proof there rested on the fact that K, L, and D are on a straight line, which is the very thing the converse wants to prove. Hultsch's and Ver Eecke's reading implies a petitio principii for Prop. 13's converse. There must be a better solution. Apparently, the converse just assumes parallelity. It can certainly do so, if we assume that DC ‖ AH was part of the protasis in the proposition itself. Then we can also explain why CD is drawn, and AH ‖ DC is shown in the apodeixis of the proposition itself. The only drawback of this reading is that it must assume that the ekthesis of Prop. 13 is "muddled," specifically that either Pappus or someone between the original author and Pappus changed the text from a version where the decisive parallelity was part of the protasis into a version where only those features that will be used in Props. 15 and 17 explicitly are mentioned there. Perhaps the change of tense within the ekthesis (cf. translation) lends some support to this assumption. It is for the logical and structural reasons given above, however, that I have decided to read Prop. 13 the way proposed in the proof protocol. I am not claiming that the text should be changed.

4.3 Prop. 14: Technical Theorem

As said in the introduction, Prop. 14 is the most complex lemma in the group. It provides the technical results needed in the arbelos theorem. It is proved as a complete lemma, in full generality, for all three possible configurations, although only the first and third configuration will come into play in what follows, and although Props. 15–17 mostly rely on an intermediate result within the argument of Prop. 14 rather than the proposition itself. The importance and role of the intermediate result is not emphasized within Prop. 14 (it has been marked out by me to facilitate reference). Within Prop. 14, it is presented in its appropriate place with respect to the theoretical content of Prop. 14 itself. All these features contrast to the style of exposition in Apollonius' works, and are reminiscent of Archimedes's monographs (cf. #6). Prop. 14 operates with abstract proportions, in the sense of V (cf. #7). The first step in Prop. 14 implicitly uses a lemma proved in proposition I of the *Lib. ass.*, which is associated with Archimedes in the tradition. A second possibility for this step is reference to a more extended version of Prop. 13, indicating, perhaps, that the arbelos treatise may originally have contained a more substantial first part dealing with tangencies and points of similarity. The topic cannot be pursued here, cf. the bibliographical references in the footnotes to Prop. 13. A third possible route of justification is appeal to lemmata on tangent circles for which Pappus provides a proof in *Coll.* VII, Prop. 102 ff., and this is the route taken in this translation and commentary. Finally, *Lib. ass.* IV provides a lemma that is a simplified version of Prop. 14. This means the historical reception of the arbelos associates Archimedes, the *Lib. ass.*, and the arbelos theorems in *Coll.* IV (cf. #8).

---

[1] 211, #1 Hu, (Ver Eecke 1933b, p. 160, #7).

4.3.1   Proof Protocol Prop. 14

1. Protasis/ekthesis
   Starting point: semicircles over BD, BC, with B, D, C on a straight line, BC > BD.
   We get two possible configurations for this first construction stage:

   1. B–D–C (D inside BC, cf. configuration 1)
   2. C–B–D (D outside BC, cf. configuration 2 and 3)

   Then, a circle with center A is constructed. It touches both semicircles. We get
   a total of three possible configurations:
   Configuration 1: circle(A) must lie inside the semicircle over BC.
   Configuration 2: circle(A) comprises both given semicircles.
   Configuration 3: circle(A) touches both semicircles from outside.

   Draw the perpendicular AM from A onto BC, and the parallel to BC through A,
   marking the radius AZ, and the diameter TZ.
   Then BM:AZ is uniquely determined. Specifically:
   Configuration 1
   BM:AZ = (BC + BD):(BC − BD) [= (BC + BD):CD]
   Configurations 2, 3
   BM:AZ = (BC − BD):(BC + BD) [= (BC − BD):CD]

2. Apodeixis
   2.1. TZ ∥ BC by construction
   H, T, B, and H, Z, C lie on a straight line,
   and Z, E, B, and T, E, D lie on a straight line as well.
   [elementary lemmata on parallel chords in tangent circles][1]
   ### 2.2. * CB × BK = DB × BL
   [similar triangles, VI, 4; VI, 16; III, 36]
   [if the configuration entails D = L, we get: CB × BK = BD^2; this limiting case
   is used in Prop. 17, and may have been added in 14 for that reason]
   2.3. Prop. 14 now follows via V, 16–18, V, 22

Additions

Add. 1: BK × LC = AM^2
        This follows from * in 2.2 for Prop. 14,
        considering triangles BTK, ZLC

---

[1] E.g. Prop. 104 in *Coll.* VII, p. 828 Hu, (Jones 1986a, p. 234 # 166). Note the connection to Prop. 8,
and recall that the lemma invoked here was not presented in the source for Prop. 8, but in Pappus'
commentary to Apollonius' *Tangencies*. The geometrical situation for the arbelos theorem is
connected to the Apollonian problem and its theoretical framework. Hultsch's explanation (cf. above,
translation) involves an auxiliary construction, and reference to Prop. 13, converse. Configurations
1, 2 could alternatively appeal to *Lib. ass.* I, cf. Archimedes, *Opera Omnia* II, p. 510–512
Heiberg.

Add. 2a: BC × KL = BC × diameter of circle (A) = BL × DC
   [BC:BD = BL:BK; proportions; CD:BD = KL:BK; VI, 16]
Add. 2b: BD × KL = BD × diameter of circle (A) = BK × DC
   [BC:BD = BL:BK; proportions; CD:BD = KL:BK; VI, 16]

4.4 Prop. 15: Lemma for Induction

As said in the introduction, Prop. 15 is a technical lemma that provides the basis for
the (complete) induction in 16. Separating it out and presenting it beforehand has
the effect that the central theorem Prop.16 becomes slim and elegant, free of any
technical ballast.[1] Prop. 15 uses equation * from Prop. 14, and the converse and
addition for Prop. 13.

Prop. 15 holds also in the case that BC is a straight line, tangent in B. A proof
for this case is not given in the transmitted text. It may very well have been part of
the original source. For the manuscript A has a figure for it, but not the argument.
For a proof cf. Ver Eecke (1933b, pp. 170–171, #4, or pp. 1227–1228 the appendix
to Hultsch's edition). This case is used in Prop. 16, Addition 1.

4.4.1 Proof Protocol Prop. 15

1. Protasis/ekthesis
Starting from the three circles in Prop. 14, with their three configurations, we
add a fourth touching circle, with center P.[2] With diameter of circle(A): = d(A)
(and analogously for any circle(X) with d(X)), we get, for all three possible
configurations: (AM + d(A)): d(A) = PN:d(P).
2. Apodeixis
   2.1. With Z as intersection of AP and perpendicular in B,
   AT:TP = AZ:ZP   [Prop. 14, VI, 2; V, 16]

   2.2. BZ = ZT   [converse and addition to Prop. 13]

---

[1] Compare *QP* 22and *QP* 23 in relation to *QP* 24.

[2] In Hultsch's edition, the sequence of the resulting configurations is permutated: configuration 2
in Prop. 14 yields configuration 3 in Prop. 15. This re-numbering is, of course, of no consequence
for the mathematical content of Prop. 15. The manuscript A has three diagrams. The first one
concerns configuration 1, building on configuration 1 from Prop. 14, the second concerns the limit
case when the second semicircle is replaced by a tangent to the first one (see appendix Hu
p. 1227f.), and the third concerns configuration 2. There is no diagram for configuration 3 in A.
See part I, text and translation, with notes.

2.3. B–O–T–S [ΔSAT ~ ΔOPT, ΔBZT ~ ΔOPT, Prop. 13]

2.4. Prop. 15 follows [ΔBKM ~ ΔBPN, and ΔBKS ~ ΔBPO]

4.5   Prop. 16: Arbelos Theorem

Prop. 16 argues via induction. In fact, it gives what would in modern terminology be called a complete induction. In the context of ancient geometry, this proof strategy is very rare, and can be linked to Archimedes (cf. #3). It is perhaps worth noting that Pappus uses the technical term "apodeikhthesetai." In the arbelos configuration, the ratios of perpendiculars to diameters turn out to be expressible in numbers. The geometrical magnitudes are commensurable, and they are so according to a surprisingly simple pattern (sequence of natural numbers in ratio, cf. #2).[1] In Archimedes' treatises, we encounter an analogous phenomenon: likewise, all the technical detail work is done ahead of time, so that the main theorems become slim and straightforward (cf. #1, #2, #3, and #6).

An appendix to Prop. 16 explores the consequences when the outer, and when the smaller inner semicircle of the arbelos configuration degenerate into tangent straight lines (Additions 1 and 2), and when the starting configuration for the sequence of inscribed circles contains only one instead of two semicircles. Because of Prop. 15, the first ratio, AM:d(A) is decisive. In turn, it is directly related to the division of BC by D, and this fact is the content of Addition 2. The proof of Addition 2 rests on a lemma that is given afterward, as Prop. 17. In Addition 2, ratios of magnitudes are almost identified with ratios of numbers (compare the translation). Addition 1 uses the limiting case for Prop. 15, for which our manuscripts still provide the figure, but not the actual argument. These observations indicate, once again, that the Arbelos treatise probably comes from a source that was originally more extended, and that several stages of transmission lie between the original and the version given in *Coll.* IV.

4.5.1   Proof Protocol Prop. 16

1. Protasis/ekthesis

In the arbelos configuration, one has:

AM = d(A), PN = 2d(P), OS = 3d(O),

and generally: the perpendicular of the nth arbelos circle = n times its diameter.

---

[1] The fact that the configuration and the theorem relate to the theory of points of similarity means that Prop. 16 encapsulates at least the potential for a very deep insight, even if we do no longer have direct access to the actual mathematical context for such discussions in antiquity.

2. Apodeixis
    2.1. AM = d(A) [Prop. 14, *; V, 16 and 17 and VI, 16]
    2. 2. Induction, from $n = 1$ to $n = 2$ and from $n = 2$ to $n = 3$ [Prop. 15]
    2.3. The induction step via Prop. 15, used in 2.2, can be repeated indefinitely.
       Prop.16 follows.

Additions
Add. 1: When BC is a perpendicular BT to BD in B, the same proposition holds as
       in Prop. 16 [Obviously, AM = d(A); Prop. 15 for induction]

Add. 2: When the semicircle over DC is replaced with a tangent DZ, one has: AM
       ~ d(A) ⇔ BC:CD is expressible as a ratio of two square numbers.
Apodeixis for Addition 2
    1. DZ = AM.
    2. BC:CD = DZ^2:TZ^2 [Prop. 17]
       ⇒ BC:CD = AM^2:$d(A)^2$.
    3. X, 9: Iff BC:CD is expressible as a ratio of two square numbers, AM and d (A)
       will be commensurable in length.
       Example: If D is chosen on BC so that BD = 4CD,
       one gets: AM:d(A) = 2:1, and the perpendiculars will follow the sequence
       of the natural numbers from 2 on.

4.6   *Prop. 17: Supplementary Auxiliary Lemma for Prop. 16, Corollary 2*

Prop. 17 is an auxiliary lemma for Prop. 16, Addition 2. Unlike the lemmata for the
main proposition, which were placed before Prop. 16, the lemma for the addition
comes as a kind of afterthought. The reason for this difference in presentation is
probably the different theoretical status of the addition in comparison to the main
theorem. Prop. 17 does not take over the notations from Addition 2. Thus, it appears
as an independent lemma.[1] Like Hultsch and Ver Eecke, I tend to think that Prop.
17 is not by the same author as 13–16. Perhaps Pappus himself is its author. It may
be a replacement that became necessary when an originally more extended treatise
was "downsized."

    Hultsch's Latin commentary explains the crucial step (2 in the protocol below)
with a reference to duplicate ratios and uses V, def. 10, together with VIII, 11.
There is no explicit exposition on the handling of duplicate ratios in Euclid.
Archimedes seems to use them freely. Thus, duplicate ratios may very well have

---

[1] It is perhaps worth noting that even though the proof as transmitted explicitly appeals to Prop.
14, *, Prop. 17 could be independent from 13–16, because the result from within Prop. 14 could
easily be proved ad locum. Also, Prop. 17 uses a special case that appears to have been added in
within Prop. 14 precisely with a view to Prop. 17. For it is not used anywhere else within the
arbelos treatise.

been what the author of Prop. 17 had in mind. In the translation, I have nevertheless preferred to take a route that does not appeal to a theorem from the arithmetical books to explain the intermediate step about geometrical magnitudes.

4.6.1   Proof Protocol Prop. 17

1. $BC:CD = BD:TZ = DA:TA$ [*Coll.* VII, 104^1; Prop. 14, * for $D = L$; proportions]
2. $DA:TA = DA^2:AZ^2$ [$\Delta BAD \sim \Delta ZAT$ and $\Delta DAZ \sim \Delta TAZ$; VI, 8, VI, 16, VI, 1]
3. $DA:AZ = DZ:TZ \Rightarrow DA^2:AZ^2 = DZ^2:TZ^2 \Rightarrow BC:CD = DZ^2:TZ^2$

4.7   Prop. 18: Analogue to the Arbelos Theorem When the Second Inner Semicircle Is Missing

Prop. 18 is introduced by the following phrase: "the following, too, has been established through investigation by the lemmas written down above." The occurrence of this phrase lends support to the thesis that the treatise which Pappus had could not have come directly from Archimedes and showed signs of several stages of "work-over," additions for which Pappus did not know the source and date. Together with his description of the arbelos group as an "ancient theorem" deriving from "some books," without mentioning an author, this might induce us to refrain from ascribing the group as we have it to Archimedes. Still, one can find a number of "Archimedean" traits in the little treatise, as we have seen. Therefore, the group of theorems on the arbelos can be described as plane geometry, Archimedean style.

4.7.1   Proof Protocol Prop. 18

Protasis/ekthesis

When instead of the arbelos configuration, we have a configuration with two semicircles, one within the other, and a sequence of inscribed touching circles, the perpendiculars have to the radii a ratio that follows the sequence of the odd numbers (1:1, 3:1, 5:1 etc.).

---

[1] P. 828 Hu, (Jones 1986a, p. 234, # 166); Hultsch and Ver Eecke prefer here, as in Prop. 14, step 1, a reduction to an extended version of the converse for Prop. 13.

Apodeixis

1. The proposition is obviously true for the first circle:
   $perp(Z) = r(Z)$

2. $n = 2$, $n = 3$: induction steps
   $perp.(H):2r(H) = (perp.(Z) + 2r(Z)):2r(Z) = 3:2$ [Prop. 15]
   $\Rightarrow perp.(H):r(H) = 3:1$
   $perp.(T):2\,r(T) = (perp.(H) + 2r(H)):2r(H) = 5:2$ [Prop. 15]
   $\Rightarrow perp.(T):r(T) = 5:1$

3. The induction steps as illustrated in step 2 can clearly be continued indefinitely. Therefore, the perpendicular of the nth circle is the $(2n - 1)$-fold of its radius.

II, 5 Motion Curves and *Symptoma*-Mathematics

5 Props. 19–30: Motion Curves and *Symptoma*-Mathematics

5.1 General Observations on Props. 19–30

Props. 19–30 (as well as 35–41) deal with lines and curves that are different both from the circles and straight lines of Euclidean geometry, and from the conic sections. They are generated from moving points, where a rule is given which regulates the "motions" involved. They will be called motion curves here. An example would be the plane spiral of Archimedes, where a point moves along the radius of a circle in uniform speed, and is at the same time carried along on that radius as it rotates the full circle, also in uniform speed. The point describes a spiral line in the process. Another example, though this is not used in ancient geometry, would be the generation of a circle as the "motion curve" described by the endpoint of a radius as the radius rotates a full 360°. In order to study the mathematical properties of such curves, one has to come to a quantifiable characterization, as a proportion, or an equality that applies to all the points on the curve and only to them. All mathematical properties have to be derived from, or related back to, this original characterizing property. It is called the *symptoma* of the curve. It ultimately rests on the motions used to generate the curve, but as they do not appear in the mathematical discourse, the mathematics develops out of the *symptoma* itself as the starting point. I will call this type of mathematics *symptoma*-mathematics. The conchoid of Nicomedes,[1] e.g., has the *symptoma* that all lines drawn from a point of the curve to the pole have a definite *neusis* property: the segment cut off on it between the canon and the point on the curve has a fixed length. The curve itself is viewed as the locus for this property, and this is how it is employed in mathematical argumentation. An analogy would be to view the circle as the locus of all points that have a fixed distance to a given point. Arguably this could even be seen as the Euclidean *symptoma* of the circle. The case of the conics is somewhat similar: they could be viewed (and some scholars think they were) as the *symptoma*-curves for certain equalities expressible via application of areas, and whether this is their true definition or not, they were often employed this way in mathematical investigation.

The motion curves discussed in *Coll.* IV are: Archimedean plane spiral (Prop. 19), Nicomedean conchoid (Prop. 23, though defined as quasi-*symptoma*-curve), quadratrix (Prop. 26, also defined as a *symptoma*-curve via analysis of loci on surfaces in Props. 28 and 29), Archimedean spherical spiral (Prop. 30) and Apollonian helix (used, not defined in Prop. 28)

---

[1] For this curve, and for the technical terms, *neusis*, pole, canon, see the translation and commentary on Props. 23–25.

H. Sefrin-Weis, *Pappus of Alexandria: Book 4 of the Collection*,
Sources and Studies in the History of Mathematics and Physical Sciences,
DOI 10.1007/978-1-84996-005-2, © Springer-Verlag London Limited 2010

The account given by Pappus suggests a certain developmental line, which has, on the whole, been tacitly accepted by most scholars, even if they do not think highly of Pappus as a mathematician (e.g., Knorr 1986). For Props. 19–30 are our main source for this type of "higher" ancient geometry, the basis for its reconstruction.[1] Generally, there are two types of motion curves, developing from curves like the Archimedean spiral and the quadratrix. They can be associated with two strategies for dealing with the problem of finding a mathematically acceptable "definition" of the curves.

(a) "Archimedean" track: motion approach, meta-mechanics (or: quasi-mechanics), with *symptoma*-mathematics on properties derived from synchronized abstract motions.

The generation (*genesis*) of the basic curves takes place via abstractly conceived uniform motions. The description of the *genesis* needs to be free of conceptual contradictions. This means that for use of synchronized motions, the speeds must be in a definite ratio (cf. spiral, below; the quadratrix was subject to objections along those lines). The curve is then characterized via a (resulting) mathematically describable *symptoma* (proportion or equality), which is directly read off of the *genesis*. Geometry with, or on, the curve considers the curve as the locus corresponding to the *symptoma*. One might compare the way geometry operates with circles and straight lines. *Coll.* IV, Props. 19–22 discuss a version of the Archimedean spiral and present it as the starting point of a research trend. The problem with this version of the Archimedean spiral (inscribed into a circle given beforehand) is that the motions used have to be synchronized according to the ratio of circumference to radius (π, essentially), and that ratio is unknown. The version in *SL* avoids this dilemma: it uses synchronized motions in a given and fixed ratio, and circumscribes a circle afterward. *SL* separates the motion lemmata from the rest of the treatise, as quasi-postulates and emphasizes that separation in addition by positioning the definitions for the treatise between these lemmata and the main treatise. Both versions, even the one with the problematical *genesis*, result in a curve that can be grasped exactly via its *symptoma*, and through it can be subjected to mathematical treatment. Pappus does not reject either the theorems on the spiral (Props. 20–22), or the theorems on the quadratrix (Props. 26 and 27, 35–41), although the original *genesis* of the quadratrix encounters similar conceptual difficulties, because it implicitly involves π. For the quadratrix, he emphasizes in fact that the theorem on it is much more acceptable than the *genesis* of the curve itself. The foundations of this type of motion curves, treated as "lambanomena," and resting on the *genesis*, do remain somewhat hypothetical, however, and perhaps there remains some uneasiness about them: you have to operate with something for which you must restrict your attention on certain conceptualizable aspects, though the object itself cannot be fully grasped conceptually. Yet, one could take the view that this is the case for all geometrical objects. Already in Aristotle's *Analytica Posteriora*, where he is abstracting from the mathematical practice of his day, it is stated that "postulates" (aitemata, not to be confused with axioms) for a science can be hypothetical, that does not detract from their scientific character. Prop. 30, which uses uniform motions with velocities in the ratio 1:4 for

---

[1] Additional examples: Archimedes, *SL* and *Eutocius in Arch., Sph. et Cyl.* II, pp. 54–110.

the *genesis*, and is based on the genetic properties of the curve, is completely uncontroversial as a theorem in *symptoma*-mathematics on a motion curve. No treatise in which Prop. 30 was incorporated survives.

(b) The second track of mathematics on motion curves operates with analysis of loci on surfaces, leading to a *symptoma* that can be captured in an exact mathematical relation to other features of the given configuration.[1]

The mathematics is restricted to such properties as follow for the curve qua observing the *symptoma*. This second track appears to have started in the generation after Archimedes. Nicomedes was one of its founders, and one of the major figures in this field – at least that is how it appears from Pappus' portrait. This line of approach was developed further by other Hellenistic geometers, as is apparent from Pappus' list in the meta-theoretical passage (between Props. 30 and 31). A substantial body of mathematical works in this area must once have existed. Unfortunately, no works of this branch of ancient "higher" geometry survive today, and we owe the little knowledge we have of them mostly to Pappus. This second, analysis-dominated way of looking at the motion curves has features that connect it to the treatment of the conics. According to the view taken by most scholars, the conics were found and treated, per analysis-synthesis, as locus curves, characterized via *symptomata* in a specified configuration: cf. Menaechmus, Aristaeus, and Euclid on solid loci.[2] Attested in *Coll.* IV are the following examples of analytically determined *symptoma*-curves. In Props. 28 and 29 (one of which may very well be by Nicomedes), the quadratrix is reduced via analysis to loci on surfaces created in dependence from the Archimedean spiral or the Apollonian helix and shown to be determined by it. The mathematics on it is *symptoma*-mathematics, again. The *symptoma* is used, the *genesis* of the assumed curves is left out of the mathematical consideration.[3] One might compare this procedure to Nicomedes' characterization of the conchoid (between Props. 22 and 23); it appears to be a transitional step in that he does generate the conchoid via motions, but does not read the *symptoma* off of the motions used. Rather, he determines it via pointwise characterization of a conchoid already drawn: the conchoid is the locus for a *neusis* property.[4]

---

[1] Here the trait of ancient geometrical analysis as essentially an analysis of configurations is craftily exploited. See below, on the use of analysis in Props. 28 and 29.

[2] A notable exception to this view of the ancients' understanding of the conic section is S. Unguru. The reader may wish to consult his contributions for a different account. Since the "dominant" view (conics treated as locus curves, so to speak, with heavy reliance on a characterization via their *symptoma*) concurs very well with Pappus' presentation in *Coll.* IV, I have opted, in the present work, not to raise and discuss this general question. Perhaps it could be discussed again, with profit, by taking into account the "solid" arguments in Pappus (Props. 31–34 and 42–44) in addition to the now extant Apollonian arguments from the *Konika* (in Eutocius' revised, purely synthetic edition).

[3] Props. 28 and 29: Chasles (1875) explores the connections to the differential geometry of higher curves.

[4] My assessment, as given here, differs from Knorr's, cf. Knorr (1989, p. 31). Knorr objects to Pappus' portrait of the conchoid as a *neusis* curve and insists that the conchoid was not intended for the *neusis*, but as a replacement for it.

Of these two alternative strategies, the former, Archimedean approach, seems to have been the original one. Nicomedes, for example, picked up on Archimedes' suggestions. His description and treatment of the quadratrix shows close connections to the Archimedean plane spiral. He does, however, develop this branch in a different direction (analysis of loci, getting rid of the mechanical metaphors). And it seems that it was the second path, not the Archimedean "quasi-mechanics" that was pursued by the Hellenistic mathematicians after him. In both strategies, one avoids having to pronounce on what a curve is in itself, instead defining it as a locus of points that fulfill a given relation. What remains somewhat unclear and vague is not the status of the mathematical argumentation, but the status of the curves themselves. Philosophers were much interested in an answer to questions like this, especially the Platonists. Already in Plato's time, it seems that the mathematicians chose to remain vague, and silent, about the exact ontological status of their objects. Aristotle's reflections on the status of mathematical objects, which, though apparently espousing some kind of abstractionist view, remain uncharacteristically vague as to the actual status of the mathematical objects, and appear to do justice to the historical facts: the status of mathematical objects remained undecided among the mathematicians (and perhaps still is so today). That the ancients did not reject the *symptoma*-mathematics of motion curves just because motions are involved, and that Descartes (to whom part of our modern prejudice toward ancient mathematics in this regard may very well go back) was mistaken, or misleading, in his explicit assessment of the ancient mathematics of higher curves as portrayed in Pappus (in the *Géométrie*), was already argued convincingly by Molland (1976). Descartes went on to use an elaboration of this constructed dichotomy between mechanics and geometry for his own program. We cannot follow up Descartes' considerations, and their possible connections to his reading of *Coll.* IV in detail here.[1] It has to be noted, however, that his picture of the ancient mathematicians' views on the curves of the third kind is somewhat skewed. For Pappus, the *symptoma*-mathematics of lines of the third kind was a branch of geometry. What was not settled, and here Descartes may have picked out a tension implicit in Pappus' account, was the status of the lines themselves. Part of the efforts of the ancient mathematicians was directed at an account of the *genesis*, eliminating conceptual problems for the motion *genesis* or reducing the *genesis* to an analysis of loci on surfaces (which have to be *given*). Their ontological status remained undecided. A few summary remarks on the problem of the ancients' views on mechanics versus geometry may be useful. A comprehensive solution is at present not in sight. Perhaps the material offered in the present edition, complete and within its originally intended context, could be usefully brought to bear on further discussions.

---

[1] It is not implausible that Descartes' reading of Pappus, *Coll.* IV had an impact on his own view of the definition of higher curves via controlled synchronized motions *and* algebraic- analytical characterization, not fully harmonized in his account. The issue cannot be pursued here in detail, but might be worthy of further investigation. See also below, footnotes to the passage on mechanics versus geometry.

5.1.1 Mechanics Versus Geometry and the Problem of Motions

Are motions allowed in ancient geometry, if only for the *genesis* of curves? Or are they banned? There seems to be no simple, straightforward answer. On the one hand, Aristotle and Plato seem to suggest they are excluded from geometry proper. Yet Plato and later Platonists use and accept abstract motions (viewed as generating into intelligible matter) for the generation of geometrical lines and objects without qualms (cf. Proclus). Euclid I, 4 notoriously uses superposition, defines congruence by appeal to moving geometrical objects so that they fit onto each other. Mechanics and astronomy, e.g., deal with objects in motion, and abstract from them. What becomes of the motions? One might wonder what role they play for the objects treated in theoretical mechanics. Popular definitions of straight lines and circles via flowing points existed. Euclid avoids that, as does Aristotle. Still we encounter such definitions as late as Proclus. Tracing a line (as the conchoid, or the spiral, or the circle) could be seen as a motion, and if so, one might again wonder how such "motions" figure in mathematics. Mechanics, in antiquity, was a mixed science. As such, it had a theoretical aspect: geometry, and a physical aspect. The latter did have to do with practical and technical devices (instruments and their operation).[1] To call an argument, or a curve, "mechanical" can mean a number of things; it may simply mean that idealized motions are involved; e.g., Nicomedes' *genesis* of the conchoid in Prop. 23, and Archimedes' *genesis* of the spherical spiral in Prop. 30 are mechanical in this sense. The use of such motions for the generation of curves is not, as such, viewed as problematical. The "mechanical" *genesis* of the conchoid (via tracing) is not criticized by Pappus, and the curve is fully accepted as a mathematical curve, because the motions are well-defined and can be grasped in thought. The motions used for the quadratrix are not objected to as such, but because they involve a logical inconsistency (see below). That motions are not explicitly objected to for the *genesis* of lines does not mean, however, that they are seen as part of the mathematical discourse. Whether well-defined or not, the generating motions themselves will never turn up in the *symptoma*-mathematics of the respective curves. They are used to read off the *symptoma*, and only the latter becomes part of the argument.[2] Where does that

---

[1] Cf. also Pappus on mechanics versus geometry in *Coll.* VIII. The passages there would, in my view, reward scholarly attention for a clarification of the ancient view on the relation of geometry and (theoretical) mechanics. See also Ver Eecke (1933a)

[2] It seems to me at least to be a plausible hypothesis that Descartes, when formulating his view on permissible geometrical curves in the *Géométrie* was strongly influenced by the examples in *Coll.* IV. For what he presents as his dual view on the characterization of geometrical curves, as an improvement on the ancients, is, at least in nuce, already to be found here in *Coll.* IV. Descartes decided that motions are permissible, and lead to geometrical curves in his sense, if the motions and their synthesis can be controlled at all times – as they can for his parabolas of higher degrees, for the conchoid, and for the spherical spiral, but not for the quadratrix (Prop. 26) and the plane spiral as in Prop. 19. Under Descartes' conception of permissible modes of generating geometrical curves, both the Archimedean spiral (in 19–22), and the quadratrix, are non-mathematical. The mathematics about them is no real geometry. As said repeatedly: for Pappus, Props. 20–22, and Props. 26–29 do qualify. In this respect, he differs from Descartes.

leave the motions with respect to the field covered by mathematics? Discussion existed, at least among philosophers of mathematics.[1] For lack of sources, it can at present not be fully reconstructed. What can be given, and what the present edition attempts to provide by presenting the relevant source material in its actual context, is a glimpse at the way in which mathematicians handled the motion curves (when no alternative *genesis* was available), and how they dealt with problems of defining them mathematically, so they could do mathematics about/of them. The salient point, which has so far not been emphasized enough in secondary literature, is the *symptoma*. If a curve as such cannot be grasped fully as an object, its defining properties perhaps can, and that is what mathematics concentrates on in any case. The ancient mathematicians may have deliberately pushed aside the question of the ontological and epistemological status of the curves themselves. The assessment in Molland (1976) – motions as such were not the problem; but the relation of geometry and mechanics is unclear – still holds today.

5.1.2   "Mechanical" Versus "Instrumental"

There is, however, a related issue, on which I need to comment here. "Mechanical" can, but does not need to, relate to the use of instruments and devices, physical objects. It would be a mistake, and has led to a misreading of the mathematics of higher curves in Pappus, to take the two notions "mechanical" and "instrumental" as virtually equivalent.

As said above, the label "mechanical" can refer to the use of abstract motions, and to metaphorical usage of terms taken from mechanics, e.g., weight of a triangle (cf. *Ephodos*, but also *QP* by Archimedes). Such language does not imply actual physical motions, accomplished by physical objects such as instruments. This is why Pappus, after describing the conchoid as resulting from a motion (mechanically; this would be called "mechanikos") mentions that it can also be generated by means of an instrument devised by Nicomedes ("instrumentally," organikos). The use of specified permissible instruments is not the focus of mathematical construction for the ancients. Euclid, for example, never uses the terms for ruler and compass. His mathematical constructions are by means of circles and straight lines, and they

---

[1] Aristotle's position on the status of mathematical objects, their relation to natural objects is much discussed and notoriously problematic. It may not be completely consistent. Most likely, however, it is a kind of abstractionism, where the mathematical objects are essentially idealized properties of physical objects, depending on them ontologically, while having some degree of epistemological priority insofar as they pick out essential features. Geometry in relation to mechanics could, with some plausibility, be construed as meta-mechanics, as that theory which gives the "why" – explanations within mechanics. Pappus' discussion of the roles of geometry versus mechanics in *Coll.* VIII (cf. footnote above) appears to be drawing on Aristotelian conceptions. This issue cannot be pursued here. It is not inconceivable that a study of Pappus' views on mechanics versus geometry in *Coll.* VIII, in comparison with the Aristotelian theory of science, could help understand Pappus' stance on the motion curves in *Coll.* IV.

are not necessarily viewed as tools, i.e., instruments. The focus is certainly on construction, but not quite as this is commonly understood. The material aspects of how one creates the circles and straight lines used in the arguments is irrelevant in Euclidean geometry.[1] The use of instruments, precise or otherwise, as such, does not make an argument mathematical or non-mathematical, in fact, it is irrelevant to the mathematical content concerned. Therefore, "mechanical" does not mean, in a mathematical context: "created by a physical device"[2] or "designed for actual practical production." I have decided to avoid the label "mechanical" for the motion curves, because it may lead to the wrong associations.

5.1.3 Survey of Props. 19–30

(a) Props. 19–22: *genesis* and *symptoma* of the Archimedean plane spiral (inscribed in a circle), two *symptoma*-theorems on it: spiral area, in relation to the circle into which it is inscribed; spiral sectors in relation to cubes over maximum radii (Archimedes)

(b) Props. 23–25: *genesis* and *symptoma* of the conchoid, further properties, *symptoma*-theorem on the conchoid: two mean proportionals, cube multiplication (Nicomedes)

(c) Props. 26–29: *genesis* and *symptoma* of the quadratrix; criticism of the *genesis*, *symptoma*-theorem on the quadratrix: rectification of the circumference of a circle and squaring of the circle; analytical reduction of the *symptoma*-description of the quadratrix to loci on surfaces (Dinostratus, Nicomedes, Apollonius (?), Sporus, Pappus)

(d) Prop. 30: *genesis* and *symptoma* of the spherical spiral (inscribed), *symptoma*-theorem on the spherical spiral: quadrature of a curved surface (Archimedes)

Obviously, there is a common structure to the presentation of mathematics of "higher" curves in Pappus: *genesis* – *symptoma* – mathematics treating curves as locus for the *symptoma* (cf. handling of conic sections – which were defined as sections of cones, but predominantly employed for arguments under the perspective of *symptoma*-curves). Also, there appears to be at least the trace of a developmental line for this branch of Greek geometry, and Pappus' meta-theoretical remarks, which follow immediately after Prop. 30, reinforce that impression. The fact that the examples for *symptoma*-mathematics fit extraordinarily well with Pappus' meta-theoretical passage in this regard need not mean that Pappus is representative of the way mathematicians viewed their work. We may see a rational reconstruction here,

---

[1] Cf. Netz (1999) on the use of highly schematized/standardized diagrams, set in stone and completed already at the outset in geometrical argumentation/instruction.

[2] While concern with devices was not an emphasis in ancient mathematics, it was of interest to Descartes, who included in the *Géométrie* a description of a mesolabum-compass, and an instrument for generating higher degree parabolas (also: device for creating ellipses etc. in the *Optics*).

didactically polished, by a well-informed commentator looking back at history and trying to make sense of it. Even under that precaution, however, Pappus' views are well-grounded and deserve attention and respect in their own right. In order to arrive at a balanced view on how, and how far, Pappus' image corresponds to a general understanding of the mathematical tradition, one should take the whole of Pappus' account as a basis and look for discrepancies and similarities in direct and indirect sources outside Pappus. This has not been done so far, because Pappus' overall position has not yet been taken seriously as a reflected and meaningful one. Perhaps the present documentation of Pappus' account as a whole can serve as a basis for further research.

As said above, Pappus, *Coll.* IV, 19–30 (and 35–41), is a valuable source on this branch of ancient mathematics, the only extensive one in existence now, apart from Archimedes, *SL*.[1] The curves presented by Pappus attracted much attention in the seventeenth century: conchoid, spiral, and quadratrix were recurring subjects of discussion. It would be an interesting topic to investigate in more detail the way in which texts like Pappus' here were used in the seventeenth century to negotiate a shift in perspective, a new program in geometry. Examples for such a reception include: Descartes (1637), Hudde and Heuraet on the conchoid in Descartes (1659), and de Sluse (1668) on the conchoid; Witt/Schooten in Descartes (1659) on conics, Newton *Arithmetica Universalis* in Whiteside (1972) V, pp. 420–490. Perhaps Prop. 21 played a role for the investigation of quadratures via indivisibles.[2] Props. 28 and 29, for some reason, did not receive as much attention in this discussion, even though these two passages do focus on method, and show how analysis, especially: *resolutio*, can be used in this area.[3] Props. 19–30 are also very interesting in themselves, and they deserve scholarly attention.

5.2   Props. 19–22: Archimedean Plane Spiral, "Heuristic" Version

5.2.1   Relation of Props. 19–22 to *SL*

For a discussion of this topic see especially Knorr (1978a). Knorr argues that Props. 19–22 document the heuristic version of Archimedes's inquiry into spiral lines, probably connected to research on the quadrature of the circle.

---

[1] Because Archimedes is such a brilliant mathematician, it is perhaps tempting to view his contributions as exemplary. However, it is also possible, and precisely for the same reason, that he was extraordinary, non-typical, as far as his research into non-explored terrain in mathematics is concerned. The issue cannot be investigated here.

[2] The discussions on spiral lines were, however, based mostly on the version in *SL* (e.g., in Jacob Bernoulli's work), not on the version in Props. 19–22.

[3] But see Pascal, *De la dimension d'une dolide forme par le moyen d'une spirale autour d'une conique*, related to Prop. 29, and Chasles (1875, p. 30) on Prop. 28 (quoted according to Ver Eecke (1933b) p. XXXII, with #4.

The curve as defined in Prop. 19 differs from the version in *SL*. Prop. 19 introduces a spiral inscribed in a given circle, whereas in *SL*, one has a spiral and circumscribes the circle. Only the version in *Coll*. IV can yield the quadrature of the circle. In the *Coll*.'s version, the *genesis* via two synchronized motions implies π (not so in *SL*, and that is exactly why *SL* 18 does not yield a constructive rectification of the circle). That said, and with the appropriate caution, Prop 19 corresponds to *SL* 14 with *SL* 2, and Prop. 20 corresponds to *SL*12 with *SL* 1.[1] *SL* 24 is an area theorem on the spiral with circumscribed circle. Though closely related to Prop. 21 in *Coll*. IV, it is really a different theorem. The true parallel to *SL* 24 comes as an appendix (corollary) of Prop. 21. Also, the proof strategies differ considerably. Prop. 21 uses a parallel auxiliary figure with rotation solids inscribed, a process of continuous division, and quasi-indivisibles, whereas *SL* 24 (relying on *SL* 21, *SL* 12, *SL* 10) gives a proof via double reductio, no indivisibles are used, the parallel auxiliary figure involves a single circle, and no progression of inscribed figures into that circle is envisaged. A proof protocol of *SL* 24 will be given below for convenient comparison with Prop. 21.

5.2.2   Heuristic Method in Props. 19–22

The full scope and machinery of Archimedes' heuristic method was described by himself in the *Ephodos*, "mechanical theorem method." The work was lost until rediscovered by Heiberg in the early twentieth century (Heiberg 1906). Thus, the mathematical tradition had no direct full access to this aspect of Archimedes' work. For his monographs do not show any traces of the method of discovery. Part of what makes Prop. 21 (and, to a lesser degree, 30) so valuable and interesting is the fact that it does exhibit some of the characteristic traits of that method, and it was indeed accessible to the mathematicians in the early modern period studying Archimedean mathematics. Because Prop. 21 is a vital part of what the early modern mathematicians, in going toward the calculus, knew about Archimedean heuristic procedures for finding quadratures, it is obviously a very important source for historians of mathematics. It deserves closer attention and analysis with a view to its reception in early modern times. Specifically: the method employed in Prop. 21 may very well have influenced Cavalieri's treatment of area and volume theorems via indivisibles (in addition to the *QP*), as well as other discussions of similar problems during that time. Quite possibly, the mathematicians in the sixteenth and seventeenth centuries consciously made use of this limited glimpse of Archimedes at work. The issue can, of course, not be pursued in detail here. What will be given is a documentation of the actual text in Pappus, as a basis for further investigation. For a comprehensive survey of Archimedes' mechanical method, cf. Dijksterhuis (1987).[2]

---

[1] *SL* 12 is used in *SL* 24; Prop. 20 is used only in the addition to Prop. 22.

[2] Incidentally: Archimedes does not use the term "methodos," which would have implied a scientific character for the procedure, but rather the word "*Ephodos*," i.e., "attack, approach," emphasizing the strategic aspect of it, and its direction toward success in the form of a concrete result.

According to Dijksterhuis, the strategy described in the *Ephodos* relies on the meta-phorical use of mechanical features and operates with (a) levers and centers of gravity and (b) indivisibles. Only (b) can be observed in Prop. 21 (more broadly: in 19–22).[1] To some degree, (a) can be observed in *QP*, but Archimedes' procedure there differs slightly from the version of the same result as given in the *Ephodos*.

5.2.3   Survey of Props. 19–22

context: geometrical properties of the Archimedean spiral, quadrature of the circle, motion curves, and *symptoma*-mathematics.
source: an otherwise lost, and unattested, work by Archimedes.
means: V, VI, exhaustion (XII), quasi-indivisibles.
method: synthesis, limit process argument via indivisibles.
format: theorems.
historical significance/reception: addition/alternative to Archimedes' *SL*, Archimedes' heuristic method exemplified partially, possible influence on arguments via indivisibles in the sixteenth and seventeenth centuries.
embedding in *Coll. IV*: presentation of mathematics on higher curves in the order *genesis-symptoma*-theorems: same as in Props. 23–25, 26 and 27, and 30; motif "geometry, Archimedean style": 13–18, 30, 42–44; motif "squaring the circle": 26 and 27; connection to the curve quadratrix: 29[2]; spiral used in 35–38; insofar as the quadratrix can be derived from the spiral, 39–41 may be included here, too. In all these instances, it is the basic *symptoma* of the curve, not the theorems on it (i.e., Props. 21 and 22) that will be used.
purpose: exhibit plane spiral as a classic and basic curve for the methods of *symptoma*-mathematics of the "linear" kind (the starting point for this kind?). Props. 19–22 show the conceptual problems inherent in the *genesis* for some of the motion curves, the heuristic efficiency of the "mechanical method," and the resulting style of mathematics on motion curves.
literature: Dijksterhuis (1987, pp. 268–274), Knorr (1978a, 1978b, 1986, pp. 200–201); Knorr's work on Archimedes' treatment of the spirals is the authoritative one at present. This portion of the commentary on *Coll.* IV relies heavily on his results. There are some limitations, in my view, mostly concerning Props. 42–44. They will be pointed out ad locum. Hultsch remarks that the treatment of the spiral in Props. 19–22 is not identical with the treatment in *SL*, but is still willing to see it as parallel. Ver Eecke (1933b), through-out his commentary on Props. 19–22, erroneously assumes that the treatment is virtually identical with the one in *SL*.[3] Heath (1921, I, 230–231) is mistaken is assuming that *SL* 18 yields a constructive rectification of the circle. See also Heath (1921, II, 377–379) on Props. 19–22.

The proof protocols for 19–22 will differ in format from the protocol of "standard" mathematical expositions, e.g., Props. 13–18, reflecting the different character and style of the arguments employed here.

---

[1] Dijksterhuis defends the thesis that only (b) is problematical for geometry. Compare the above discussion of mechanics versus geometry.

[2] Knorr (1986, pp. 161–168) supports an even closer connection between spiral and quadratrix.

[3] In addition, Ver Eecke claims at XXVIII that Archimedes's exhaustion method is equivalent to the infinitesimal calculus, and that 19–21 are analytical in form. One might disagree with these judgments.

5.2.4   Prop. 19: *Genesis* and *Symptoma* of the Archimedean Spiral

Genesis: In a given circle, inscribe a spiral; use two uniform synchronized motions, a clockwise rotation of radius BA, and a linear motion of P from B to A. P is to arrive at A when BA has completed a full rotation. Obviously, the two motions have to be coordinated according to $2\pi r{:}r$.[1]

Symptoma: For any radius BC, drawn arbitrarily, intersecting the spiral in P, BP:BA = arc BC:circumference, and an analogous property holds for the angles under the arcs.[2]

This *symptoma* will be used in all places where the spiral is employed for argumentation. The *genesis* is kept out of the picture from now on.

5.2.5   Prop. 20: Progression of Spiral Radii

Directly from the *symptoma*, one can derive the following.

A progression of angles, increased by the same increment at each step, produces an arithmetical progression of spiral radii.[3]

5.2.6   Prop. 21: Area Theorem

5.2.6.1   Argument in Prop. 21

1. Protasis/ekthesis
   Assume a circle with inscribed spiral.[4] The spiral area is one third of the circle.

2. Apodeixis[5]
   2.1. extension of configuration, auxiliary construction
   circle, sector CBA (S(C)), sector ZBH (S(Z)),[6]
   rectangle KNLP, rectangle KNT'R,
   in the same ratio as the circle and S(C),
   rectangle MNT'W, rotation cylinders over NL, NY'
   (C(R), C(W))

---

[1] This constitutes a problem for the genetic definition for the curve, analogous to what will be said about the quadratrix below. Pappus is, however, silent on this issue here. That the conceptual inconsistency was brought up, and that it must have bothered Archimedes, can be seen from the fact that *SL* avoids this problem, even at the cost of no longer being able to rectify, and square the circle constructively. Compare, and contrast, Prop. 19 with *SL*, Def. 1, 2, 3, and 7.

[2] Compare *SL* 14.

[3] Compare *SL* 12 + *SL* 1.

[4] A counterclockwise motion of the generator for the spiral is understood.

[5] The argument employs an infinite inscription process and quasi-indivisibles. Still, Pappus calls this argument a proof (apodeixis); cf. Knorr (1978a), pp. 52 ff. on Prop. 21.

[6] A division in the ratio 2^n is likely.

2.2. S(C):S(Z) = C(R):C(W)

[*symptoma*, VI, V, XII, 2]

2. 3. Continue analogous parallel inscription processes in circle and rectangle, create progression of sectors and rotation cylinders, *exhausting* circle and rectangle.

In each case, you get a proportion as in 2.2.

2. 4. Summing up after a finite number of inscriptions

circle:sum of all inscribed sectors

= C(R):sum of all inscribed rotation cylinders

2. 5. Approximation from above

The analogous proportion will hold for progressions of circumscribed sectors and rotation cylinders.

2.6. Limit argument

Imagine the partition made more and more fine-grained.

The inscribed and circumscribed circle sectors approximate the spiral area from both sides, and the inscribed and circumscribed cylinders approximate the rotation cone over KN with side KL. By an (implicit) continuity argument (a transition to infinity, or an appeal to indivisibles), we infer: the above proportion will still hold in the "limit case," and thus: circle:spiral area = rotation cylinder:cone = 3:1.

5.2.6.2   Proof Protocol of *SL* 24

(for comparison with Prop. 21)

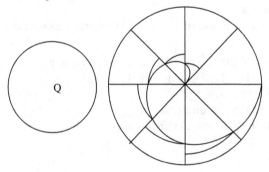

1. Protasis/Ekthesis

   When a spiral of first rotation with a circumscribed circle is given, the following proposition holds: The spiral area (S) will be one third of the circle (A).

2. Apodeixis

   2.1. Extension of configuration, auxiliary construction

   circle Q with Q = 1/3A.[1] We need to show: Q = S.

---

[1] With r: = radius of circle A, construct a line r' so that $r^2 = 3r'^2$ (e.g., by trisecting r (VI, 9), and transforming the rectangle with sides r, 1/3 r into a square (II, 14). Then the circle Q with radius r' will have the area 1/3 A (XII, 2).

2.2. Double reductio: If not, then either Q > S or Q < S

    2.2.1 Assume Q > S

    then Q – S = : f is a definite magnitude.

    Construct a sequence of similar sectors circumscribed around the spiral, so that for their sum F we get

    F – S < f (*SL* 12, 21). Then F < Q.

    On the other hand: F > 1/3A (*SL* 10, addition),

    i.e., F > Q, a contradiction.

    2.2.2 Assume, then, that Q < S

    then S – Q = : f is a definite magnitude.

    Construct a sequence of similar sectors inscribed in the spiral, so that for their sum F we get

    S – F < f. Then F > Q.

    On the other hand, *SL* 10, addition, yields F < 1/3 A, i.e., F < Q, a contradiction.

    2.2.3 Therefore, Q = S must hold.

3. Symperasma: Therefore, S = 1/3 A.

Prop. 21 is a *symptoma*-theorem on the spiral *inscribed* in a given circle. The proof uses the *symptoma*, V, VI, XII, 2, and XII, 10 and 11. The heart and core of the argument, however, is the progressive inscription, and the limit process. This is "unorthodox," non-Euclidean, and there is no analogue in *SL*. It is probably the implicit limit argument employing rotation solids that Pappus refers to in his introduction to Props. 19–22 as a "certain astonishing plan of attack."

Addition to Prop. 21

The analogue to Prop. 21 holds for spiral sectors (of the original inscribed spiral), and corresponding sectors of circumscribed circles.[1]

5.2.7   Prop. 22: Ratio of Spiral Areas and Spiral Segments[2]

5.2.7.1   Argument in Prop. 22

Protasis

    The ratio of the spiral area to spiral segment areas is equivalent to the ratio of the cube over the radius to the cubes over the corresponding spiral radii.

---

[1] Since in this addition, the secondary circles end up being circumscribed, one might (as I have done in the introduction) read the addition as the true equivalent of *SL* 24. All one needs is a continuity argument, interpreting the spiral and circumscribed full circle as a limit case for spiral and sectors of circumscribed circles.

[2] On Prop. 22, cf. Knorr (1978a, pp. 57 ff).

Ekthesis
> Start with a configuration containing a circle ACD with radius BA, and a spiral
> AZEB. Draw the arc ZT, and produce BZ to C.

Apodeixis
1. Spiral area:spiral sector area = circle ACD:sector ZBT
   [Prop. 21, with addition].

2. Circle ACD:sector ZBT =
   (circle ACD:circle ZBT) × (circle ZBT:sector ZBT)
   Circle ACD:circle ZBT = $AB^2:BZ^2$ [XII, 2]
   and circle ZBT:sector ZBT = circumference ZBT:arc ZBT
   = circumference ACD:arc CA   [VI, 33, with addition]
   = AB:BZ   [*symptoma* spiral]

3. Spiral area: spiral sector area = $(AB^2:BZ^2) × (AB:BZ)$,
   and this is equivalent to $AB^3:ZB^3$

Compound ratios are written here as quasi-products; this is somewhat problematical.
See the discussion of compound ratios and multiple (double and triple) ratios in Saito
(1986) and Heath (1926) on VI, 23. Knorr (1978a, p. 57) discusses a similar use of
ratios in the alternative proof for *Sph. et Cyl.* II, 8. It appears as though Archimedes
handled compound ratios in the way suggested here. Hu and Ver Eecke refer to XI,
33, and V, def. 10 for a justification. The triple ratio here may very well have been
seen as the analogue to VI, 23 for solids. Co p. 85 seemed to have qualms about this
proof in Prop. 22, with its implied identification of ratios of solids with composite
ratios of planes and sides; he offers an alternative via centers of gravity for solids.

5.2.7.2   Addition to Prop. 22: Areas of Spiral Quadrants[1]

1. Ekthesis
   Start with a circle, inscribed spiral, and partition of circle into quadrants using
   the end stage of the rotating radius, yielding spiral quadrants Q1–Q4.
   Posit Q1 as 1 (unit). Then: Q2 = 7, Q3 = 19, Q4 = 37.

2. Apodeixis
   Progression of spiral radii is 1:2:3:4 (*symptoma* of the spiral),
   progression of spiral area sector is 1:8:27:64 (Prop. 22),
   subtract preceding quadrants at each step, progression of quadrants is 1:7:19:37.

5.3   Props. 23–25: Conchoid of Nicomedes

5.3.1   General Observations on Props. 23–25

> context: *neuses*, motion curves, cube duplication/multiplication (two mean proportionals).
> source: Nicomedes, treatise on conchoid lines, with additions from Pappus.

---

[1] Prop. 22, Addition, is not specific as to the relative priority of circle and spiral.

means: *neusis*, I, II, V, VI.

method: analytical use of the *symptoma*, synthesis.

format: problem.

reception/historical significance: seventeenth century studies on the properties of algebraic curves: Vieta, Descartes and the ensuing development/discussion (*inter alia:* Sluse, Heuraet, Hudde, Newton).

embedding in *Coll.* IV: motif "motion curves and *symptoma*-mathematics": Props. 19, 26, 30 and meta-theoretical passage; motif "analytical determination of curves via *symptoma*": Props. 28/29; motif "*neusis*": Props. 31–33, 42–44, meta-theoretical passage; motif "author Nicomedes": quadratrix (Props. 26–29, perhaps also 35–41). The conchoid is not used again in *Coll.* IV; however, the *neusis* in 31 can easily be constructed with it.

purpose: illustrate determination of motion curves via analytical determination of *symptomata*, and illustrate the operation with such a *symptoma*-curve; the conchoid is transitional between the straightforward motion approach in Archimedes (Props. 19–22, 30), and the analytical *symptoma*-characterization in Props. 28 and 29.

literature: On Nicomedes and his mathematical achievements see Knorr (1986, pp. 219–233); on Arabic sources for cube duplication and angle trisection see Knorr (1989, pp. 63–70, 77–129, and 247–372), and on their ancient sources, see Heath (1921, I, pp. 238–240 and 260–262, and II, pp. 197–199). The most extensive ancient source on ancient cube duplications, *Eutocius in Arch Sph. et Cyl.* II, pp. 54–106 Heiberg, also contains a treatment of the conchoid and the cube duplication with it (pp. 98–104 Heiberg). It was probably more influential for the reception of the conchoid in seventeenth century mathematics than Pappus' account here, or the parallel in *Coll.* III (pp. 58–64 Hu) and *Coll.* VIII. Of special interest, because demonstrating close acquaintance with Pappus' account, are Newton's considerations in Whiteside (1972, II, pp. 196–201, V, pp. 460–465) (within a larger context of treating "solid" problems via construction of equations, pp. 420–495).[1]

5.3.1.1 Nicomedes

This mathematician lived in the third/second century BC – after Archimedes and before Apollonius. Most of his work is lost. He studied the quadratrix (Props. 26–29, perhaps also Props. 35–41), worked on the analytical justification of higher curves (Prop. 29), and wrote a treatise on the properties of conchoid lines. His treatise on the conchoids is lost, but a fragment from it, dealing with the "first conchoid," usually simply called conchoid, is preserved (Prop. 23–25, but see also the references given above). Nicomedes may have known (some version of) the conchoid of a circle, but this cannot be ascertained. From what we can see in *Coll.* IV, Nicomedes appears to have developed the *symptoma*-mathematics of motion curves along the second path mentioned in the introduction to Props. 19–30. Unlike Archimedes, who experimented with the approach using idealized motions as

---

[1] Newton explicitly refers to Pappus several times. In the above-mentioned text, he voices his preference for the conchoid for solid *neusis* constructions and attempts to portray himself as in line with the ancient geometrical tradition, as against the Cartesians. Descartes' discussion of the conchoid in Descartes (1637) does not rest on Pappus, and neither does his construction for the cube duplication and the angle trisection (for Descartes' discussion of the conchoid, and some examples from the Cartesian tradition in the seventeenth century see the references below, under the heading Conchoid).

metaphors, Nicomedes worked toward a justification of such curves via an analytical reduction of their "defining" *symptoma* to loci on surfaces.[1] Prop. 23–25 seem to move somewhat in that direction; a full-fledged example for such a quasi-definition via analysis on loci is given in Props. 28 and 29. While the author of Props. 28 and 29 cannot be identified with certainty, Nicomedes is at least a possible candidate for Prop. 29, and at any rate this is the branch of mathematics on which Nicomedes and others during that time period centered their research activities (for further names, see the meta-theoretical passage after Prop. 30). Above all, however, Nicomedes is known for his treatment of the conchoid line. The conchoid has a *neusis*-property as *symptoma*, and can be used for the duplication of the cube (indeed: multiplication in any given ratio) and the trisection of the angle. In Props. 23 and 24, Pappus presents the solution for the former. A few remarks on Nicomedes' conchoid, on cube duplication in ancient mathematics, and on *neusis* constructions may be helpful.

5.3.1.2   Conchoid

In Pappus' excerpt, the conchoid is generated via motions, but in such a way that it is really a locus-property throughout that characterizes it (see below, *genesis* of the conchoid, and the translation[2]). Nicomedes' "definition" lingers between a *genesis* via motions as in Prop. 19 ("mechanical"), and a definition as a locus curve answering to a *symptoma*. After the description, Pappus mentions that the curve can also be traced by means of an instrument, a kind of *neusis*-compass, which was devised and described by Nicomedes already.[3] The compass "materializes," as it were, the defining locus property of the curve, and the tracing implies, of course, a motion. Note that, as discussed in the introduction, generation with instruments (ὀργανικῶς) is not the same as "mechanical" (μηχανικῶς)[4] generation. In modern terminology, the conchoid is an algebraic curve of order 4, in polar coordinates: $\rho = b + a \sec \theta$. It is indeed a "higher" curve in comparison to circles

---

[1] The references in Proclus, quoted above, fit nicely with such an interpretation. Again, this issue cannot be pursued here. Perhaps it can be investigated further on the basis of the material in Pappus.

[2] For a different assessment cf. Knorr (1986, p. 31) (mentioned above).

[3] For a description of Nicomedes' compass, cf. Eutocius *in Arch., Sph. et Cyl.*, 98, 12–100, 14 Heiberg.

[4] Note also the close analogy to Descartes' discussion of generative motions in geometry in Descartes (1637, pp. 315ff) (40ff Smith/Latham). Descartes may have used the conchoid as reported in Pappus for his discussion of mechanics vs. geometry, and the use of controlled successive motions, with special focus on instruments. His evaluation suggests that Pappus rejects the conchoid and all the mathematics on it, because motions are used for generation, whereas he should have rejected only the quadratrix and spiral, as using uncontrollable composite motions. As stated in the introduction to 19–30, Pappus in fact accepts all these curves as representatives of geometry of the linear kind.

and conics, i.e., it can solve problems that cannot be solved with either of these. Pappus' position on the conchoid and its *symptoma*-mathematics is as follows. The line itself is fully accepted as a mathematical curve of the "linear" kind, and so is its characterizing *neusis*-property. Pappus even lets the *symptoma*-theorem on the two mean proportionals stand uncontradicted in *Coll.* IV – qua *symptoma*-mathematics. So far, his portrait is consistent and convincing. The curve really belongs to the "linear" kind in Pappus' classification. A problem arises, however, in connection with the use of the conchoid for the construction of the cube multiplication. The *neusis* for it is solid, it can be constructed via conics; therefore, using the conchoid for it violates Pappus' homogeneity criterion (cf. below, meta-theoretical passage). It is subject to the same kind of criticism Pappus voices against Archimedes and Apollonius in the meta-theoretical passage. They used a solid argument where a plane one would have sufficed. Yet Pappus did not object to Nicomedes' cube multiplication via the conchoid, even though he explicitly rejects the use of the conchoid for cube duplication in *Coll.* III as a violation of the homogeneity principle. Pappus may have thought the cube multiplication was a "linear" problem, because it was more general than the cube duplication, which he knew to be "solid." In Prop. 35, he will make a transition from angle trisection, characterized as "solid," to general angle division, which he labels as "linear" (correctly so). If he thought the cube multiplication was "linear," that would explain why he does not criticize the use of the conchoid for it.[1] Perhaps one should not assume an author's ignorance lightheartedly. But in this case here, the assumption of a slip, or error, on Pappus' part would yield a straightforward explanation for the omission of the conchoid's failure to meet the homogeneity requirement, in the very passage leading up to his formulation of that criterion. And even if Pappus' portrait of the *symptoma*-mathematics of the conchoid were marred by this error of putting it to use for a problem that is not appropriate, it should be pointed out that cube duplication/multiplication is used in *Coll.* IV mostly as a kind of anchor, or motivator. The focus of Props. 23–25 in fact is to give an illustration of the *symptoma*-mathematics of motion curves in the generation after Archimedes, shifting toward their characterization via analysis of loci. This move proved fruitful in the future, and the conchoid as presented in Props. 23–25 is a classic in this regard, an important step in a developmental line.

---

[1] Another possibility would be that Pappus did after all think that the mathematics of motion curves is not real mathematics, therefore its "unmathematical" use for solid problems does not make any difference any more. One might perhaps think so, because he also mentions that instead of the conchoid one could, for practical purposes, just as well use a marked ruler and proceed by trial and error until it fits (similar remarks can be found in *Coll.* III for "unmathematical" *neuses*, and in *Coll.* VIII). This might be taken to indicate that Pappus accepts neither as mathematically exact. In view of the general tone in Props. 19–30, with regard to the *symptoma*-mathematics of the motion curves, I would nevertheless refrain from that inference and resort to the hypothesis mentioned above.

Pappus mentions four types of Nicomedean conchoids, without giving any further descriptions. The most likely candidates are the other branches of the ordinary conchoid.[1] Nicomedes' work was known perhaps as late as Eutocius, although it is also possible that Eutocius did not have direct access to Nicomedes' treatise. No other reception in antiquity is documented. Proclus' portrait in *Procl. in Eucl.* passim (deriving from Geminus) suggests that Nicomedes must have been of considerable importance for the branch of mathematics dealing with *symptoma*-curves.[2]

The conchoid attracted much attention in the seventeenth century. A few examples include:

Conchoid in Vieta (1593), Supplementum geometriae[3]

Generation of curves via idealized motions and instruments, Descartes[4]

Properties of the conchoid: asymptotic behavior, tangents, points of inflection, local maximum: de Sluse (1668), appendix; also Descartes and the commentators in Descartes (1659): Hudde, Heuraet, Huygens[5]

Newton, Arithmetica universalis: operational advantage of the conchoid over conics for *neusis* – constructions[6]

LaHire and Réaumur also studied this algebraic curve.[7] In the eighteenth century, compasses for general conchoids were designed by Suardi (1752) and Gaetani, taking their queue from Nicomedes' conchoid compass.[8] An example for the discussion of the properties of the conchoid with the means of the calculus is Witte (1813). Let us now return to the problem to which the conchoid is applied in Props. 23 and 24.

---

[1] Cf. Ver Eecke (1933b, pp. 186–187), Heath (1921, II, p. 240). Other suggestions for the other conchoids include the conchoid of the circle (Cantor 1896; Knorr 1986, p. 220) and the hastaria specifically in Curtze (1874).

[2] References to Nicomedes and the conchoid are found, *inter alia* on pp. 110–113, 177, 272, and 356 Friedlein.

[3] Cf. Vieta 1970, reprint of Schooten's 1646 edition). Especially close to *Coll.* IV are propositions 19–28 and 31–38 there; see also Hofmann (1990, I, pp. 343–366) on Vieta and *neuses*.

[4] Descartes (1637, pp. 315–327); (40–58 Smith/Latham *inter alia* generalized mesolabum, shifting parabola); Descartes ed. Schooten (1659, pp. 19–25) (shifting parabola, conchoid). According to Whiteside (1972, V, p. 474, #700), Descartes' s construction is modeled on Menaechmnus', and an earlier draft from 1636 exists.

[5] Descartes (1637, pp. 351/352 (113–115 Smith/Latham), normal to the conchoid), Descartes ed. Schooten (1659, pp. 246–265) on the conchoid, contributions by Hudde et al.

[6] Within Whiteside (1972, V, pp. 420–495), see especially pp. 454/456, 426/428, 432, and 454/456, including the following remark: "constructionem per Conchoidem praefero ut multo simpliciorem et non minus geometricam & quae resolutioni aequationum a nobis propositae optime conducit." (p. 432) Newton's cube duplication p. 456 is essentially the same as Nicomedes'. Whiteside (1972, II, pp. 460–465) contains several constructions for two mean proportionals and angle trisection, one of which is closely parallel to Props. 23 and 24.

[7] See Chasles (1875). These authors investigated generalized Conchoids with the methods of analytical geometry along the lines proposed by Descartes.

[8] See von Braunmühl (1892); Suardi developed ingenious drawing devices, including one for the conchoid of a circle.

5.3.1.3   Cube Duplication: Two Mean Proportionals

To find a cube the volume of which is double the volume of a given cube was one of the three classical problems in ancient geometry (doubling the cube, trisecting the angle, and squaring the circle). Around these three problems a good deal of the development of ancient geometry can be aligned. As motivations, they guided the development of mathematical methods and theories.[1] Specifically, the cube duplication and the angle trisection were among the factors promoting the development of the theory of conic sections. Pappus, who means to give, in *Coll.* IV, a survey of classical Greek geometry from a methodological point of view (my thesis), includes these three problems, also. Angle trisection will come up in Props. 31–34 (solid problem, with generalization into the linear problem of angle division in a given ratio in Props. 35–38), and the quadrature of the circle will come up in Props. 26–29 (linear problem by nature). Cube duplication, or rather: finding two mean proportionals and constructing a cube in a given ratio, is the subject of Props. 23–25. The ancient mathematicians were able to provide a mathematically exact constructive solution for the angle trisection and the cube duplication, with a *neusis* that later turned out to be reducible to a construction via conic sections. Through Apollonius' work in analysis-dominated geometry, the nature of the problem was finally settled. Apollonius' solution via conics has not survived, but a very closely related construction is presented for the angle trisection in *Coll.* IV, Props. 31–33. In fact the same *neusis* constructed in Props. 31–33 can be used for Prop. 24, and it then yields two mean proportionals. The Apollonian "solid" construction for the cube duplication and the angle trisection was, as it were, the endpoint of a discussion that had lasted from 450 BC to ca. 150 BC. Perhaps a few names and highlights in this development are worth recalling here.[2] Hippocrates of Chios reduced the cube duplication to finding two mean proportionals.[3] Archytas made use of three-dimensional rotation figures and a torus.[4] Eudoxus developed a special (*symptoma-?*) curve in the plane, probably identical with the Hippopede, and derived from Archytas' curves in space for the solution.[5] A mechanical solution that also relies

---

[1] Knorr (1986) even tells his story of the history of ancient mathematics as a story evolving around solving these problems, and they were singled out in Heath (1921, I) as a separate chapter in his otherwise chronological account centering on authors; for the cube duplication see pp. 244–270 there.

[2] For a survey in secondary literature see Heath (1921, I, pp. 244–270) and Knorr (1986) passim. The most important ancient source is, as said above, *Eutocius in Arch. Sph. et Cyl.* pp. 54–106 Heiberg.

[3] Heath (1921, I, pp. 200/201). This description is to be taken with caution. See Netz (2004).

[4] Heath (1921, I, pp. 246–249), Knorr (1986, pp. 50–51); Eutocius pp. 84–88 Heiberg. Note the motivic connection to the attempt, in Props. 28 and 29, to determine the curve "quadratrix" from loci on surfaces that are created as intersections of rather similar surfaces in space (see below).

[5] The solution does not survive. If the curve resulted from orthogonal projection of Archytas' curves (as Tannery suggested), it would be a prototype for the curves determined analytically in Props. 28 and 29 (see below).

heavily on the manipulation of a physical instrument is ascribed to Plato. It is
not genuine, but it is in all likelihood pre-Euclidean.[1] Menaechmus used conic sections, viewed as locus curves with *symptoma*.[2] Archimedes (no solution by him
survives) seems to have used a *neusis* very much like the *neusis* in *Lib. ass.* VIII
for angle trisection and cube duplication. This connection is drawn in an Arabic
source. The *neusis* used for angle trisection in another Arabic source is in fact
closely associated with *Lib. ass.* VIII, and with the *neuses* in *SL* 5–9.[3] Eratosthenes
devised a quasi-compass, the mesolabum.[4] Nicomedes, in an attempt to improve
on Eratosthenes, solved the problem with the conchoid. Diocles' solution with
the cissoid is in some respects methodologically similar to Nicomedes' conchoid
solution.[5] Finally, Apollonius gave a systematic analytical treatment of plane versus
solid *neuses*, differentiated into kinds (gene): *neuses* for plane problems use only
circle and straight line, *neuses* for solid problems use at least one conic section in
addition.[6] The *neusis* for angle trisection and cube duplication is solid. Apollonius
also gave a construction of this solid *neusis*. Apollonius' contribution is lost.
Since *Coll.* IV, Props. 31–33 draw on Apollonius' treatment of conics, and proceed
via analysis-synthesis, Knorr has argued for a close connection between Prop. 33
and Apollonius' solution.[7] At any rate, a solution along the lines of Props. 31–33
(or one of the alternatives in Prop. 34) settled the question as far as ancient geometry
was concerned.

   The ancient geometers' sustained interest in the two problems on the cube
and on the angle was not just a fancy idea. Although the ancients may not have
known this explicitly, these two problems are really fundamental in the following
sense. All problems that can be solved via circles, lines, and conics (that is: all
solid problems) reduce to either the angle trisection or the cube duplication.

---

[1] Cf. Eutocius pp. 56–58 Heiberg.

[2] Cf. Eutocius pp. 78–84 Heiberg; note the possible connection between Menaechmus' handling
of his locus curves (they turned out to be conic sections) and the attempt to utilize more general
curves, described and describable solely through their *symptoma*. See also Jones (1986a,
pp. 573–577), Knorr (1986, pp. 61–66, 1989, pp. 77–129), and Zeuthen (1886, pp. 455–469).

[3] Cf. Knorr (1986, p. 221 f.); compare also Hogendijk (1986), and the discussion of angle trisection
below.

[4] Cf. Eutocius pp. 88–96 Heiberg; see Heath (1921, I, pp. 258–260), Zeuthen (1886, pp. 455–469),
and Knorr (1986, pp. 61–66). A discussion can also be found in *Coll.* III, 56–58 Hu. The mesolabum construction quite obviously influenced Descartes' invention of a compass for the construction of proportions in the *Géométrie* (Descartes 1637, pp. 317–319 (44–49 Smith/Latham)), and
it may very well have inspired Sluse book of the same title.

[5] Cf. Eutocius, pp. 66–70 Heiberg.

[6] Cf. Jones (1986a, pp. 527–534) on Apollonius' *neuses*.

[7] Cf. Knorr (1982, 1986, pp. 302, 305–308). In these contributions Knorr argued for Apollonius as
the direct author of Prop. 33. In Knorr (1989), he favors a somewhat less immediate connection,
which is perhaps preferable.

This was shown by Vieta (Supplementum Geometriae, pp. 240–257 of the 1646 Schooten edition[1]). It is quite possible that the ancients had an inkling that the two classical problems already exhausted all solid construction problems. But even if they did not come to formulate such a suspicion or hypothesis, they could not have but noticed that in point of fact solid problems regularly did reduce to these two.[2]

5.3.1.4 Neusis

As said above, cube duplication reduces to the finding of two mean proportionals, which in turn reduces to a *neusis*. What, then, is a *neusis*, and what is its general significance in Greek geometry[3]? A *neusis* is a construction in which one has to insert a line (usually of a given length), between two given lines (usually a straight line and a circular circumference), so that it meets, when produced, a certain point: it verges toward that point (Greek verb: neuein). Ancient examples for *neuses* include:

(i)   *Neusis* in Hippocrates' third lunula-quadrature[4]
(ii) Archimedes, *SL* 5–9, the *neusis* for angle trisection in Lib. ass. VIII, and the one used for his construction of the regular heptagon[5]
(iii) Apollonius: work on plane *neuses*, lost; a commentary by Pappus survives[6]
(iv) Examples for *neuses* in Pappus, *Coll.* IV: Props. 23 and 24 (Nicomedes), Props. 31 and 32 (probably pre-Apollonian, see below, with post-Apollonian justification via conics: Prop. 33), Prop. 34, and Props. 42–44 (Archimedes, and perhaps Aristaeus)

The ancient *neuses* that are still extant can all be constructed either with circle and straight line, or with conics. This is due to the fact that the two lines chosen for

---

[1] For his proof, Vieta made use of his new algebraic techniques. Even so, his procedure is astonishingly close to *Coll.* IV in its general set-up. The same result was shown also by Fermat, and again by Descartes (1637, pp. 389–402) (193–219 Smith/Latham). Compare also Newton's treatment in Whiteside (1972, V, pp. 420–491), mentioned above. Similar results had been found earlier by Omar Kayyam, who may have had an impact on European mathematics in the Middle Ages, and by Raffael Bombelli in the geometrical part of his Algebra. The latter remained unpublished, however, until it was rediscovered by Bortolotti in the 1920s (cf. Bortolotti 1923,1929, pp. 265–267), and cannot have had much impact on geometry and algebra in the sixteenth/seventeenth century.

[2] Compare Pappus' remarks on the analysis of the solid *neusis* in Props. 42–44; cf. also Jones (1986a, pp. 527–530) for attempts towards classification made possible through Apollonius' work.

[3] Cf. Zeuthen (1886), Heath (1921), Knorr (1986), RE Suppl. IX (1962), col. 415–461 on this topic; cf. also Jones (1986a, pp. 527–534) (on Apollonius' work with that title).

[4] On the Hippocrates fragment see Simpl. in Phys. 61–68 Diels, Heath (1921, I, pp. 183, 195–196); Knorr (1986a, pp. 32–34), Netz (2004).

[5] Knorr (1978b, 1986, pp. 178–187).

[6] *Coll.* VII, pp. 770–820 Hu (Jones 1986a, pp. 196–229, # 120–157, with comments pp. 527 ff.).

insertion are either two straight lines, or a straight line and a circle. Zeuthen's account of the role of *neuses* in Greek geometry has largely been accepted. What follows, is a brief summary of it.[1] According to Zeuthen, the *neusis*-operation historically started out as a legitimate standard argumentative device in mathematics: no construction of it was needed for justification, just like there is not one needed for circles. In actual practice, *neuses* were probably accomplished by means of a marked ruler; this procedure was in all likelihood still a valid device at the time of Archimedes, and that is why *SL* 5–9 give no argument for the construction of the *neuses*. It was Apollonius' analytical work which enabled geometers to differentiate between plane and solid *neuses*, and to construct them, as separate entities, via analysis-synthesis. Apollonius obviously did address the *neusis*-operation as something that can, and should, be constructed from simpler entities, and through his contributions, the *neuses* lost their status as simple devices. From then on, they had to be constructed explicitly, and there were distinct different types. Specifically: there were plane *neuses*, which had to be constructed by plane means, and solid *neuses*, which required at least one conic.[2] As said above, Zeuthen's account was mostly accepted, and it is indeed quite plausible, although no explicit ancient testimony exists on the status of the *neuses* before Apollonius. Assuming this rough outline of the history of *neusis* as a construction device in ancient geometry, Nicomedes' conchoid appears to hold a kind of middle ground. Nicomedes lived after Archimedes, and before Apollonius; his curve is really a *neusis*-curve, and his attempt to define the curve pointwise, analytically, could perhaps be seen as a step toward "mathematizing" the *neusis*-operation.[3]

5.3.2   Prop. 23: *Genesis* and *Symptoma* of the Conchoid

Genesis: a straight line AB with perpendicular CDE is given; C and E are *given* in position[4]; CE moves along AB, while E remains fixed; D is always the point of intersection with AB, and the segment DC remains fixed in length; C describes a conchoid

Symptoma: every straight line drawn between E and the curve cuts off a segment of the same length between AB and the curve (neusis-property).

---

[1] Cf. Zeuthen (1886, especially pp. 261–265/ 269–272).

[2] Cf. Heath (1921, I, pp. 235–241, II, pp. 65–68) (Archimedes, *SL* 5–9), 189–192 (Apollonius' work on plane *neuses*), Heath (1926, I, pp. 150/151), Dijksterhuis (1987, pp. 138/139) (Archimedes), Knorr (1986, pp. 365 ff.) and passim (essentially repeating Zeuthen's arguments). A classic contribution, addressing also a differing opinion on ruler and compass in *neuses* as early as Hippocrates, is Steele (1936). According to Jones (1986a, p. 530), the identification and exhaustive construction of all plane *neuses* may have been the very purpose of Apollonius' *Neuses*.

[3] For a comparable assessment cf. Knorr (1986, p. 303); for a different assessment, see Knorr (1989, p. 31), however.

[4] Nicomedes/Pappus is indeed using technical terminology from geometrical analysis.

The curve's *genesis* is via motions (as in Prop. 19). The *symptoma* is, however, not directly derived from the motions used (as in Prop. 20). Rather, Nicomedes' derivation is quasi-analytical, and presupposes the curve as drawn, considering it pointwise. As said above, this is a step toward "defining" higher curves analytically as *symptoma*-curves.[1]

Further information, from Nicomedes' treatise, is only listed in Pappus. The curve created in the above *genesis* is called "first conchoid"; the line AB "ruler," the point E pole, and the line segment DC "distance." The perpendicular from the curve onto AB has a maximum in CD, perpendiculars closer to CD are larger than those further away; the curve is asymptotic with respect to AB. Nicomedes discussed a total of four types of conchoids. After listing some of Nicomedes' results, Pappus also adds: the curve can be described by means of an instrument, which Nicomedes also devised. It can be used to double the cube, and to trisect the angle, as Pappus says he himself has done. The conchoid also yields the cube multiplication (a generalization of the duplication). For practical purposes, says Pappus, one might just as well use a marked ruler to find the *neusis*.

5.3.3 Prop. 24: Two Mean Proportionals via *Neusis*

There is a slight problem with the authorship for Prop 24. If we emend the text as Hultsch did, Pappus claims that Nicomedes gave only the construction, and that he himself supplied the proof. Since Nicomedes' work was still around at the time, this could not really have been an outright falsehood. Yet Eutocius (independently from Pappus?) reports the very same proof almost verbatim, and strongly suggests, though he does not say, that it is Nicomedes' own.[2] Does our proof stem, essentially, from the second century BC, or rather from Pappus, i.e., from the fourth century AD? Like most scholars, I think the essence of Prop. 24 belongs to Nicomedes and does illustrate "Nicomedean" mathematics. In what follows, I offer a brief summary of attempts to explain Pappus' introductory sentence. Perhaps Nicomedes gave only the analysis and kataskeue explicitly, leaving the synthesis to the reader, and Pappus merely spelled it out. Ver Eecke (1933b, p. 188) assumes that Eutocius was quoting from *Coll.* IV, but this leaves the authorial claim on Pappus' part unaccounted for. Jones (1986a) considers the possibility that Eutocius may be drawing not on *Coll.* IV, but on a report in *Coll.* VII. Perhaps (as suggested by Knorr 1989, pp. 65 ff.) Pappus wishes to claim authorship for certain minor intermediate steps only, those that make the connection to the *Elements* explicit. In fact, the close connection to Euclid I–VI does suggest some editorial input on Pappus' part. Another explanation results from restituting the transmitted text "monen" for "monon." Then the problematic sentence in Pappus could be understood as stating that Nicomedes provided a single construction (for both angle trisection and cube duplication), whereas he himself is going to

---

[1] Cf. Jones (1986a, p. 529), Knorr (1986, pp. 219–222).
[2] Eutocius *in Arch. Sph. et Cyl.* 104–106 Heiberg.

excerpt and present that part of it that concerns two mean proportionals for cube duplication. Then the apparent contradiction is diminished, and Nicomedes is acknowledged as the source for Prop. 24. The degree to which Pappus edited his source cannot be determined with certainty, unless one can find evidence for Eutocius' independence from Pappus. I will treat Prop. 24 as essentially Nicomedean.

The proof protocol for Prop. 24 will be given in some detail, to illustrate the fact that Nicomedes' methods, within *symptoma*-mathematics, were quite "standard," unlike, e.g., Archimedes' procedure in Prop. 21, and correspond in scope to the means employed in Apollonius' analytical works.

5.3.3.1   Proof Protocol Prop. 24

1. Protasis
   Find two mean proportionals for CL, LA

2. Ekthesis/Kataskeue
   Rectangle ABCL, construct, M, Z, K, TK
   CK, MA solve the problem, i.e.: CL:CK = CK:MA = MA:AL

3. Apodeixis:
   Show that BM:BK can be expressed in three ways
   3.1 MB:BK = CK:MA

$BK \times KC + CE^2 = EK^2$	[II, 6]
$BK \times KC + CZ^2 = KZ^2$	[I, 47]
MA:AB = ML:LK	[VI, 2 with V, 16]
ML:LK = BC:CK	[VI, 2]
MA:AB = BC:CK;	
MA:AD = HC:CK; HC:KC = ZT:TK	[VI, 2]
MA:AD = ZT TK;	
MD:DA = ZK:TK	[V, 18]
DA = TK $\Rightarrow$ MD = ZK	[V, 9]
$MD^2 = BM \times MA + DA^2$	[II, 6]
$BM \times MA + DA^2 = ZK^2 = BK \times KC + CZ^2$;	
$DA^2 = CZ^2$	
$\Rightarrow BM \times MA = BK \times KC$	
$\Rightarrow$ BM:BK = KC:MA	[VI, 16].
3.2 MB:BK = LC:CK	[VI, 4]
3.3 MB:BK = MA:AL	[VI, 4]

4. The equations 3.1–3.3 establish
   CL:CK = CK:MA = MA:AL.

5.3.4   Prop. 25: Cube Multiplication in a Given Ratio

Set the ratio out as a:b. Via Prop. 24, construct c, d so that a:c = c:d = d:b. Then $a:b = (a:c)^3 = a^3:c^3$.

Note that triple ratios are here identified with ratios of cubes. V, def. 11: $(a:c)^3 = a:b$. According to XI, 33, the cubes stand to each other in the same triplicate ratio.

5.4   Props. 26–29: Quadratrix/Squaring the Circle

5.4.1   General Observations on Props. 26–29

5.4.1.1   Structure of Props. 26–29

Genesis and *symptoma* of the quadratrix as a motion curve
Sporus' criticism of the quadratrix (specifically of the *genesis*)
Props. 26 and 27: *symptoma*-mathematics of the quadratrix: rectify and square the circle.
Props. 28 and 29: geometricize the *genesis* of the quadratrix via analysis on surfaces.

context: motion curves and *symptoma*-mathematics, squaring the circle.
sources: Nicomedes or Dinostratus on quadratrix, Sporus' *Aristotelian Wax Tablets* for criticism of the *genesis*, Nicomedes (?) for exhaustion proof in Prop. 26[1]; unknown sources for Props. 28 and 29 (Apollonius? Nicomedes?).
means: I, II,V,VI, *Circ. mens.* 1 for Props. 26 and 27;
no recourse to the *Elements* in Props. 28 and 29.
method: exhaustion proof, synthetic (Prop. 26), analysis (Props. 28 and 29).
format: non-uniform: *genesis* and *symptoma* is descriptive; Sporus' criticism is an excerpt from a philosophical refutation argument, rhetorically styled; Prop. 26 is a theorem, Prop. 27 a problem; Props. 28 and 29 give an analytical determination of a curve.
reception/historical significance: the quadratrix was much discussed in the seventeenth century, as an example for a non-geometrical, or a transcendent, curve.
embedding in *Coll.* IV: connection to the plane spiral (Prop. 19): Props. 25, 26 and 29, motif "author Nicomedes": Props. 23–25; motif "*genesis* via synchronized motions": Props. 19, 30; motif "linear problems and *symptoma*-mathematics of the quadratrix": Props. 35–41; motif "analytical interpretation of the *symptoma*": Prop. 23.
purpose: exemplary illustration of the problems, and the mathematical potential of curves of the third (linear) kind.
literature: Heath (1921, I, pp. 225–230), Knorr (1986, pp. 80–88; 226–233; 166–167) for Props. 26 and 27; on Props. 28 and 29 and its context of analysis on surface loci see Heath (1921, I, 439–440; II, pp. 380–382), Knorr (1978a, pp. 62–66, 1986, pp. 129 and 166–167), Jones (1986a, pp. 595–598), Ver Eecke (1933b, pp. 197–201), Chasles (1875, pp. 30–37) and Notes VIII. *Coll.* VII, pp. 1004–1014 Hu (Jones 1986a, pp. 362–371) contain Pappus' commentary on Euclid's work in loci on surfaces (probably conics). The work he comments on probably rested on prior contributions by Aristaeus.

5.4.1.2   Authorship for Props. 26–29

Pappus mentions both Dinostratus and Nicomedes as authors that used the quadratrix in connection with the squaring of the circle. For biographical information on

---

[1] Pappus himself can be excluded as the author of Prop. 26, because he explicitly says he is reporting.

Nicomedes see above, introduction to Props. 23–25. Dinostratus was the brother of Menaechmus (inventor of conic sections). He lived ca. 350 BC and may very well have been a pupil of Eudoxus. We know almost nothing about his mathematical work outside of our passage in *Coll.* IV, so what will be said here is to some degree speculative.

Pappus clearly associates both him and Nicomedes with the use of the quadratrix for squaring the circle. Since the fifth century BC sophist Hippias is mentioned elsewhere as the inventor of the curve itself, perhaps Hippias used the curve for the angle trisection (indeed: arbitrary division), while Dinostratus discovered its rectification property, and possibly proved it with the Eudoxean method of exhaustion (not Prop. 26, however). After Archimedes, due to *Circ. mens.* I, the quadratrix could then have been employed by Nicomedes to square the circle (Prop. 27). Perhaps it was Nicomedes, also, who is responsible for the proof of the rectification property in its present form. The proof of Prop. 26 as given in Pappus is almost certainly post-Archimedean, because it relies implicitly on a theorem that is equivalent to Arch., *Circ. mens.* 1 (see proof protocol below). Furthermore, Nicomedes could have pursued the analytical *symptoma*-approach to the properties of the curve, and be at least partially responsible for the analytical reduction of the quadratrix, viewed as *symptoma*-curve, to Archimedes' spiral (Prop. 29). In analogy to the conchoid, which is determined pointwise as a kind of *neusis*-curve, the quadratrix can be seen as a curve corresponding, at each point, to the correlation of the very same rotation + linear motion used in the plane spiral (when it is inscribed in a circle), and such a perspective leads to the way the quadratrix is characterized in Prop. 29. On this view, the contribution of Dinostratus is substantial, but Nicomedes would be the one who developed the *symptoma*-mathematics of the curve theoretically, as a member of the class of higher curves.

This is, as it were, the maximum option for Dinostratus' and Nicomedes' achievements in relation to the quadratrix. I am putting it forth tentatively. It has the advantage of fitting well with a sympathetic reading of Pappus' text, and it can account for the fact that in most later sources, it is only Nicomedes that is associated with the study of the properties of the quadratrix as a higher curve.[1] In what follows, I will briefly sketch two alternative views.

Heath (1921) would like to ascribe much of the mathematics on the quadratrix to Hippias already, including the discovery of the rectification property and its proof via exhaustion. Perhaps this is a little too optimistic for Hippias.[2] It would place a considerable amount of mathematical theory and expertise already in the fifth century BC, and is therefore somewhat unconvincing. Knorr (1986) has argued a different view. He denies that Hippias could have had anything to do with the

---

[1] Cf. Iamblichus apud Simpl. in Cat. 192 Kalbfleisch, 645 b Brandis. Procl. in Eucl. 272 Friedlein mentions only Nicomedes as well, but not Dinostratus, although Proclus must have known about him. His name had been mentioned earlier, in the catalog of mathematicians derived from Eudemus.

[2] Hippias would be credited with an expertise in handling the exhaustion method that is in this form usually associated with Eudoxus, who lived considerably later than Hippias, and this reading also would leave no room, as it were, for Dinostratus.

curve, because it involves a considerable degree of mathematical sophistication and expertise. Rather, according to Knorr, Dinostratus may have invented the curve and discovered its rather obvious angle division property (i.e., he ascribes to Dinostratus what others had ascribed to Hippias). Because the *genesis* of the curve via motions has strong affinities with the Archimedean plane spiral (inscribed version), Knorr believes that in the generation after Archimedes this connection was made use of, by Nicomedes probably, to formulate, and to prove the rectification property of the curve (Prop. 26 entirely), whence the quadrature of the circle follows as a corollary thanks to *Circ. mens*. I. The rectification property is implicit in the curve, but unlike the angle section property, it cannot be read off directly. It seems plausible to assume that the latter was discovered after the former. As for Props. 28 and 29, Knorr envisages Apollonius as a possible source for Prop. 28 (because of the central role of the Apollonian helix therein), and associates Prop. 29 with Archimedes.

Perhaps this is a bit too pessimistic with regard to Hippias and Nicomedes. There seems to be no compelling reason to discard the unequivocal testimony that Hippias invented the curve itself. How much he knew about it may be uncertain, but the angle section property is indeed easily deduced. Archimedes seems unlikely as a source for a predominantly analytical investigation of motion curves as in Prop. 29. In his other works, and in his heuristic method, Archimedes shows no preference for the analytic approach. His interests point rather in the direction of quasi-mechanical methods, and perhaps infinitesimals. Prop. 29 appears not to be Archimedean in style. More plausible is the connection between Prop. 28 and Apollonius, because Apollonius did in fact favor, and develop, the analytical approach in geometry, and is said in Iamblichus to have called the quadratrix "sister of the cochlias." This some-what enigmatic statement could be read as a description of Prop. 28, and then Apollonius could be its author.[1] Finally, Apollonius may have written a treatise on the helix. The evidence on such a treatise is slim, however. I prefer to refrain from ascribing the substance of Prop. 28 to him directly, while supporting the claim that Prop. 28 is well in line with higher mathematics, Apollonian style, i.e., in a tradition developing an approach to mathematics that is exemplified in Apollonius. As said above, I am inclined to assign to Nicomedes the leading role in the shaping of Props. 26 and 27, and to consider a substantial contribution to Prop. 29 on his part as a distinct possibility. Furthermore, I am of the opinion that the upcoming minor results on the *symptoma*-mathematics of the quadratrix in Props. 35–41 may derive from his treatise on the quadratrix as well.

5.4.1.3   Quadratrix

As outlined above, the quadratrix is usually associated with Hippias of Elis (fifth century BC) as its inventor. It was at first used to trisect the angle; in fact it can

---

[1] Iambl. apud Simpl. in Cat. 192, 19–24 Kalbfleisch. Heath (1921, I. p. 225) supports a different view.

divide an angle in any ratio. Later on, it was discovered that it can also be used to rectify the circle, and thus to solve the problem of squaring the circle. This solution via the quadratrix was not accepted as a constructive solution, because in setting out the quadratrix one already has to assume access to the ratio of diameter and circumference of the circle (see below). Nevertheless, Pappus was willing to fully accept the mathematics on, or about, the quadratrix as an example of geometry of what he called the "linear" kind. This applies to Props. 26–29, and to Props. 35–41. The status of the curve itself was left somewhat in limbo by him, and his ambivalent portrait probably contributed to the fact that, in the seventeenth century, the quadratrix was used as one of the primary examples of curves that did not fit the bill of Descartes' definition of a proper geometrical curve,[1] and might therefore be used as either a vantage point to enlarge, or a counter-example in the attempt to delimit the horizon of geometry and analysis.[2]

Props. 26 and 27, with their prefatory detailed description of the quadratrix as a motion curve are our only surviving evidence on this curve (and the squaring of the circle with it) from antiquity. Props. 35–41 show us the use of the quadratrix for the general angle division, and results derived from it, as well as further properties following from the rectification property. Again, those are the only such sources extant from antiquity. Finally, Props. 28 and 29 are our only extant detailed examples for an analysis of loci on surfaces (used here for the geometrical justification/description of the *genesis* of the quadratrix). Obviously, this makes Props. 26–29 (to a lesser extent: Props. 35–41) a document of the highest importance for the history of ancient mathematics. However, the fact that no other ancient source gives such a detailed insight into the discussion of this curve and its geometrical properties, and that there are no traces of parallel accounts on other higher curves, also entails, unfortunately, that we have no context in which to set, and from which to evaluate, Pappus' portrait of the curve and its mathematics. There seems to have been a mathematical community (or just a small group of mathematicians?) who pursued this kind of mathematics for some time (how long? just one generation, or 100 years?). Pappus and others list names. What did these mathematicians think they were doing? What was their view on the status of curves like the quadratrix, and on the *symptoma*-mathematics on them? What was the mainstream view (if such a view existed) on this collection of mathematical treatises? Is Pappus' rational reconstruction of the quadratrix and the degree to which it can be "geometricized" representative, and if so: representative of what? Wherever his pronouncements

---

[1] In Descartes ed. Schooten (1659, pp. 18 and 38), for example, the quadratrix and the spiral appear as the primary examples for non-geometrical curves.

[2] I am not aware of any systematic study of the evidence. Such a project would seem to me to be rather promising, because the amount of available source texts is rather extensive and widespread. The quadratrix does turn up, e.g., in Jacob Bernoulli's papers (he was also interested in spiral lines), and also in Leibniz's mathematical manuscripts. Leibniz borrowed the name and applied it to a more general type of curve with "quadrature" properties. For the Cartesians, the quadratrix was a classic example for a non-permissible curve.

remain vague, should we conclude that he is uncertain or does not understand, or that the discussion had not reached a consensus, or that there was no discussion and the project just died out? Perhaps a detailed comprehensive comparative study of Props. 19–30 and 35–41 in connection with *SL*, taking into account also the scattered summary remarks in other authors (especially Proclus and the commentators on Aristotle) could shed some new light on this issue. It cannot be pursued here. What is given is a documentation of the ancient evidence on the mathematics of the quadratrix and on Pappus' evaluation of it, as far as the full text in *Coll.* IV attests it. Perhaps this material could be the basis for further investigations on a broader scope.

5.4.1.4 Squaring the Circle

The squaring of the circle, i.e., the problem of finding a geometrical construction to transform a given circle into a square, caught the attention of the Greeks very early on. Already in the fifth century BC, they appear to have found a (very elementary) way of transforming any given polygon into a square – II, 14 in the *Elements*, resting on I, 44 and I, 45. It seemed that an analogous procedure to do the same with the circle should be possible. The question captured the imagination of mathematicians and non-mathematicians alike, and it sparked the development of new methods and new theories in geometry, with the goal to become able, among other things, to solve this problem with the new mathematics. Already in Aristotle's time doubts arose as to whether the problem was solvable at all. But the discussion, and the search, continued nevertheless, and it continued beyond antiquity. In a sense, the matter was finally settled only with Lindemann's proof of the transcendence of π in 1882. No construction with means that are equivalent to the solution of an algebraic equation with rational coefficients is possible. One needs infinitesimal methods, or else a curve like the quadratrix. In what follows, a survey of the attested ancient attempts at solving the problem, and a selective list of later attempts and judgments is given.[1]

Hippocrates of Chios, in the fifth century BC succeeded in constructing three out of the five quadratures of lunulae that are possible within plane geometry. One of them is located over a semicircle, one over a segment that is larger than the semicircle, and one over a segment that is smaller. He also squared a figure composed

---

[1] For further information on the ancient quadratures cf. Heath (1921, I, pp. 183–201, 220–235) and Knorr (1986) passim. Knorr is perhaps not always careful in demarcating textual/historical evidence from his own reconstructions. Tropfke (1923, pp. 195–238) provides a survey of attempts to square the circle, focusing on contributions known in Western Europe. It is still valuable for its numerous bibliographical references. Also worth reading, though in some respects outdated, is Rudio (1892), a monograph on the measurement of the circle that prints the major contributions by Archimedes, Huygens, Lambert, and Legendre in full, and also contains a survey on the history of the quadrature.

of a circle and a lunula. His constructions probably were intended as steps toward squaring the circle. As Aristotle points out, they were fully valid mathematical arguments, but they do not square the circle.[1] Still in the fifth century, the sophists Antiphon and Bryson presented arguments that they held to be solutions to the problem of squaring the circle.[2] Both started with an inscribed square and constructed a sequence of polygons approximating the circle more closely at each step. Bryson used, in addition, a corresponding sequence of circumscribed polygons. Both assumed that in the process, the circle is exhausted.[3] Antiphon in effect assumed that the process would be finite, and that the circle coincides with a polygon having very small sides. This polygon can then be squared, via the equivalent of II, 14. His argument was considered as invalid mathematically, because it rests on a non-geometrical concept of a circle.[4] Bryson did not identify the circle with a polygon. He argued that since all the "inside" polygons are smaller than the circle, and all the "outside" ones are larger, there must exist a polygon that has the same area as the circle. Such a polygon could, again, be squared via the equivalent of II, 14. Bryson claimed thus to have squared the circle. His argument rests on the idea that "area," taken abstractly, is a continuous quantity. Though not invalid in itself, it was taken by Aristotle not to be a geometrical argument at all, because it violates his homogeneity criterion.[5] Whether or not the mathematicians would have shared Aristotle's opinion here, it is clear that the argument does not amount to a geometrical solution to the (construction!) problem of squaring the circle. The required square – though Bryson's argument may reassure us that it exists – cannot actually be produced from this argument. Bryson did not square the circle. The problem thus was still unsolved in Aristotle's times, and he uses it frequently as an illustration for failed or as yet unsuccessful attempts in scientific inquiry. In his writings, one

---

[1] Heath (1921, I, pp. 183–201); cf. Simpl. in Phys. 56–68 Diels, Rudio (1907). See also Aristotle on Hippocrates' quadratures in Heath (1970, originally: 1949). The above judgment is taken from Aristotle, but there is no reason to assume that the mathematicians would not have shared his opinion.

[2] Cf. Heath (1921, I, pp. 221–225).

[3] Note that this does not mean that they used the Eudoxean method, the so-called "method of exhaustion." That method is, in its essence, a double reductio argument. Even though it was often used in connection with area and volume theorems, and in a context where a process of approximation is assumed, such a process is not essential to the method as such. That is: although it was used for arguments that we translate into limit arguments – and for others, too – it was in itself not a concealed limit argument. Heath is mistaken in assuming that Antiphon's and Bryson's "quadratures" contain the nucleus of the famous Eudoxean method, even though they may have anticipated infinitesimal procedures to some degree.

[4] This argument, found in Aristotle, again, would in all likelihood have been shared by the mathematicians.

[5] *Anal. Post.* I, 9. Aristotle differentiates between Antiphon's and Bryson's attempts. It is perhaps interesting that he does not reject Bryson's argument in itself as invalid (as he does in Antiphon's case), but rejects it as involving a "katabasis eis allo genos," as being ungeometrical, because of its failure to differentiate between geometrical and other continuous quantities. For Aristotle, arguments in geometry have to address geometrical entities qua geometrical, and not qua something else that they may also be.

also finds the suggestion that the squaring of the circle may be impossible in principle, because straight line and circle are generically different – even though it's clear, on the basis of continuity assumptions, that a straight line with a square equal to the circle must exist.[1]

Alongside the squaring, rectification became an issue soon. Is the ratio of circumference to diameter expressible as a ratio of numbers? With Archimedes's work (*Circ. mens.* I), it became apparent that and how the rectification and the quadrature entail each other.[2] Archimedes' investigation of the plane spiral in the heuristic version could, taken together with a theorem like *SL* 18[3] and *Circ. mens.* I, be used for the squaring of the circle. Archimedes never presented such an argument, and in effect replaced the *genesis* of the spiral with one that avoids the problems the quadratrix and the original (inscribed) spiral have. With the spiral in this description, one can indeed no longer square the circle. Archimedes also provided, by means of logistics, two different approximations for π[4] (*Circ. mens.* II, III). Apollonius, Sporus, and Ptolemy gave further approximations for π, closer in numerical value than Archimedes'.[5] Dinostratus/Nicomedes used the quadratrix for circle rectification, and Nicomedes applied *Circ. mens.* I to obtain a "quadrature" (Props. 26 and 27). As noted above, this is not a constructive quadrature in the sense required, either, because the setting out of the quadratrix involves the ratio of circumference and diameter (essentially π), and that was the equivalent to what was sought in rectification. The squaring of the circle with the quadratrix, insofar as it is geometrical, is purely *symptoma*tic.

5.4.1.5 Circle Quadrature Through the Ages

The following selective list of examples is intended to give an impression of the different results, perspectives, and methods developed within the horizon of

---

[1] Descartes would later repeat that same general statement in Descartes (1637, pp. 340/341) (90/91 Smith/Latham). He took it for granted that the circle cannot be squared, because circle and straight line belong to different, incomparable kinds.

[2] *Circ. mens.* I implies that the problem of squaring the circle can be reduced to the problem of rectifying the circumference; cf. Knorr (1986, p. 159).

[3] *SL* 18 is a *symptoma*-theorem on the spiral with circumscribed circle. It shows that in such a situation, the circumference is equal to the subtangent of the spiral at the endpoint of the first rotation. It does not yield a constructive rectification of the circle. It also does not provide a constructive solution for finding the tangent to a spiral of first rotation (cf. Vieta, *Varia responsa*, including an approximate construction for such a tangent).

[4] The name π was not used by the ancients. It was coined in the early seventeenth century by Ludolph van Ceulen (*Fundamenta geometrica* ed. W. Snel, Würzburg 1615). According to Tropfke (1923, p. 232), its first occurrence is even later: 1706, in Jones' *Synopsis palmariorum matheseos*. The label refers to a number, in modern terms. The ancients had a very different view on numbers and ratios. I am using π merely as an abbreviation here.

[5] Cf. Heath (1921, II, pp. 232–235) and Knorr (1986, pp. 155–159) for Archimedes' and other approximations; see also Heath (1921, I, pp. 180–189).

Western European culture over the centuries in connection with the squaring of the circle. It is based on the above-mentioned Tropfke (1923, pp. 198–238). The Indian mathematician Aryabhata, fifth century AD gave an astonishingly close approximate value for π.[1] His methods are interesting because of their combination of geometry, algebra, and what would in Greek terminology be called logistics. There must have been substantial contributions in Islamic culture, but they are not accessible to me. They did have an influence on Fibonacci's contribution.[2] During the Middle Ages, Archimedes' approximations were widely used, and they came to be regarded as exact by many (especially the simpler one: 22/7). Exact geometry, as a demonstrative science, did not receive much attention during the Middle Ages in Europe. An interesting use of mathematical motifs for philosophical purposes (in connection with infinity) including a treatment of the circle and π can be found in Cusanus.[3] In the Renaissance, Leonardo squared the circle by rolling up a cylinder with a base equal to the circle that is to be squared, and appealing to Arch., *Circ. mens.* I.[4] Stifel devised a mechanism using levers and scales to "weigh" π.[5] In the modern era, when algebraic and improved calculation methods became available, we find approximations of π by Vieta and Huygens, still operating within the framework of classical geometry and logistics. They developed Archimedes' basic approach via inscribed polygons, and refined the limits for the approximations.[6] Gregory, Newton, and Leibniz employed infinite series,[7] and Wallis used his arithmetic of infinites to characterize π. Leibniz called his result an "arithmetical quadrature of the circle."[8] In the eighteenth century (1766), Lambert used continuous fractions, and showed that π was an irrational number. Euler studied both π and e in connection with trigonometric functions.[9] Toward the end of the nineteenth

---

[1] Cf. Elfering (1975). The work has been translated into English.

[2] Cf. Tropfke (1923, pp. 211/212) for references; Fibonacci's value for π is 864/275, ca. 3.141818.

[3] Cf. Tropfke (1923, pp. 213/214). On Cusanus' studies in connection with the quadrature of the circle see also Hofmann (1990, I, pp. 47–77, II, pp. 179–192, 351–395).

[4] Cf. Tropfke (1923, p. 214), Cantor II², pp. 301–302.

[5] Cf. Tropfke (1923, p. 215); on Stifel see also Hofmann (1990, II, pp. 78–109).

[6] Cf. Tropfke (1923, pp. 215–216, 218–219); e.g., Vieta, in *Variorum de rebus mathematicis liber VIII*, p. 392 in the 1646 Schooten edition gives a value for π that is exact in the first nine digits, and uses the "Archimedean" approach via inscribed polygons. For Huygens's contribution, see the appendix to his *Theorenata de quadratura hyperboles, ellipseos et circuli ex Data portionum gravitatis centro*, Leiden 1651, and his *De circuli magnitudine inventa*, Leiden (1654), both in Ch. Huygens, *Varia Opera*, Leiden (1724, pp. 328–340, 351–387); see also Rudio (1892). The latter work also contains a full German translation of Archimedes's, Lambert's, and Legendre's treatment.

[7] Gregory's double series is equivalent to an approximation via arctanx, Leibniz's series for $\pi/4$ converges very slowly, so that it only has theoretical value. Newton used the series for arcsinx, Brouncker developed Wallis' solution into an infinite fraction.

[8] Cf. Tropfke (1923, pp. 223–230).

[9] Cf. Tropfke (1923, p. 229).

century, Lindemann proved the transcendence of π.[1] As said above, Lindemann's result means that a constructive quadrature is impossible with circle and straight line, with conics, with any curve expressible as a polynom with rational coefficients. One needs a curve like the quadratrix. In the end, Pappus' Prop. 26 and Prop. 27 are as good as it gets.

5.4.2 *Genesis* and *Symptoma* of the Quadratrix

Genesis: Start with a quadrant BAD in a square ABCD (clockwise) over the radius; use two motions: of BC along BA, parallel to AD, and of AB along the arc BD, synchronized so that they both reach the position of AD at the same time; during the process they create an intersection line BH, the quadratrix.[2]

Symptoma: As can be seen from the *genesis*, for any line AZE drawn to the curve and extended to the circumference, we get:

arc BD:arc ED = BA:ZT.

5.4.3 Criticism of the *Genesis* by Sporus

Sporus of Nicaea, ca. 200 AD was not a mathematician. Rather, he seems to have been a philosopher interested in epistemology and theory of science. Of his work *Aristotelian Wax Tablets* only fragments remain. They contain reflections on mathematical arguments from the standpoint of an Aristotelian theory of science.[3] His criticism of the *genesis* of the quadratrix is as follows.[4]

[1] Cf. (Tropfke 1923, pp. 231–232). Lindemann's proof for the transcendence of π (Lindemann 1882) is modeled on Hermite's proof for the transcendence of e. For ensuing improvements and simplifications of this proof cf. Tropfke loc. cit.

[2] Note the close connection to the *genesis* of the spiral as given in Prop. 19.

[3] On Sporus cf. Tannery (1912, I, pp. 178–184); the main source for our information on Sporus outside this passage in Pappus and the one mentioned below, in Eutocius, are the scholia on Aratus' *Phaenomena*. A further example for a criticism by Sporus, also in connection with the squaring of the circle, is found in *Eutoc. in Arch. Circ. mens.* III, 258–259 Heiberg. Sporus insists that Archimedes' approximate values are not exact, discusses the decisive difference between an exact and an approximate value, and produces a closer approximation than Archimedes' to show that it was not the exact value. Apparently around 200 AD already the nature of an approximation was not properly understood by some, who believed that a value – a ratio in numbers – must be true, i.e., correct.

[4] As mentioned in the survey table of Props. 26–29, the style of this passage is decidedly different from the rest of *Coll.* IV. We are clearly dealing with an argument from a philosophical work, in polemical style, one in which objections were raised against another position on epistemological grounds. Note, e.g., the rhetorical questions and the device of a ficticious dialogue; cf. translation.

(i)   The definition via synchronized motions contains a petitio principii: to coordinate the rotation and the linear motion, you need the ratio of arc BD to AB – essentially π – the very thing the quadratrix was supposed to provide.[1]

(ii) Even objection (i) aside, the genetic definition does not capture the endpoint of the curve, because the intersection stops right when the moving lines coincide with AD. This endpoint, however, was needed for the rectification of the quadrant (Prop. 26, see below). Infinitely many other points on the curve are in fact constructible, e.g., all points that one would get by successive division of angle and radius in half. But the endpoint is not among them.

(iii) The endpoint of the curve cannot be interpolated by extending the line in the manner of producing a straight line, because the curve does not have a fixed direction (as the straight line does). In fact, the quadratrix does not even have a constant curvature.

Sporus concludes: as long as the ratio of circle and radius is unknown, or not given, the curve cannot be accepted. Pappus will in effect pick up right here in Props. 28 and 29, and show that it is *given* in the specific sense of geometrical analysis, if a helix, or the spiral, is granted. Whereas Hultsch (and others around 1900) dismissed Sporus' objections, most modern interpreters accept them as valid.[2] The curve is not well-defined. Note that the reason for Sporus' objections is not the use of motions as such, but the conceptual inconsistencies involved in this particular motion description. These inconsistencies will have to be circumvented, or abolished, if Sporus' objections to the mathematical use of the curve are to be met. And Pappus explicitly agrees with Sporus' reasons for rejecting the curve under the motion description (under the description as "mechanical"). He uses the word "eulogos" (with good reason). On the other hand, he insists that the argument about the quadratrix – Props. 26 and 27, the *symptoma*-quadrature – is "much more acceptable" mathematically. In Props. 28 and 29, Pappus will provide a geometrical analysis for the generation of the curve, via analysis of loci on surfaces. It is intended to meet, or rather perhaps to circumvent, the objections raised by Sporus, so as to "geometricize" the curve as a basis for valid *symptoma*-mathematics (see below).

5.4.4   Prop. 26: Rectification Property of the Quadratrix

5.4.4.1   Proof Protocol Prop. 26

This proof protocol is given in detail, because its content is a "classic," and also because it is the only example of a full-fledged argument via double reductio in *Coll.* IV (the other example in Prop. 13 is much less complex).

---

[1] As in the case of the spiral in the version given in Prop. 19, you need π to determine the speeds involved.

[2] Cf. Heath (1921, I, 229/230), Knorr (1978a, 1986, p. 230), Jones (1986a, pp. 596–598).

1. Protasis/Ekthesis

Start with a square ABCD,[1] circular arc BD (K1 in what follows), and quadratrix BT. arc BD:BC = BC:CT. (CT is the third proportional to arc BD and BC).

2. Apodeixis (by double reductio: "exhaustion"[2])

If not, then either arc BD:BC = BC:CK, CK > CT
or arc BD:BC = BC:CK, CK < CT

    2.1 Assume arc BD:BC = BC:CK, CK > CT

        2.1.1. Auxiliary construction
        circle THKZ, center C (K2),
        perpendicular HL, draw CHE

2.1.2 arc BD:BC = BC:CK = CD:CK	[assumption]
CD:CK = arc BD:arc ZK	[see argument *]

* argument for this (not in *Coll.* IV): similar arcs in the ratio of the radii (or diameters)[3]

$K1:K2 = CB^2:CK^2 = CD^2:CK^2$	[XII, 2]
$K1:K2 = (U1 \times CD):(U2 \times CK)$	[*Circ. mens.* I]
$CD^2:CK^2 = (U1 \times CD):(U2 \times CK)$	
$=> CD^2:(U1 \times CD) = CK^2:(U2 \times CK)$	[V, 16]
$=> CD:U1 = CK:U2$	[VI, 1]
$=> CD:CK = U1:U2$	[V, 16]
$=> CD:CK = $ arc BD:arc ZK	[V, 15]*

Thus, BC = arc ZK	[V, 9]
2.1.3 arc BD:arc ED = BC:HL	[*symptoma*]
= arc ZK:arc HK	
	[equal parts of quadrants]
=> HL = arc HK	[V, 9]

This is impossible.

    2.2 Assume, then, that arc BD:BC = BC:CK, CK < CT

        2.2.1 Auxiliary construction:
        circle ZMK, center C;
        perpendicular KH, draw CHE
        2.2.2 As in 2.1.1, we see: arc BD:BC = CD:CK,

---

[1] The labeling of corner points for the square in the starting configuration is now counterclockwise, as opposed to the original *genesis*. Perhaps this is an indication that the author of Prop. 26 is different from the source for the *genesis* and *symptoma*.

[2] On Prop. 26 see also Heath (1921, I, pp. 226–229) and Knorr (1986, pp. 226–230). Knorr's account contains some interesting speculative remarks on the study of tangents, subtangents etc.

[3] The theorem that circumferences have to one another the ratio of the respective diameters is used repeatedly in *Coll.* IV, cf. Props. 26, 30, 36, 39, and 40. A proof is given by Pappus in *Coll.* V, 11 and VIII, 22.

and CD:CK = arc BD:arc ZK,

$\Rightarrow$ BC = arc ZK

[use an argument analogous to argument *].

2.2.3 As in 2.1.3, we see that

arc BD:arc ED = BC:HL   [*symptoma*]

arc BD:arc ED = arc ZK:arc MK

[equal parts of quadrants]

$\Rightarrow$ HL = arc MK   [V, 9].

This is impossible.

3. Symperasma: Therefore, arc BD:BC = BC:CT must hold.

Corollary

By constructing a line a with CT:BC = BC:a, and finding 4a, one has rectified the circle. For a = arc BD.

This means that the quadratrix has a rectification property, which can be derived from its *symptoma*. Further results, directly from the *symptoma*, or from the rectification property, can be found in Props. 35–41. They are much less spectacular than this one here.

5.4.5   Prop. 27: Squaring the Circle

After rectifying the circle, one can apply Archimedes, *Circ. mens.* I, and construct a triangle that has the same area as the given circle: base 4a, with a as in Prop. 26, appendix, height BC. This triangle can then be transformed into a square via II, 14.

5.4.6   Prop. 28: Geometrical Analysis, Linking the Quadratrix to Loci on Surfaces Through a Cylindrical Helix

5.4.6.1   Outline of the Analysis in Prop. 28

Start with a quadrant ABC, radius BD, E on BD, perpendicular EZ, assume that EZ:arc DC is *given.*[1]

Then E lies on a uniquely determined curve.

---

[1] Note that this is a response to Sporus's demand after criticizing the definition of the quadratrix via motions. He had demanded that a crucial ratio be given. In Prop. 28, it is taken as *given* in the sense of geometrical analysis.

Analysis

1. Extension of the configuration
Cylinder-segment over ABC; in it, take an Apollonian helix as *given* in position L,
T, I as in the figure create a garland-shaped surface, determined by the helix

2. *Resolutio*
   2.1 I lies on a uniquely determined plane
   (a plane *given* in position[1]), through BC and ZI
   (or perhaps EZ and ZI); here the *given* ratio is used;
   2.2 It also lies on the plectoid surface created by the helix
   [use the *symptoma* of the helix[2]].
   Since the helix is also *given* in position, I lies on an intersection curve of surfaces, which is also *given* in position.
   2.3 Project this line onto the plane of the original quadrant.
   By construction, E will lie on this uniquely determined line.

3. Specification
When the *given* ratio EZ:arc DC = AB:arc AC, this line will be the quadratrix.

5.4.6.2   Intersection Plane in Step 2.1: Through EZ or BC?

Pappus' description is not sufficiently precise. In addition, there are several illegible
spots in the main manuscripts for this part of the text, and they were already there
when the minor manuscripts were copied. With Knorr, I favor the reading according
to which the intersection plane is the one through BC and ZI, for it is obviously
given, i.e., constructible, at this stage of the analysis (assuming that one has the
helix). BC is *given* in the starting configuration, and the inclination of the sought
plane toward the underlying plane is determined by the *given* ratio EZ:EI. The
drawback is that with this intersection plane, the endpoint Z for the intersection
curve in space is not uniquely determined. Neither will the endpoint of the resulting
special case quadratrix be. If one opts for the plane through EZ and ZI, as Hultsch,
and apparently Treweek did, one has to assume that EI and EZ are *given* in position.
It is not clear, at this stage of the analysis, that they are.

Ver Eecke assumed that the segments ZE and EI are *given*, because they go
through *given* points. One might object that if Z and E were *given*, there would be no
need for further argumentation at this point. It is unclear how the points can be seen
to be *given* at this stage. Ver Eecke also assumed that the intersection plane on which
I lies goes through LT. One might find this objectionable, too.

Even if we cannot decide with certainty which plane is used in the analysis in
Prop. 28, the main thrust of the argument is clear: it provides a conceptual connection

---

[1] My reading of Prop. 28 differs considerably from the one given in Ver Eecke (1933b, p. 199, #2);
it is compatible, however, with the discussion in Knorr (1986); compare also the following notes
on the crucial intermediate step 2.1.

[2] On the helix and its *symptoma* cf. *Procl in Eucl.* 105, 271 Friedlein, Knorr (1986, p. 295/296).
The ratio of height and rotation angle is a constant, i.e. *given* with the curve.

between the quadratrix of Dinostratus and a locus created on a curved surface in space, in dependence from the Apollonian helix, and that is its purpose.[1]

5.4.7   Prop. 29: Geometrical Analysis, Linking the Quadratrix to Loci on Surfaces with Spiral

Start with a circular sector (not necessarily a quadrant) ABC, *given* in position, a radius BD, point E on it, and a perpendicular EZ, where EZ:arc DC is *given*, and EZ:arc DC = AB:arcAC (spiral-creating ratio). Assume that a spiral BHC is inscribed in the sector ABC.

Then E lies on a uniquely determined line.

5.4.7.1   Outline of the Analysis in Prop. 29

1. Extension of configuration
Cylindroid over spiral, height BH;
BH = EZ   [construction],
EZ:arc DC = AB:arc AC = BH:arc DC   [*symptoma*]
right cone, vertex B, generating line at an angle of $\pi/4$ with respect to the underlying plane

2. *Resolutio*
   2.1 Analytical determination of a locus for K
   K on HK, perpendicular to the plane, KH = BH
   HK is *given* in position
   K lies on the cylindroid surface,
   and on the surface of the cone
   $\Rightarrow$ K on the intersection line created by those two surfaces:
   a conic spiral that is *given* in position.[2]

---

[1] I agree with Ver Eecke's summarizing statement: "En exposant ce premier mode de construction géométrique de la quadratrice au moyen des Lieux à la Surface, la proposition démontre donc, sans l'énoncer explicitement, une propriété remarquable de la surface de la vis a filet carré à axe vertical, a savoir que, si l'on coupe une surface hélicoide rampante (y = x tang (2π z/h)) par un plan passant par une de ses génératrices rectilignes (z = my) [I opted for BC, Ver Eecke for LT], et si l'on projette orthogonalement, sur un plan perpendiculaire à l'axe de cette surface la courbe détermineée comme section, on obtient une quadratrice de Dinostrate" (Ver Eecke 1933a, p. 199, #4).

[2] Note that the conic spiral used in Prop. 29 is not automatically accepted, as the helix in Prop. 28, and the spiral in Prop. 29 were. It must be reduced to the spiral in order to be revealed as *given* in position. This could be an indication that the Archimedean spiral and the Apollonian helix were viewed as privileged basic curves for the analytical determination of other motion curves by Pappus (cf. Molland 1976). If so: was this the case just for Pappus, or: more generally? Were these curves perhaps seen as basic for the *symptoma*-definition of higher curves, as Prop. 29, but also Prop. 28 seem to suggest? In the meta-theoretical passage, Pappus significantly speaks of quadratrices and spirals as exemplary curves for the third kind. Recall also Apollonius' claim on his helix as a basic curve, on a par with circle and straight line, reported in Proclus on authority of Geminus (*Procl. in Eucl.* 251 Friedlein). The issue cannot be pursued here.

2.2. Analytical determination of a locus for I and E
    2.2.1 Extension of configuration, second part:
    analogous to the "garland" in Prop. 28, create a plectoid surface, derivable
    from the original spiral; use BL, and the conic spiral; both are *given* in
    position: LI moves along the spiral and BL, parallel to the underlying
    plane, creating a twisted surface in space that is *given* in position.
        2.2.2 I lies on that surface.
        2.2.3 I also lies on a uniquely determined plane [through BC and ZI;
        use the *symptoma* of the spiral].
        ⇒ I lies on the intersection curve created by those surfaces.
        2.2.4 Project this curve onto the underlying plane.
        By construction, E lies on this projection, on a uniquely determined
        line.

3. Specification:
When the sector ABC posited in this analysis is a quadrant, this line is the
quadratrix.

5.4.7.2   Lines, Planes, and Surfaces in Prop. 29

Whereas Prop. 28 used a cylindrical helix from the start, Prop. 29 starts with a plane
curve, the spiral, and constructs a curve in space from it as a first step: a conical
spiral. Most commentator agree that the conical spiral is created by erecting a cylindroid over the given spiral and intersecting it with a right cone with axis BL,
inclined at 45° toward the underlying plane. The point K lies on it. This much
seems uncontroversial, and for this reason I have used a diagram for Prop. 29 that
shows the cylindroid surface and the point K.
    Different interpretations have been offered for the second part of the construction
in Prop. 29. The reading offered here is minimalist, and modeled on Prop. 28. One
draws the parallel LKI to BE, leaving the exact location of I open, i.e., reserving the
possibility to extend KI if needed. The generator BZ with flexible endpoint,
adjusted between BL and the conical spiral, creates a "plectoid," garland-shaped
surface. It is intersected with the plane through BC and ZI, analogous to Prop. 28,
and projected orthogonally onto the plane. This reading is only tentative. Its advantage over some other ones is that they all assume that LKI is extended to the circumference, and that the same cylindrical helix, and the same garland as in Prop.
28 is created. The text of Prop. 29 does, however, not mention the helix and seems
to propose the analysis in terms of the plane spiral as an alternative to the one using
a helix. It is not to be excluded that the original author of the argument in Prop. 29
did intend to show, with his analysis, how the plane spiral, the conical spiral, the
cylindrical helix, and the quadratrix are all connected. The text as reported by
Pappus does not explicitly say as much, though. Therefore, I opted for the minimalist reading (and accordingly, a very reduced diagram). If one accepts the presence
of the helix, perhaps a reading along the lines of Commandino is the most straightforward one. Commandino does assume a cylinder in addition to the spiral-induced,

cylindroid, extension of LKI to H on the cylinder surface, creation of a helix in dependence from the conical helix, with a garland-shaped surface for I. His diagram (Co p. 91) shows all these features. Commandino then assumes the creation of an intersection curve in space, and orthogonal projection onto the underlying plane as in Prop. 28. There are some problems with his reading in detail, for which see the translation. Hultsch ad locum refers to Chasles and Bretschneider, and does not offer an interpretation. His diagram is also minimal. For Ver Eecke's reading see Ver Eecke (1933a, p. 200f). Knorr (1989, p. 166f) offers an explorative interpretation of the material in Prop. 29, drawing a connection to Archimedes's study of tangent problems on the plane spiral. It is very interesting in itself; I am somewhat diffident, however, that it works well as an explanation of Prop. 29 as given in Pappus' text. Therefore, I have restricted my presentation of the content of Prop. 29 to the information as given in the text for the most part. For further clarifications and alternatives, the reader is referred to the literature mentioned above.

5.4.8   Additional Comments on Props. 28 and 29

5.4.8.1   Loci on Surfaces

As noted before,[1] Props. 28 and 29 are our only explicit sources on analysis of loci on surfaces. This means that observations drawn from them provide only limited knowledge of the discipline for which they are an example. There is a danger of over-interpretation, because we lack a context to check our reading against. The following observations on Props. 28 and 29 may nevertheless capture some representative features for this kind of mathematical approach.

1. The dominant method of investigation, and the method for determining the basic objects of study, is geometrical analysis in the technical sense.
2. Certain spiral-type curves have a privileged role, others are determined relative to them.
3. We operate with surfaces in space, created by rotation, by a motion that combines a linear progression and a rotational motion in synchrony (twisted surfaces, controlled "motions"), or by establishing cylindroid surfaces over a plane figure, and intersecting them with each other, and with planes.
4. The created curves in space are in the end projected onto the plane.
5. Because we are using analysis, the result is not a constructive solution, or a constructive *genesis* of the curve. This is also not intended. The content really is a mathematical analysis of the *genesis*, establishing unique determinateness for the "target curve" inside a configuration.

---

[1] For bibliographical references, see the literature given at the beginning of the chapter on Props. 26–29.

A Potential Context for the Analysis of Surface Loci: Analysis of loci
and Conic Sections

Consider the parallel between items 3 and 4 above and Archytas' solution for the
cube duplication with Eudoxus' procedure for his curve devised for cube duplication.[1] According to Zeuthen (1886, pp. 460–461), this procedure was taken over by
Menaechmus as a model for the conic sections, viewed as analytically determined
loci (cf. item 1.), as plane *symptoma*-curves.[2] Even after the conics were discovered
to be sections of cones, and their definition was in terms of this *genesis*, the actual
handling continued to be focused on the *symptoma*-characterization. Consideration
of those aspects of the Apollonian treatment of conics that might be viewed as
analogous to *symptoma*-mathematics – and there are quite a few examples (see
again also Zeuthen 1886 passim) – might help to reconstruct a context for the
symptoma-mathematics of the third kind, by studying the analogue in *symptoma*-
mathematics of the second kind. Perhaps even the reduction of the conic sections
as plane curves to the intersection of a plane and a surface in space (i.e., the surface
of a cone) could be seen as somewhat of a model for the reductions we see in Props.
28 and 29. In addition, the analytical Euclidean work on loci on surfaces, on which
Pappus comments in *Coll.* VII, and which is based on related work by Aristaeus,
might be considered.[3] Perhaps the outlines of a context for *symptoma*-mathematics
become visible here. The issue is worth exploring. A decisive difference, even if parallels
can be found and brought to bear, would be the fact that conics can be viewed as
essentially defined, although *symptomatically* handled; the higher curves cannot.[4]

Use of Analysis for the "Definition," or Determination of Curves

Without drawing far-reaching conclusions from our scarce evidence in Props. 28
and 29, one thing can nevertheless be said, and it has been somewhat overlooked in
secondary literature on the propositions. The propositions have a clearly analytical
character, with analysis taken in the full technical sense of the word. And it seems
plausible to assume that this feature would have been typical of the geometry of the
third kind. Specifically, geometrical analysis (*resolutio*) is used here, not to (only)

---

[1] Compare the remarks on cube duplication and conics in the commentary on Props. 23–25. For a
hypothetical reconstruction of Eudoxus' curve cf. Tannery (1912, I, pp. 53–61). It seems plausible
to assume that Eudoxus projected the space curves, created in Archytas' solution, onto the plane,
creating a curve with which he could solve the cube duplication. It would have to be defined by
deriving the characterizing properties from the properties inherent in the space curves. For the
symptoma of the helix in Proclus cf. pp. 105 and 271 Friedlein; cf. also Knorr (1986, pp.
295–296).
[2] Cf. Knorr (1986, pp. 50–66, 112).
[3] *Coll.* VII, pp. 1004–1014 Hu (Jones 1986a, pp. 363–371, see also pp. 503–507, 591–599).
[4] Cf. Zeuthen (1886, pp. 459 ff.), Knorr (1986, pp. 61–66, 112) on the combination of essential,
genetic definition, and operation with the *symptoma* in the theory of conic sections.

solve problems but to "mathematize" motion curves as *symptoma*-curves, by reducing
them to properties of other curves that are taken as *given*.

Given in 28:
Per hypothesis: sector ABC, quadrant (as in quadratrix), radius BD with perpendicular; EZ, and ratio EZ:arc DC (this ratio is not necessarily the one used in the
quadratrix). An Apollonian helix with a *given* progression ratio for angle:height (connected to EZ:arc DC).
    Entailed: each such configuration determines, i.e., turns into a *given*, a certain
unique projection curve, in direct dependence from the ratio that is embodied in the
helix: a quadratrix-like curve. We get a family of curves. The quadratrix is the one
where the given ratio is the same as AB:arc BC.

Given in 29:
Per hypothesis: sector ABC (not necessarily a quadrant, unlike quadratrix), radius
BD, perpendicular EZ with ratio EZ:arc DC = BA:arc AC (ratio as in quadratrix)
an inscribed spiral (embodies the ratio BA:arc AC)
Entailed:
(i)   A conical spiral, as an intersection curve in space.
(ii) An intersection curve between two curved surfaces in space.
(iii) A certain unique projection curve, in direct dependence from the ratio embodied
      in the spiral: a quadratrix-like curve. We get, again, a (different) family of
      curves; the quadratrix is the one where the given sector is a quadrant.

The analysis in Props. 28 and 29 is restricted to the *resolutio*-phase: the phase
where that which we need or want to establish is shown to be *given*, if certain other
features (theorems, prior results, etc.) are posited. The arguments show that the
curves in question are uniquely determined in a hypothetical sense: We cannot
derive them from essential properties rooted in the archai and the principle objects
of our discipline, but the properties we focus on in mathematical argumentation can
be put in an exact, conceptualizable relation to properties of other entities in a specific spatial configuration (ultimately the *symptoma* of a privileged curve). The
latter we just assume and posit – much like we posit the straight line and circle. This
much one can assert. We will have to leave it undecided, because that is what
Pappus does as well, whether this determination "saves" the curves completely, so
that *symptoma*-curves were taken to be just as solidly defined as the archai of the
plane and solid kind, even though the *symptoma*-approach operationalizes, lets the
curve itself disappear and replaces it by a kind of relation/equation. We will not try
to determine, at this point, what it means that Pappus asserts and supports the fully
mathematical character of arguments about the curves, but is hesitant about the
status of the curves themselves (cf. also meta-theoretical passage, where a similar
ambivalence shows).[1]

---

[1] See the excursus below for some speculative remarks in this regard.

What does Pappus achieve with Props. 28 and 29? The reading suggested here is a sympathetic one. Pappus does not achieve, and does not believe he has achieved, the quadrature of the circle. He has not "saved" the *genesis* of the quadratrix in the sense that the curve can now be constructed geometrically, and he does not claim to have "solved" the problem. He succeeds in partially circumventing Sporus' objections, i.e., he interprets the demand that the crucial ratio be given before the curve can be accepted by showing in what sense, and to what degree, the ratio can be seen, via geometrical analysis, to be *given* in the technical sense of the word. He gets an analytical characterization, not a constructive definition. And he is explicit about that. Even so, the analytical determination has achieved something. Its effect is that the quadratrix, although not constructible, can be investigated geometrically, without conceptual inconsistencies, qua locus curve for a certain *symptoma*. It is well-defined, uniquely determined. The geometry on it is true geometry, geometry of the third kind. Its results are geometrically demonstrable properties of the curve as *symptoma*-locus. As long as we only had the genetic definition, which was conceptually inconsistent, such geometry did not have a satisfactory basis.

Even so, the issue of the quadratrix's foundation is not completely settled. As noted above, we cannot be sure just how solid the analytical basis is. In the metatheoretical passage, where Pappus will, once again, classify this kind of mathematics as legitimate mathematics, alongside plane and solid mathematics, he also does say that the curves have a somewhat forced *genesis*, and he shows a certain hesitancy with respect to the third kind of mathematics. Also, the analytical approach leaves a gap: *symptoma* – analysis cannot guarantee that a more elementary construction is impossible for a curve thus characterized (e.g., that in certain specific cases, it might reduce to a locus of the second kind).

Most interpreters so far have not given Pappus a sympathetic hearing. One basic error, which is rather pervasive, is that they read Pappus' statement that he will provide a geometrical analysis (analuesthai) as actually saying that he claims to "solve" the problem (the quadrature) geometrically (equivalent to luesthai). As has been pointed out in the notes to the translation, this is a serious misunderstanding, for Pappus does indeed provide an analysis for the *genesis* of the curve, and he does not provide a solution of the problem. Jones (1986a, p. 598), e.g., seems to believe that Pappus is trying to give a construction of the quadratrix and remarks that, as constructions, they do not meet Sporus' objections. Similar attributions of confusion to Pappus can be found in Knorr (1978a, 1986). Knorr also offers, however, the consideration that Pappus may, after all, have tried not to meet Sporus' objections, but to circumvent them. In this respect, his reading concurs with mine.

5.4.9 Excursus: Speculative Remarks on the Potential of Analysis-Based *Symptoma*-Characterization of Higher Curves

The decisive difference between using the circle mathematically by focusing on its *symptoma*, and using the helix and other curves solely accessible through their *symptoma* is somewhat like this: We think (perhaps) we know what the circle is, essentially, and the properties we use in mathematics are seen as properties of that object, for which we can posit some kind of epistemological or ontological priority. Of the Apollonian helix we do not have such a direct grasp. It has to be constructed in thought as the thing which has the decisive property. The helix itself disappears, as it were, behind its *symptoma* in a way the circle does not. So what is the epistemological, or ontological, grounding of such curves, when they are viewed exclusively as loci for a *symptoma*?

In view of the complete absence of statements from ancient mathematicians on this question, and the deplorable lack of evidence on their actual practice in this area, it is perhaps fruitless to try and establish what the commonplace opinion among them would have been on that question. The following, speculative remarks should be taken as an elucidation of the potential impact of this question, its horizon of potential for future developments in the intellectual history of mathematics. Specifically, I have the sixteenth and seventeenth century readers in mind, and "anchor points" for the routes they took to transform mathematical investigations toward algebraization on the one hand, and infinitesimal calculus on the other. Could the ancient mathematicians, in defiance of the essentialist view on science and explanation – whether Aristotelian or Platonist – have taken the view that circle, helix, and spiral are really equivalent, because all mathematics is *symptoma*-mathematics and does not really care about the ontological status of the objects the *symptomata* of which it studies? That the circle, e.g., is treated as a locus curve just as the helix is? That what is mathematically interesting about it, its property, can equally well be seen as stemming directly from the motion generating it? Apollonius for one argued that the helix should be placed alongside circle and straight line as a basic, unanalyzed principle in geometrical argumentation. Does this imply an anti-essentialist thrust, a turn toward making locus-properties, i.e., relations, the final objects of mathematics? Are the basic items all loci, as it were, characterized as such via "defining" relations? That would make Apollonius a forerunner of the paradigm shift toward algebra that occurred in the seventeenth century. It cannot be ruled out.[1]

If such was the case, and there was an Apollonian programme to implement a new paradigm for mathematics, one in which operationalism, and the manipulation of relations are key ideas, we would have to say that the programme did not carry the day in antiquity, and the ancient research project of *symptoma*-mathematics might have died out precisely for that reason: re-channeling into the mainstream essentialist approach. What we see in *Coll.* IV, and what the seventeenth century readers saw as well, would then be like the remnants of a large-scale re-orientation project for mathematics which was abandoned, with the remnants still bearing the traces of the revolutionary ideas behind them, of this push toward operationalizing geometry into a proto-algebraic discipline. Such an ideological clash, an unsuccessful

---

[1] The fact that Apollonius apparently wrote a work called "katholou pragmateia" (universal treatise, attested in Marinus (Eucl. Op. 6, p. 234 according to Jones (1986a, p. 530/531), and the fact that the remarks attested in Proclus seem to point towards an attempt at radically reorganizing the foundations of Euclid's *Elements*, do invite speculations in this direction.

frontal attack against the ruling paradigm, which in turn was backed by non-negotiable essentialist convictions and preconceptions, and which in the end prevailed, would explain why Apollonius' minor analytical works were lost, why his *Konika* were stripped of their analysis-parts, which in the original must have been dominant (Pappus groups the *Konika* with the analytical works), and recast by Eutocius in purely synthetic form, and also why no works of the authors who worked on the analysis of loci on surfaces are preserved. The essentialists in the field of epistemology/theory of science would have won the day, and forced the continuation of the old paradigm.

Such a speculation is tempting. But it is equally possible that the mathematicians, including Apollonius, went along with the essentialist views on the nature of science and explanation, or – and that is perhaps the most likely option – that they did not reflect on such questions at all and just went ahead doing their mathematics of *symptoma*-curves. After all, even in the orthodox Aristotelian paradigm, any science is entitled to positing its principles and does not have to go beyond, justifying them, so that in the end, any science can do its job while focusing in on the rigorous development of arguments about *symptomata*. On the whole, we cannot get beyond the observation that geometrical analysis in the technical sense was applied to derive a hypothetical definition, or characterization, of motion curves through their *symptoma*. This was not just a side thought, since a considerable amount of sophistication and argumentation is needed to perform this task. It must have been of some importance, and served a serious purpose. Whether the result was that these curves were then seen as on a par, epistemologically, with objects like circles, straight lines, or conics, must be left undecided. Also, the details of this kind of mathematical argumentation are at present opaque to us, and certainly were so for the sixteenth and seventeenth century readers as well. This may be part of the reason why so much effort was spent on reconstructing the analytical works, and the analytical strategies of the ancients. Still, the material presented in Pappus is suggestive toward a new perspective on what mathematics essentially is, one in which analysis and operation with relations are central. One could pick up here; in a way that needs to be explored and spelled out in more detail, Vieta, Descartes, Fermat and others did.

5.5 Prop. 30: *Area Theorem on the Spherical Spiral*

context: Archimedes on spiral lines, motion curves, quadratures.
source: lost text of Archimedes.
means: I, III, V, XII, *Sph. et Cyl.* I, 33, I, 35, I, 42.
method: synthesis; infinite inscription process, quasi-infinitesimals, limit argument.
format: *genesis*-description, *symptoma*-theorem and corollary.
reception/significance: no reception, the only related extant treatise is *Sph. et Cyl.*; the addition to Prop. 30 is the first example for a quadrature of a curved surface in space.
embedding in *Coll.* IV: motif "Archimedes": Props. 13–18, 19–22, 35–38, 42–44; Archimedes, with his "mechanical" approach "frames" the treatment of motion curves; motif "spiral lines": Props. 19, 20, 26, 29, 35–38; motif "area theorem": Prop. 21; the content of Prop. 30 is not picked up again in *Coll.* IV.

<u>purpose:</u> illustrate the "mechanical" path for the treatment of motion curves: *symptoma*-mathematics as meta-mechanics.
<u>literature:</u> Heath (1921, II, 382–385), Knorr (1978a, 59–62, 1986, 162–163), Ver Eecke (1933b, 206, #2).

Prop. 30 is the first known example for the quadrature of a curved surface in space. Methodologically, it picks up the first, "Archimedean" path for dealing with motion curves. Although no author is named for Prop. 30, the theorem is usually ascribed to Archimedes. The parallels to the argumentative style and the structure of Prop. 21, especially the use of indivisibles, and the infinite inscription process, as well as the parallel argument using two figures with parallel division processes, are very compelling indeed. The spherical spiral is created by motions in the ratio 1:4. Unlike the plane spiral inscribed in a circle, the spherical spiral described here is conceptually well-defined in its *genesis* via motions.[1] The *symptoma* is directly read off from the coordinated idealized motions.[2] The theorem on the spiral uses aspects of Archimedes' "mechanical method," namely indivisibles. Mathematics appears as meta-mechanics, where mechanics itself is already highly abstract. We do not have a context for Prop. 30. It may have been part of a larger work.[3] No applications outside *Coll.* IV are attested. The only surviving Archimedean complete monograph on the *symptomata* of a motion curve is *SL*, and its argumentative method and style differ significantly from the quasi-mechanical approach attested in Props. 21 and 30. As in the case of the analytical branch of *symptoma*-mathematics, the lack of a context makes it impossible to draw far-reaching conclusions on the status of the mathematics of the third kind "Archimedean style," which, I think, is represented in Prop. 30 (see Knorr (1986) for an interesting, if perhaps sometimes speculative, evaluation of the possible development of motion curves in the generation after Archimedes). Certainly plausible is Knorr's assumption that the "mechanical" approach was picked up and put to use for analytical (*symptoma*-) mathematics, and this assumption also agrees with Pappus' statements on the geometry of the third kind in the meta-theoretical passage, as well as with his developmental story in Props. 19–30. Knorr also points out that the approach via infinitesimals and indivisibles was not pursued further. Pappus voices no objection to the result in Prop. 30, and obviously treats it as valid.

---

[1] Polar coordinate description for the spherical spiral: $\rho = 1/4\omega$; compare the analogous equations for the plane spiral in Prop. 19: $\rho = 1/(2\pi)\varphi$, and for the spiral as used in *SL*: $\rho = a\ \varphi$, where a is arbitrary, but fixed. Both Prop. 30 and *SL* avoid having to take recourse to π.

[2] No instruments are involved, the verb used for the *genesis* via motions is noein. We deal with abstract motions. The verb kinein, used in 19, is absent; no application context for Prop. 30 can be envisaged. Its "mechanical" character is purely theoretical.

[3] A comparison of the argumentative means in Props. 30 and 21 shows: Prop. 30 adds *Sph. et Cyl.* to V, XII, which were already used in Prop. 21. Knorr (1978a, pp. 59–62) argues that the material in Prop. 30 belonged to the heuristic version of *Sph. et Cyl.* The connections are clearly there. I doubt, however, that they are sufficient for postulating an immediate and precise relation such as the one postulated by Knorr.

Genesis of the spherical spiral: In a hemisphere, rotate the arc TNK of a quarter-circle through the pole along the base circle (arc KLM). At the same time, let a point N travel from the pole toward the base, and assume that it completes the quarter-arc at the same time in which the rotating quarter- arc completes the full circle. The traveling point describes a spherical spiral.

Symptoma: If one draws an arbitrary quarter-arc TOL, with O on the spiral, arc TL:arc TO = circumference: arc KL.

5.5.1   Proof Protocol Prop. 30

1. Protasis/ekthesis
Assume a hemisphere with pole T, surface A, and spherical spiral TOIK (area above: ASp), a quadrant ABCD of a maximum circle (area Q), and a segment ABC (area ASg).
Then A:ASp = Q:ASg.
2. Apodeixis
   2.1 Extension of the configuration and transformation of the protasis
   Construct sector AEZC (area S); show that S = Q[1]
   The protasis has now become A:ASp = S:ASg
   2.2 Auxiliary construction
   (set-up for the "exhaustion process")
   On the hemisphere, cut off a sector LTK (area: AL),
   describe a circle on the surface through O, center T,
   cutting off the surface OTN (area A(O)),
   with a sector cut off in it by KT, KL (area A'(O));
   cut off from arc ZA the arc ZE,
   as the same part as KL is of a maximum circle,
   cutting off from S the sector EZC (area: A(E));
   in it, cut off sector BHC (area: A(B))
   2.3 Lemma for the "exhaustion process"[2]
      2.3.1 arc ZE:arc ZA = arc BC:arc AC
      2.3.2 arc TO = arc BC
      2.3.3 AL:A'O = A:AO
      A = area of circle with radius TL
      [*Sph. et Cyl.* I, 33[3]]

---

[1] S is 1/8 of the circle with radius CA, Q is 1/4 of the circle with radius AD, $CA^2 = 2 AD^2$.

[2] Compare Prop. 21: inscribe a sector into the spiral; then compare sector and spiral sector on the one hand, and rotation cylinder and cone-related rotation cylinder on the other.

[3] The reference to *Sph. et Cyl.* is, of course, anachronistic. The material in Prop. 30 probably predates the treatise. Archimedes must have been aware and convinced of these theorems independently of his theoretical work. It seems plausible to assume that he found the results in the context of his pre-formal research activity, using his heuristic method.

A(O) = area of circle with radius TO

[*Sph. et Cyl.* I, 42]

A:A(O) = TL^2:TO^2 [XII, 2]

$\Rightarrow$ AL:A'(O) = TL^2:TO^2

2.3. 4. TO = BC   [III, 29]

TL = AC = EC by construction

$\Rightarrow$ AL:A'(O) = EC^2:BC^2

2.3.5. EC^2:BC^2 = A(E):A(B)   [XII, 2; V, 15]

$\Rightarrow$ AL:A'(O) = A(E):A(B)

2.4 "exhaustion from above"[1]:

Iterate the process described in 2.3, and sum up;

A: sum of all circumscribed (spherical) spiral sectors =

S: sum of all circumscribed partition-induced plane sectors

2.5 "exhaustion from below"

The analogous proposition will hold for inscription instead of circumscription.

2.6 limit process

Imagine the partitions more and more fine-grained.

The inscribed and circumscribed spherical sectors approximate the spiral surface from both sides, and the inscribed and circumscribed plane sectors approximate the segment. The same propositions will always hold. By an implicit continuity argument (a transition to infinity, or an appeal to indivisibles[2]), we infer: they still hold in the limit case, and thus: A:ASp = S:ASg = Q:ASg

Addition: Quadrature of a Spiral-Induced Surfaces on the Hemisphere

Since A = 8Q by [*Sph. et Cyl.* I, 33], we can derive

(a) For the area above the spiral: ASp = 8 ASg

(b) For the area below the spiral

A – ASp = 8Q – 8 ASg = 8(Q – ASg) = 8 triangles ABC,

and triangle ABC = 1/2(1/2d)^2 = 1/8d^2

---

[1] Compare Prop. 21. There the exhaustion from "within," i.e., "below" was discussed at length, and the other case glossed.

[2] An analogous limit argument was used in Prop. 21.

II, 6 Meta-theoretical Passage

6 Meta-theoretical Passage

This passage is a locus classicus for methodology in ancient mathematics. It is perhaps the best-known passage in *Coll.* IV. A doublet can be found in *Coll.* III, and a shorter version in *Coll.* VII. There are to be three kinds of mathematics: plane, solid, linear, corresponding to three kinds of basic curves. In addition, a homogeneity requirement holds: only arguments that use means from the mathematical kind to which the problem belongs are fully valid mathematically. The passage is referred to in Descartes' *Géométrie* (Descartes 1637, p. 315, pp. 40/41 Smith/Latham). Newton also quotes it with approval, and employs it against the Cartesian program in geometry. Up until relatively recently, it was taken to be the communis opinio for mathematics throughout antiquity, and quoted or referred to in secondary literature in this way. In fact, it is, at least in this generality, only to be found in Pappus. For him, it is obviously important. He is committed to this view in the following sense: he uses it to structure his material to give a representative survey of ancient mathematics, to give a coherent methodologically oriented picture of the geometrical tradition. It is not certain, and in fact not all that relevant for the understanding of *Coll.* IV itself, whether this meta-theoretical position was shared, in this full generality, by the mathematicians. Pappus may very well be generalizing a feature to be found in Apollonius' analytical works on locus problems: separate plane problems from solid ones.[1] Still, he is well-informed, competent, and manages to tell a reasonably coherent story. It should be appreciated as a whole. An extensive discussion will not be given here (for the full text, see the translation in Part I). In the present edition, I have taken this passage quite literally, and propose a reading of the whole of *Coll.* IV in light of it. In what follows, I will comment on the two main items in the passage: the mathematical kinds, and the homogeneity criterion, and briefly indicate how the different parts of *Coll.* IV relate to remarks in the passage.

6.1 The Three Kinds of Geometry According to Pappus

There are three kinds of mathematical problems, generalized to three kinds of geometry, according to the means needed to solve the problem or demonstrate features.

---

[1] Cf. Jones (1986a, p. 530, 540/541), Knorr (1989, p. 34) for a similar assessment (Pappus generalizing a trend to be found in Apollonius' plane analytical works); e.g.,: "Pappus is our only explicit authority on this mathematical pigeon-holing, and he says nothing about how it developed and when. However, it is difficult not to see Apollonius' two books on *Neuses* as inspired by the constraints of method imposed on the geometer.... The only conceivable use for such a work would be as a reference useful for identifying 'plane' problems." (Jones 1986a, p. 530). On p. 530f., Jones also voices the opinion that Apollonius may have had a similar purpose in the *Plane Loci* and the *Tangencies*.

H. Sefrin-Weis, *Pappus of Alexandria: Book 4 of the Collection*,
Sources and Studies in the History of Mathematics and Physical Sciences,
DOI 10.1007/978-1-84996-005-2, © Springer-Verlag London Limited 2010

1. Circle and straight line only[1]: *genesis* of these two is in the plane
2. In addition, one or several conics: *genesis* from solid figures (cones or cylinders)
3. Even more complex lines

"Plane" geometry operates with circles and straight lines only. Euclid's *Elements* and *Data*, and anything that can be proved or constructed with these means, would fall into this kind. Within *Coll*. IV, plane geometry is illustrated in Props. 1–18. It becomes apparent that geometry of this kind is not uniform. It allows for a spectrum of styles and approaches. Prop. 1, for example, is directly modeled on the Pythagorean theorem and uses classical synthesis, Props. 4–12 illustrate different facets of the method of analysis within analysis-synthesis, and Props. 13–18 illustrate a monographic style of exposition within synthetic plane geometry. Note that not all problems or curves in the plane are "plane" in Pappus' sense. For example, the conic sections, the conchoid, the quadratrix, and the Archimedean spiral, are drawn in the plane, but they do not belong to the first kind. Neither do the problems that can be addressed with them. The quadrature of the circle and the trisection of the angle are problems set out in the plane, but they are not "plane." *Neusis* constructions, even those formulated for configurations with circles and straight lines, can be either "plane" or "solid." Perhaps a modern reader might think that the separation of circles and conics into different kinds is somewhat artificial. Both these lines can be defined by a mathematical equation specifying defining distance relations in the plane. Pappus, however, thinks the conics are essentially connected, for what they are as objects, to the cone, a three-dimensional figure, whereas the circle is not. In this respect, Pappus may represent a communis opinio among the ancients. For even Apollonius, who favored an operationalist approach to geometry and works with the *symptomata* mostly, does define the conics as sections of cones, and derives their *symptomata* from this essential definition.

The second kind of geometry, encompassing everything that can be successfully treated by employing circles, straight lines, and conic sections, is represented in *Coll*. IV by Props. 31–34 and 42–44. Apparently, Pappus was of the opinion that the geometry of this kind is predominantly analytic-synthetic and that it aims at creating typified configurations. Even the edition of the *Konika* to which he refers (in Prop. 33) was analytic-synthetic. The purely synthetic edition that survives today is due to a revision by Eutocius (sixth century AD). Pappus' portrait suggests that "solid" geometry arose in the context of unsuccessful attempts to solve certain problems, notably the cube duplication and the angle trisection, with "plane" means. The meta-theoretical passage in *Coll*. IV singles out the angle trisection in this regard.

Apparently, Pappus is drawing on criteria from Aristotle's theory of science for the conceptual definition of his kinds. In the *Posterior Analytics*, sciences are defined and determined by the kinds of objects they treat. The methods of the corresponding science must be "akin" to these objects. Pappus' first two "kinds" of geometry are

---

[1] The following slight misreading, already to be found in Descartes, is rather common: restrictions are viewed as pertaining to instruments: third class only mechanical, first class only compass and ruler; Pappus says nothing about instruments, and he certainly counts the third class as full mathematics.

compatible with Aristotle's description of kinds as subject matters. The third one is not. It does not have a positive description, a clear characterization via predicative content. It looks very much like "all the rest." In fact, one might call it unconvincing and unsatisfactory from an Aristotelian perspective. Nevertheless, Pappus will formulate a global homogeneity requirement in the spirit of Aristotle's homogeneity requirement in the *Posterior Analytics* (see the following sections).

For the basic curves of the third kind, Pappus offers a description that is not quite uniform. Two types/approaches can be made out: generation by "varied", "forced" motions and "twisting" of surfaces (quasi-mechanical), and "finding" via the intersection of surfaces in space that are "less structured" than the cones and cylinders used for conics (analytical). Loci on surfaces play a major role for the second type, and for all of these curves it is the "astonishing *symptomata*" that are in focus. Quadratrix-type lines and spiral lines are singled out as basic curves. Pappus lists works and authors, and it does appear that there once was a substantial corpus of treatises in this area, by authors that came after Archimedes, extending into late Hellenistic times and even beyond. Unfortunately, those works are lost. The only complete extant full treatise with geometry of the third kind is Archimedes's *Spiral Lines*. In Props. 19–30, Pappus has discussed exemplary contributions to the geometry of the third kind, depicting a developmental line, as well as typical and crucial problems in exposition and foundation (plane spiral in Props. 19–22, conchoid in 23–25, quadratrix in 26–29, and spherical spiral in 30). He strongly suggests that there were two types, a quasi-mechanical one (cf. Props. 21 and 30, with their informal limit processes) and one that relied heavily on analysis (cf. Props. 28 and 29 in particular). His portrait agrees well with what he says here in the meta-theoretical passage.

Mathematical problems of the third kind can arise out of plane or solid problems by generalization. Props. 35–38 are examples. The *symptoma* of quadratrix and spiral is used for general angle division, and problems that reduce to it can thus be solved. Solid problems form a bridge between the first and the third kind. Other problems of the third kind cannot be related thus directly to problems in the "lower" kinds, because they target properties that cannot be captured by algebraic curves (in modern notation). Circle rectification is a case in point. Props. 39–41 draw out some consequences of the quadratrix's rectification property. Such theorems belong to the third kind "by nature," as it were. As stated repeatedly, the lack of comparable sources creates problems for the evaluation of Pappus' classification. It clearly serves a purpose in *Coll.* IV. To this extent, it is valid and meaningful. Whether it is representative cannot be decided, and should not be inferred (nor denied) from Pappus' relative success at telling a coherent story.

6.2 The Homogeneity Requirement

In analogy to Apollonius' separation of plane and solid locus problems, where problems were differentiated into "classes" according to the minimal means needed to solve them (e.g., identification of plane *neuses*), and where it was required that you use those minimal means only in order to have a geometrically valid argument,[1] Pappus puts forth a generalized homogeneity requirement: All geometrical argumentation must use argumentative devices from the appropriate kind. His formulation of the criterion borrows from Aristotle's theory of science again.

The question immediately arises, of course, as to how one is to decide whether the criterion has been met, i.e., as to how one can, with mathematical means, decide whether a given argument belongs to the plane, the solid, or the linear kind. Also, one might wonder whether this strong meta-theoretical claim represents the mathematicians' perspective on the arguments in their discipline. Pappus is, in all likelihood, drawing, at least to the following degree, on an inner-mathematical discussion that took its starting point from Apollonius' work.[2] Not merely locus problems, but also already existing theorems were scrutinized, via analysis, in order to determine whether they met the requirement. An attempt was made to instrumentalize geometrical analysis to demarcate plane from solid arguments quite generally. This is another systematic technical use of Greek geometrical analysis that has been underestimated in secondary literature thus far. With the help of analysis of loci, Archimedes' *neuses* in *SL* 5–9, e.g., were identified as "solid," and it was argued by some, apparently, that he should have done with a plane argument for *SL* 18 (see below, Props. 42–44). Another example for a proposition criticized in this vein is a construction by Apollonius in *Konika* V. Pappus probably refers to the construction of a normal to the parabola in V, 62. This solution proceeds via conics, in analogy to the case for the hyperbola and ellipse; but since the problem is solvable (once we take the parabola itself as given) by plane means only, Apollonius was criticized for failing to meet the homogeneity requirement.[3]

While the scarce evidence we have suggests that a widening of the discussion on systematic discrimination between plane and solid arguments took place within mathematics, the same cannot be said with regard to linear versus solid problems. Demarcation upward is obviously possible here as well: if you can show, via analysis of loci, that your problem/theorem can be solved via conics, you are done. There are no traces of a systematic attempt to use analysis/*diorismos* to identify conditions under which a "linear" problem or theorem becomes solid. To that degree, Pappus' general homogeneity requirement was not fully developed, or integrated, into

---

[1] In Apollonius this may have been simply a pragmatic device, in line with his operationalist approach, casewise, from simplest to most complex, always with the minimum amount of machinery added.

[2] Before Apollonius' analytical works, such a differentiation, and the corresponding homogeneity requirement, would not have been possible. For Archimedes, or for the pre-Euclidean geometers, it was probably not valid, not even a consideration.

[3] Unfortunately, Pappus does not discuss the Apollonian argument within the preserved text of *Coll.* IV. We may perhaps assume that his argument that Apollonius missed the mark could have taken the form either of explicitly providing a plane argument (as he does in the plane case of the angle trisection), or by showing that the locus used in Apollonius reduces to a plane locus under the specifying conditions in *Con.* V, 62 (an argument like this, not cited by Pappus in *Coll.* IV, was provided for the plane case of the angle trisection by one Heraclius cf. *Coll.* VII, # 72 Hu).

ancient geometry, it seems. Pappus' meta-theory claims more than the practice, or the theory, could do. It should perhaps also be noted, however, that although the criteria for determining homogeneity were not watertight (cf. Props. 42–44), all of Pappus' classificatory judgments in the upcoming third part of *Coll.* IV are correct:

31–34: angle trisection is solid.
35–38: general angle division is linear.
39–41: arc rectification is linear by nature.
42–44: the analyses are correct, and the Archimedean *neusis* is solid.

In Pappus' overall scheme, the geometry of the second kind has the position of a bridge between plane and higher geometry. This may be one of the reasons why Pappus presents his portrait of solid geometry after his discussion of higher curves, and the meta-theoretical passage here (Props. 31 ff.). In addition, it was in connection with the establishment of solid geometry, in differentiation from plane geometry, that the idea of compartmentalizing mathematics along the lines pursued and generalized by Pappus arose. And so, the last part of *Coll.* IV contains solid arguments, a transition to linear problems, and an example of how analysis of solid loci was used to determine the accurate "level" of a problem.

II, 7 Angle Trisection

7 Props. 31–34: Trisecting the Angle

Props. 31–34 are our only sources for the trisection of the angle via conics/solid loci in antiquity. Following up on the introduction of the problem in the meta-theoretical passage, Pappus uses the trisection as an exemplary argument to illustrate mathematics of the second, the solid kind.[1] In his methodological portrait, it looks as though the dominant mode of argumentation in this field was analysis-synthesis, focusing on loci, and that Apollonius was the culminating figure for this discipline, although his work rested on earlier achievements (Aristaeus, *inter alia*) and did not completely supersede them. The arguments in Props. 31–34 are, together with Props. 42–44, Menelaus' cube duplication as reported by Eutocius,[2] and selected arguments from *Konika* V also our only surviving examples for a treatment of solid locus problems, and Props. 31–34 and 42–44 are the only analysis-based ones. As in the case of the *symptoma*-mathematics of motion curves, this uniqueness obviously makes Props. 31–34, presented here in their original context, most valuable sources for historians of mathematics, while also creating the problem that we cannot decide to what degree Pappus' portrait, drawn up with a visible program in mind, is representative of the actual mathematical practice. His portrait should be carefully evaluated on its own terms and as a whole. As in the case of the *symptoma*-mathematics of motion curves, Pappus implicitly traces a developmental line, from the pre-Euclidean treatment of "solid" problems down to Apollonius and his reception. For the portrait of "solid" geometry, the majority of modern commentators agree with Pappus' reconstruction, i.e., their assessment of the development of the ancient analytic treatment of conic sections is congruent with Pappus' account. In addition, there is general agreement on the character of the historical layers detectable in Pappus' report. What has not received enough scholarly attention thus far, and this is, again, parallel to the case of Props. 28 and 29, is the methodological emphasis. The portrait in Pappus stresses the practice of the technique of Greek geometrical analysis for solid loci, as a method of argumentation, in Props. 31, 33, and 34.

> context: trisecting the angle, doubling the cube, arguments of the "solid" kind.
> sources: anonymous pre-Apollonian source with *neusis* in Props. 31 and 32, reshaped with Apollonian theory (Apollonius? Pappus?) in Prop. 33, Prop. 34b draws on an argument

---

[1] The angle trisection, though it may look like a very special isolated question, is indeed rather typical, even exemplary, for problems that can be solved via conics. As pointed out in the introduction to Props. 23–25, the two problems of trisecting the angle and doubling the cube already exhaust solid geometry, in the sense that any problem that can be solved by means of conics reduces to one of these two basic construction problems. The angle trisection is thus a fitting topic for an exemplary illustration of geometry of the solid kind.

[2] Cf. Eutocius *In Arch. Sph. et cyl.* 78–84 Heiberg.

H. Sefrin-Weis, *Pappus of Alexandria: Book 4 of the Collection*,
Sources and Studies in the History of Mathematics and Physical Sciences,
DOI 10.1007/978-1-84996-005-2, © Springer-Verlag London Limited 2010

ultimately based on Aristaeus, via Euclid's *Solid loci*, with at least one intermediate layer;
Prop. 34a is based on Prop. 34b, perhaps by Pappus.

<u>means</u>: II, III, V, VI, *Con.* I, II: an analytic-synthetic version of these books (now lost).

<u>method:</u> analysis-synthesis (Prop. 32 in isolation: synthesis only).

<u>format</u>: problem.

<u>reception/historical significance</u>: no reception in antiquity is attested; reception in Islamic culture; significance for historical scholarship as (the main) source for ancient angle trisection via conics, and as a source for the analytical treatment of solid loci.

<u>embedding in *Coll.* IV</u>: classification of problems: meta-theoretical passage; motif "Apollonius": Props. 8–10, motif "(solid) neusis": Props. 23–25, 42 - 44; motif "analysis as primary investigation method": Props. 4–12, 35–41, 42–44; motif "angle division": Props. 35–38.

<u>purpose</u>: illustrate mathematics of the second kind (solid).

<u>literature</u>: Heath (1921, I, pp. 235–244, II, 119–121), Jones (1986a, pp. 363–371, 573–577, 582–584), Knorr (1986, pp. 128–137, 302–308, 324–327,[1] 1989, pp. 213–224, 316–324),[2] Zeuthen (1886, pp. 210–215, 267–268). Hogendijk's (1981) study of Arabic sources on the angle trisection which contains the same argument as Props. 31 and 32, via *neusis*, but without the reduction to the Apollonian theory of conics, independently corroborates the communis opinio on Props. 31–33. For Prop. 34a and Prop. 34b there is a consensus for the factual content: older layer (older concept of conics, going back to Aristaeus, reshaped partially by using Apollonian theory), but some disagreement remains on the authorship of those re-arrangements and overlays (see below).

7.1 Angle Trisection Through the Ages

With one exception (*Lib. ass.* VIII), all direct testimonies on ancient trisections actually derive from *Coll.* IV (details see below). The problem of trisecting the angle consists in the task of constructing an angle that divides a given angle into three equal parts. This problem, generally, the problem of dividing an angle in a given ratio, arose in the context of constructing regular polygons and inscribing them in a circle. Prop. 38 will point out that one consequence of dividing the angle in a given ratio is that we can now inscribe a regular polygon with any prescribed number of sides into the circle.[3] While bisection of an angle is an easy plane construction (I, 9), trisection resists attempts to solve it with elementary means, as Pappus has pointed out in the meta-theoretical passage.

---

[1] Specifically: Knorr (1986, pp. 128/129) on Prop. 34b; Knorr (1986, pp. 272–276) on Props. 31–33; Knorr (1986, pp. 282–284) on Prop. 34a; Knorr (1986, pp. 303/312) on *neusis*, Apollonius, and Nicomedes; Knorr (1986, pp. 321–328) on Apollonius and Aristaeus as contributors to 34a/34b.

[2] This passage addresses the testimony of Al-Sijzi and Al-Quhi. See Knorr (1989, pp. 247–372) for a comprehensive presentation of Arabic sources with connection to ancient angle trisections.

[3] Cf. Heath (1921, I, p. 235).

7.1.1 Attested Ancient Solutions

Hippias of Elis is usually credited with the invention of the quadratrix, ca. 430 BC.[1] This curve can divide any acute or right angle in any given ratio (cf. Props. 35–38).[2] At some point, the problem of the angle trisection was reduced to a *neusis*, and solved from there, apparently without the use of conics at first (see Props. 31 and 32 and remarks[3]). Archimedes has been suggested as a possible author by Knorr, but a pre-Euclidean origin cannot be excluded. Nicomedes (second century BC) used his conchoid to solve the problem via *neusis*. His argument is not preserved, but the conchoid as described in Prop. 23 can construct the *neusis* in Prop. 31 (see remarks on Props. 31, 23–25).[4] Further indirect evidence for a pre-Apollonian *neusis* – construction which is now lost can be gathered from *Lib. ass.* VIII,[5] together with an Arabic source in the Banu Musa. *Lib. ass.* VIII leads to a *neusis* that can be used for the trisection. This *neusis* is closely related to the Archimedean *neuses* in *SL* 5–9.[6] It is not at all unlikely that this lost ancient solution was by Archimedes. Another possible connection for this particular *neusis* is to the conchoid of a circle, which may have been known to Nicomedes.[7] But the actual ancient constructions do not survive. Apollonian theory made it possible to construct the *neusis* via conics (as in Props. 31 with Prop. 33). Prop. 33 could be by Apollonius, or else by Pappus, on the basis of an argument using the analytical-synthetical version of the *Konika*.[8] Aristaeus, a predecessor of Euclid, probably was the author of a *neusis*-free trisection underlying Prop. 34b. The argument uses the focus-directrix property of the hyperbola and operates with the pre-Apollonian names of the conic sections. In its present form, it is partly reshaped with the help

---

[1] Cf. introduction to Props. 26–29.

[2] Heath (1921, I, pp. 226–227).

[3] Cf. Heath (1921, I, pp. 235–237).

[4] Cf. Heath (1921, I, pp. 239–240); Procl. in Eucl. 272, 3–7 Friedlein; Cantor (1900, I, pp. 335–337).

[5] Cf. Heath (1921, I, pp. 240–241); see also Hogendijk (1981), and the remarks on Props. 31/32.

[6] A *neusis* of this type is subjected to analysis in Props. 42–44, to show that it is "solid." Knorr (1986, pp. 186–187) draws the connection between the trisection via *Lib. ass.* VIII and Archimedes; so does Heath (1921, I, pp. 240, 241).

[7] Cantor (1900) draws the connection to the conchoid of the circle; Knorr (1986, 221ff.) argues that it is plausible that Nicomedes worked with conchoids, including the one on a circle, for his angle trisection. If Nicomedes indeed investigated the conchoid of the circle, it is a tempting possibility to speculate that Archimedes may have experimented with this curve and its properties as well. For Nicomedes seems, in general, to have taken Archimedean contributions as a basis for his own, analytically based contributions. But at present, we do not have enough "hard evidence" for such a thesis.

[8] This argument is not reported in Heath (1921), cf. Prop. 33.

of Apollonian theory.[1] Prop. 34a is the simplest, and the latest of the solutions in
Coll. IV, it uses the same hyperbola as 34b, Apollonian description and techniques,
and is based on Prop. 34b.[2] Its author may very well be Pappus (see below, remarks
before the proof protocols for Props. 34a and 34b).

7.1.2   Islamic Middle Ages (Selective[3])

Hogendijk (1981) analyzed an angle trisection that is closely parallel to *Coll.* IV,
Props. 31 and 32, and avoids the use of conic sections and solid loci. He was able to
show that the Islamic author worked from the original Greek argument, which in
Pappus is overlaid with Apollonian theory, thus corroborating the thesis about the
existence of a now lost ancient Greek source, with a *neusis*, but without conics.[4]
Thabit ibn Qurra's angle trisection was derived from a Greek source as well. Knorr
(1989, pp. 218f.) argues that this source is not Pappus, *Coll.* IV.[5] *Lib. ass.* VIII,
according to Knorr,[6] goes back to the Banu Musa. For their Arabic sources, Knorr
refers to Al-Sijzi and Al-Quhi.[7] Al-Sijzi lists all the trisections known to him, and one
of them is similar to *Lib. ass.* VIII. The same *neusis* is used, according to Knorr, by
Al-Biruni. Omar Kayyam (cf. Katz 1993) made an attempt at systematizing cubic
equations into types, and solving them by geometrical construction. The angle trisection was included. Omar Kayyam's contribution could be interpreted as a precursor
of the seventeenth and eighteenth century project of constructing equations.[8]

7.1.3   Occidental Middle Ages (Selective)

Jordanus Nemorarius and Campanus treated the angle trisection and were probably
influenced by Arabic sources.[9] According to Cantor, Jordanus' argument is parallel

---

[1] Cf. Heath (1921, I, pp. 243–244).

[2] Cf. Heath (1921, I, pp. 242–243); see also Jones (1986a, pp. 582–584).

[3] Cf. Sezgin (1974), and Knorr (1989, pp. 247–372) for information on Islamic mathematics,
specifically on cube duplication and angle trisection, as well as Knorr (1983a, 1989, pp. 216–224)
on the transmission of ancient angle trisections into Islamic culture.

[4] Cf. (Knorr 1989, pp. 267–275) (angle trisection according to Ahmed ibn Musa).

[5] Cf. Knorr (1989, pp. 277–291) for Ibn Qurra's angle trisection.

[6] Cf. Knorr (1986, 197, #107).

[7] Cf. Knorr (1986, 185, #106); on Al-Sijzi's and Al-Quhi's trisections cf. also Knorr (1989,
pp. 293–309).

[8] Cf. Bos (1984, 2001) for a history of this project.

[9] It is doubtful whether *Coll.* IV could have been known in the Middle Ages to any significant
degree. Unguru (1974) argues that a passage from Witelo's *Optics* betrays knowledge of a substantial passage from *Coll.* VI. Commandino p. 95 C provides a plane argument, drawn from
Witelo, in connection with the *neuses* discussed in Props. 42–44. Perhaps this is an indication that
Witelo looked at *Coll.* IV as well, though other explanations are possible, also.

to Al-Sijzi's,[1] connected in terms of mathematical content to a *neusis* with conchoid of a circle. Cantor's reconstruction of Jordanus' source is essentially the same as the one Knorr ascribes to Nicomedes. Also according to Cantor,[2] Campanus' argument is analogous to the earlier one by Jordanus.

7.1.4 Some Examples from Renaissance and Early Modern Times

Bombelli, in the sixteenth century, discovered the connection between the irreducible case of the equation of third degree and the angle trisection.[3] He made similar observations for the cube duplication. Bombelli's investigations on the geometrical interpretation of algebraic results were not published until the 1920s. Therefore, their impact on the development of analytical geometry and the construction of equations was probably minimal. As stated above, in the discussion of the problem of cube duplication (Props. 23–25), Vieta and Descartes, working on an algebraically based approach to geometrical analysis, both proved that all "solid" problems in Pappus' sense reduce to either the angle trisection or the cube duplication.[4] They studied algebraic equations of the third and fourth degree, derived from geometrical configurations. If such an equation is reducible, a geometrical construction with circle and straight line (with ruler and compass) is possible. In the irreducible case, the construction can be accomplished either with the trisection or with two mean proportionals. Newton, also, discusses the angle trisection, cube duplication, and the use of the conchoid for the *neusis* required in several places in the *Arithmetica Universalis*, obviously in close connection to Pappus' text in *Coll.* IV. He uses the new algebraic techniques as well, and contrasts solutions via conics with solutions via *neusis*, polemicizing against the Cartesians.[5]

7.1.5 Nineteenth Century

Azemar/Garnier (in 1809) constructed a trisection curve, generating and discussing it with the means of the theory of functions.[6] A trisection compass is described in Dyck (1892, pp. 225–226).

---

[1] Cf. Cantor (1900, II, pp. 81–82).

[2] Cf. Cantor (1900, II, p. 104/105).

[3] Bombelli ed. Bortolotti (1923, 1929, pp. 265–267).

[4] Vieta's argument can be found in Vieta ed. Schooten (1646, pp. 240–257); it was first formulated in Vieta's *Supplementum Geometriae* from 1593. Descartes (1637, pp. 396/397) (206–209 Latham/Smith) uses parabola and circle for the angle trisection. He also developed an instrument, a kind of compass, for the trisection. See also Descartes (1659, pp. 178 ff).

[5] Cf. Whiteside (1972, V, 426/428) (conchoid for angle trisection); 428–432 (*neusis* reduced to construction via conics, close connection to Prop. 31–33 and Props. 23/24); 458–464 (angle trisection, with explicit reference to Pappus (Prop. 32)); cf. also the solution of cubic equations via neusis in 432 ff. (closely connected to the Archimedean *neusis* from *lib. ass,* VIII), and the summary remarks on solid *neusis* constructions pp. 454/456 and 474.

[6] According to Ver Eecke (1933b, XXXVIII, #1).

Finally, we return to the question that originally motivated the quest for the angle trisection, and the general angle division: the problem of constructing a regular polygon in a given circle. Book IV of the *Elements* shows that quite a few cases can be constructed with circle and straight line (e.g., square, and therefore all polygons with 2^n, $n > 1$ sides, similar constructions from triangle, hexagon, pentagon). With the angle trisection via conics (*Coll.* IV), one will get, in addition, any polygon that entails an angle division that can be composed of divisions by three, and constructions possible with IV. Archimedes gave a construction of the regular heptagon via circle, straight line, and conics. In the ancient sources, no attempt is attested at trying to determine which cases would be plane, which solid, and which linear. The available analytical techniques apparently were not strong enough to determine under which conditions the generally linear problem will become solid, or even plane. Still, the ancients managed to capture almost all of the constructively interesting plane cases, in the following sense. Many centuries later, in the *Disquisitiones Arithmeticae* (Gauss 1801, p. 449 in the 1889 edition), Gauss showed that a regular polygon with n sides is constructible with circle and straight lines (ruler and compass), when n is a Fermatian prime number (in addition to the cases noted above). Gauss only gave this as a sufficient condition. According to Knorr (1986, p. 373), Wentzel, in 1837, was then able to show that the condition is also necessary: no other prime number will do.[1] Fermatian prime numbers tend to become very large soon, so that an actual construction via circle and straight lines becomes uninteresting. Feasible constructions reduce to the cases that are contained in *Elements* IV, or can be gotten from there by simple additional bisection or angle trisection, plus Gauss' construction of the regular Heptadecagon.

7.2   Analysis in Props. 31–34

Analysis is the dominant method for problem solving in "solid" mathematic, according to Pappus' portrait here. This analysis has certain specific features, some of which have so far not received the attention they deserve.

1. Analysis in Prop. 31 and Prop. 34 (also in Props. 42–44) is not deductive throughout. The decisive step is engineered so that the reverse step, used for synthesis, is deductive. Reversibility, not deduction, is clearly the focus. This is relevant in light of a long-standing debate about the nature of Greek analysis. Many scholars claimed that Greek geometrical analysis, because it is essentially reductive, is a purely deductive strategy, and that the mathematicians just counted on convertibility at each step. This would mean that the synthesis and proof would be out of focus for the analysis. Analysis would be a largely independent corroboration strategy on its own. Others have argued that the analysis, even if largely deductive, nevertheless is to be viewed as an "upward" procedure, one that in essence

---

[1] Ver Eecke (1933b, XXXIX, #1), presents the matter somewhat differently.

looks for grounds to start from for a synthesis as its completion. The examples in Props. 31–34 can help decide this issue. They provide evidence that analysis was viewed as not necessarily deductive, and not as independent, but vying, as it were, for a successful "way back." The crucial non-deductive steps in the analyses in Prop. 31, and in Prop. 34 (also in Props. 42–43), make it clear that such steps were fully valid in analysis, and viewed as sufficient, if the converse, used in the synthesis, could be deduced from a valid theorem. This is another aspect in which the mathematical material in *Coll.* IV can add to our information about Greek geometrical analysis in practice, and therefore also to our understanding of the nature of the method. Since the arguments in Props. 31, 34, and 42–44 may very well be severely edited by Pappus, perhaps even originally by him, I should add that they only provide the strong evidence I have suggested if we assume that he was a competent practitioner of the method. The concurrence, or at least compatibility, between Pappus' solid loci arguments and the few examples from Apollonius that we have might contribute to an optimistic estimate.

2. Analysis in Props. 31 and 34 (and Props. 42–44) moves toward stereotyped, typical situations. A single crucial point is focused on, and the analysis shows that it can be constructed as the intersection of solid loci. This strategy has a parallel already in Menaechmus' construction of the cube duplication via solid loci, and is also to be found in Apollonius' analytical work on plane loci. There, too, a single point is identified as crucial, and constructed as the intersection of two plane loci.[1] If this was indeed a typical feature, it seems to make sense. For since all solid problems de facto reduce to either the angle trisection or the cube duplication, it must have been noticed that standard "catalog" examples could regularly be found for solving a given solid problem. And singling out a single point on which the successful analysis, and ensuing construction, hinges is really as "primitive" as it can get. The use and the availability of such standard examples would facilitate the use of analysis for the determination of a problem level, and help make analysis-synthesis arguments in this field partially algorithmic (once you have reduced a problem to a standard locus, you can go through the motions by analogy).

3. In the analysis-arguments of the solid type, conics are seen as loci, characterized by their *symptoma*. This provides a connection to the *symptoma*-mathematics of higher curves in the analytical vein.

4. In Prop. 33, Pappus refers explicitly to an analysis in the first book of Apollonius' *Konika*, and very likely to another analytical argument in the second book. This means that the edition of the *Konika* he worked with was an analytic-synthetic one, not the one we have today (edited by Eutocius considerably later). The original work by Apollonius was in all likelihood dominated by analysis. Prop. 33 is "Apollonian" in character; on the connection between Apollonius and Prop. 33 see the remarks after the proof protocol for Prop. 33.

---

[1] Cf. Jones (1986a, pp. 540–541) for the Apollonian construction of plane loci, and pp. 573–577 as well as Knorr (1989, pp. 94–100) on Menaechmus' cube duplication via solid loci.

The proof protocols for Props. 31–34 will be given in some more detail, including the syntheses, which are only sketched, or even left out in Pappus. Solid locus-arguments are non-elementary and less common, and it may help to have a summary of the steps, even if this means repetition of items covered in the translation already.

7.3 Props. 31–33: Angle Trisection via Neusis

7.3.1 Proof Protocol Prop. 31

1. Protasis/Ekthesis
Start with a rectangle ABCD (clockwise), BC produced.
Task: to draw AZ with EZ = m, m given.

2. Analysis
    2.1 Assumption: problem solved
    2.2 *Apagoge*
    Problem reduces to finding H (parallelogram DHZE).
    2.3 *Resolutio*
    H is *given*.
    [*diorismos*, not explicitly stated: DHZE is to become a parallelogram]
        2.3.1. H lies on a *given* circle (center D, radius EZ).
        2.3.2. H lies on a *given* hyperbola.
        Rectangle BCD *given*; it is equal to BZ × ED [I, 43],
        and BZ × ED is equal to rectangle BZH.
        Appeal to *Con.* II,12, "converse," yields here, in the analysis[1]: H lies on the hyperbola through D with asymptotes AB, BC.
        2.3.3. H is given.

3. Synthesis
    3.1 Kataskeue and Ekthesis
    Through D, draw the hyperbola DHT with asymptotes AB, BC [cf. Prop. 33].
    Draw the circle HK with center D and radius m.
    It intersects the hyperbola in H.
    From H, draw HL ∥ BC, and HZ ⊥ BC; join AZ.
    Then EZ = m.
    3.2. Apodeixis
    Rectangle ZHL, i.e., rectangle BZH, is equal to CD × DA [*Con.* II, 12]

---

[1] This is the decisive, non-deductive step in the analysis. *Con.* II, 12 actually states the reverse: all points H on the hyperbola through D with asymptotes AB, BC will fulfill the above conditions for rectangles/parallelograms. Because we know this, thanks to *Con.* II, 12, we can *conclude*, for the purpose of the analysis, that H must lie on this hyperbola. This is not a logical derivation, but a prospective argument, if you will. We can conclude this way, because we know that the reverse, in the upcoming synthesis, will give us a valid deduction.

$\Rightarrow$ ZH = DE [VI, 2/4; V, 16/18]

ZH $\parallel$ DE $\Rightarrow$ DEZH is a parallelogram, and EZ = DH = m.

The *neusis* in Prop. 31 could easily be constructed with Nicomedes' *first conchoid*, if we choose A as the pole, CD as the canon, and m as the distance. Perhaps this was what Nicomedes did.[1] The Greek text for Prop. 31 shows a separate figure for the analysis, and some inconsistencies in lettering. This observation is quite compatible with the following hypothesis (also in accordance with Hogendjik's findings on the existence of a trisection similar to Props. 31 and 32, but without conics): The oldest layer of Prop. 31 contained just the *neusis*, to be used in Prop. 32, without construction via hyperbola (probably via ruler manipulation). Later on, the *neusis* was constructed by means of a hyperbola, with the analysis either added in later (three layers), or the synthesis of a complete analysis-synthesis adapted to Prop. 32, while the analysis was left as it was. We do not know for certain when and by whom revisions of the older argument were put in place. The older argument may very well be pre-Euclidean. The revisions in the form presented here presuppose Apollonius' version of the theory of conic sections. Apollonius could have been the author of the argument or a similar one. It is also possible, however, that a post-Apollonian author, maybe Pappus himself, added the analysis-synthesis via Prop. 33, or that a post-Apollonian author gave a synthetic argument for Prop. 31, based on Apollonius' analytical-synthetical solution, which was then re-edited by Pappus in such a way that an added analysis made the methodological bias in favor of analysis, and the connection to Apollonius' solution explicit. Further research would be needed to decide upon this question. What can be said safely is that Prop. 31, as presented by Pappus, clearly draws attention to the analytical emphasis in the successful solution of the angle trisection as a problem of the second kind.

7.3.2   Proof Protocol Prop. 32

Task: trisect $\angle$ABC.

Diorismos:

We have three cases[2].

First case: $\angle$ABC is acute.

Extend the configuration:

Create a rectangle BCAZ, produce ZA

The *neusis* from Prop. 31, with m = 2AB = EB solves the problem; the resulting line EB forms with BC an angle that is one third of $\angle$ABC.

[proof via consideration of $\triangle$ BAH, where H is midpoint of DE; use III, 20/31].

---

[1] Other, more complex and sophisticated constructions, using reconstructions of other possible Nicomedean conchoids have been suggested *inter alia* by Knorr (1986, 220ff).

[2] This is the only explicit *diorismos* in *Coll.* IV.

Second case ∠ABC is a right angle.
In this case, we do not need the *neusis*.
We simply construct an equilateral ΔBCD, and bisect ∠DBC.

Third case: ∠ABC is obtuse.
Draw a perpendicular, dividing ∠ABC into a right angle and an acute angle.
Apply cases 1 and 2, and combine the results.

Cases 1 and 3 are in general solid; case 2 is plane, as the constructions indicate. Pappus' homogeneity criterion is essentially met (except for angles that are the 2^n-th part of the right angle). It is worth noting, however, that Pappus did not choose to show, by means of analysis, that in the second case the *neusis* from Prop. 31 becomes itself plane, i.e., constructible with circles and straight lines. He misses out, as it were, on a chance to illustrate the power of analysis as an instrument to demarcate "downward" via *diorismos*. From *Coll.* VII, Prop. 72 Hu we know that this option was open to him in principle. For there Pappus reports a plane analytic-synthetic construction by a certain Heraclitus (or: Heraclius, otherwise unknown) for the *neusis* in this case.[1] Perhaps he thought that to show that a case is plane it suffices to give a plane construction, and the one he gives in Prop. 32 is certainly the simplest one possible. But that leaves open the question of other possible plane cases, still hiding, as it were, under case 1 (e.g., angle of 45° could be trisected via plane means, etc.). Pappus might have opened himself up for further questioning, had he admitted that the *neusis* itself allows for cases which he cannot fully capture with analysis. Even in his standard example, the demarcation between plane and solid cases is not clear-cut and complete "downward" and would not have been even if he had invoked Heraclius' plane construction for the *neusis*. And this has consequences for the evaluation of Pappus' use of analysis as a criterion for determining the level of an argument, a topic that will be taken up in Props. 42–44.

7.3.3 Proof Protocol Prop. 33

1. Protasis/Ekthesis
Start with ∠ABC, point D in interior.
Task: describe the hyperbola through D with asymptotes AB, BC.
2. Analysis
    2.1. Assumption: assume the hyperbola has been described.
    2.2 *Apagoge*
    Draw the tangent A–D–C, diameter HD, DT ∥ BC.
    2.3 *Resolutio*

[1] Cf. Zeuthen (1886, pp. 280–282) for a reconstruction of a possible *diorismos* in terms of analysis of loci, *Coll.* VII, Prop. 72 (pp. 780–782 Hu, Jones (1986a, pp. 202–208), Heath (1921, II, pp. 412 and 413), and Knorr (1986, pp. 298–300) on Heraclius' argument itself; Descartes (1637, pp. 387–389) (188–193 Smith/Latham) discusses the same problem, as a case where a problem with a cubic equation ("solid-looking") can be reduced, with explicit reference to Pappus.

HD, DT *given* in position, T *given.*
AD = DC, because AC is tangent, AB and BC asymptotes.
$AD^2 = DC^2 = 1/4HD \times k$, where k = latus rectum [*Con.* II, 3].
CD = DA $\Rightarrow$ BT = TA.
BT *given* $\Rightarrow$ TA *given*; A *given* $\Rightarrow$ A–D–C *given* in position.
AC *given* in length => AC^2 *given.*
$AC^2 = HD \times k$ => $HD \times k$ *given.*
HD *given* => k *given.*

The problem has been reduced to the following situation: with HD, k *given* in position and length, describe the hyperbola with diameter HD, latus rectum k, and ordinates parallel to AC, which is *given* in position. Pappus refers to an analysis in the first book of the *Konika*. The extant *Konika* do not contain analyses, but cf. *Con.* I, 54 and 55 for a synthetic argument.

3. Synthesis
   3.1. Kataskeue
   Draw DT ‖ BC; construct TA = BT.
   Join AD, produce to C; produce DB, BH = BD.
   Construct k with HD $\times$ k = AC^2 [I, 45].
   Describe hyperbola EDZ with diameter HD, latus rectum k,
   and ordinates parallel to AC [*Con.* I, 54/55].
   This hyperbola solves the problem.
   3.2 Apodeixis
   AC is tangent to the hyperbola EDZ [*Con.* I, 32]
   AD = DC, because BT = TA
   $AD^2 = AC^2 = 1/4HD \times k$
   $\Rightarrow$ AB, BC are asymptotes to the hyperbola EDZ
   [*Con.* II, 1, 2].

The authorship for this very interesting theorem has been the subject of some discussion. Knorr defended the thesis that Prop. 33 is essentially by Apollonius in Knorr (1982). He even argued that the material now found in *Con.* II in this regard is by Eutocius, while Apollonius' own argument is Prop. 33. I would rather agree with his later judgment (Knorr 1989, p. 215) and refrain from a specific ascription, while acknowledging the general Apollonian character of Prop. 33. The fact that many intermediate steps seem to appeal to the *Data* (cf. translation) points to Pappus as the one mainly responsible for Prop. 33 in its present form. So does the explicit appeal to the *Konika* by title. As said in the introduction, Prop. 33 attests that an analytical version of the *Konika* must have existed. This material was used both for Prop.33, and for the analysis-synthesis overlay in Prop. 31 over an older *neusis*. As Hultsch points out in his notes to the Latin translation, a shorter, purely synthetic solution to Prop. 33 could have been given by means of *Con.* II, 4 (as presented in Eutocius' edition of the *Konika*). It would certainly have been accessible to Pappus as well. Apparently, Pappus goes out of his way to illustrate that the methods for geometry of the second kind are essentially connected to the method of analysis (even if, afterward, one might be able to give a shorter, synthetic solution).

7.4   Prop. 34: Angle Trisection Without Neusis

Props. 34a and 34b discuss the trisection of an arc over AC, corresponding to the crucial case "acute angle" in Prop. 32. Only the analysis is given in full; a synthesis is sketched for Prop. 34a, left to the reader for Prop. 34b. The effect of this on Pappus' readers is, of course, that analysis is emphasized as the decisive method of problem solving. In the proof protocols below, I have added a reconstruction of the synthesis so as to illustrate that the analysis indeed carries the burden, as Pappus suggests.[1] As pointed out in the introduction, Prop. 34a is the simplest of all the trisections discussed in *Coll.* IV, and it rests on Prop. 34b. Its style is very close to Props. 42–44. Perhaps Pappus is the author of all four of them. For he claims authorship for Props. 42–44 as presented in *Coll.* IV.

Prop. 34b uses a hyperbola characterized via focus and directrix. The same conditions as the ones discussed in the analysis in Prop. 34b appear in Pappus' commentary on an analytical work by Euclid in *Coll.* VII (# 237 Hu, Jones (1986a, pp. 365–369, # 316/317, with commentary pp. 503–507[2]). There they appear as *symptomata* of a hyperbola in connection with Euclid's loci on surfaces, an argument that in turn seems to be targeting an argument on solid loci by Aristaeus. We encounter in Prop. 34 b an older version of angle trisection via conics as locus curves, one that has been "worked over" in several stages, while the core of the oldest layer, i.e., Aristaeus' consideration of solid loci, was preserved. The final "work-over" is by means of the Apollonian theory of conics. This state of affairs is somewhat similar to Props. 31–33. Apparently, in Pappus' view, the mathematics of the solid kind developed around solid problems, handled analytically, and Apollonius was the culmination of a working tradition, without completely superseding the older contributions.[3] Unfortunately, we cannot identify the Aristaean, Euclidean, Apollonian, and Pappian contributions to Prop. 34b in detail. We lack sources for comparison (e.g., Aristaeus' solid loci; in fact Pappus, in *Coll.* VII, and perhaps this proposition here in *Coll.* IV, is our main source), and too many layers are involved, as it were. Even so, a sufficiently clear and coherent global portrait of the methods of "solid" geometry emerges in outline. In my opinion, Pappus' reconstruction, in Props. 31–34, and Props. 42–44, deserves closer investigation in itself.

---

[1] Commandino (Co 102–103 E and 103–104 E) also provides a synthesis. It covers all possible cases.

[2] According to Jones, there are quite a few problems with the argument as presented by Pappus in *Coll.* VII; the lemma seems to contain several errors. This makes the task of reconstructing the original "Aristaeus" from here all the more difficult. For literature on Prop. 34a/b see the list given at the beginning of the exposition on Props. 31–34, and the footnotes to the section on attested ancient solutions.

[3] The image created in *Coll.* IV by the way Pappus presents the geometry of the solid kind agrees to a large degree with the portrait given by Zeuthen (1886).

7.4.1 Proof Protocol Prop. 34a

Task: trisect the arc AC over chord AC (arc AC smaller than semicircle).

1. Analysis
   1.1 Assumption: problem solved, B divides arc AC in ratio 2:1
   in triangle ACB over fixed AC, $\angle ACB = 2\angle CAB$. We need to show that B lies
   on a uniquely determined hyperbola.
   1.2 *Apagoge*: extension of configuration: points D, E, Z, H
   1.3 *Resolutio*
   BE = AE
   $BD^2 = 3AD \times DH$
   Appeal to *Con.* I, 21 (converse):
   B lies on a uniquely determined hyperbola.[1]
2. Synthesis (only sketched in Pappus' text)
   2.1 Kataskeue/Ekthesis
   Divide AC in H in ratio 2:1; AH = 2HC, AC = 3CH   [VI, 9].
   Describe, through H, the hyperbola with axis AH,
   latus transversum 3AH   [*Con.* I, 54/55].
   It intersects the given arc AC in a point B.
   The resulting triangle ACB has the property
   $\angle ACB = 2\angle BAC$, and B divides arc AC in the ratio 2:1.
   2.2 Apodeixis
      2.2.1 Auxiliary constructions
      Draw perpendicular BD onto AC, D on AC.
      Construct E, Z on AC with DE = DC = EZ   [VI, 9].
      Draw BZ, BE.
      2.2.2 Apodeixis proper
         2.2.2.1. $BD^2 = DA \times 3DH$ [*Con.* I, 21]
         by construction: 3DH = AZ
         $BD^2 = DA \times AZ$
         2.2.2.2. $DA \times AZ = AE^2 - EZ^2$ [II, 6]
         $\Rightarrow BD^2 + EZ^2 = AE^2$
         $BD^2 + ED^2 = BE^2$ [I, 47]
         and ED = EZ by construction
         $\Rightarrow BE^2 = AE^2$, i.e., BE = AE
         2.2.2.3. Consider $\triangle$ AEB; it is isosceles
         i.e., $\angle BAE = \angle ABE$
         $\angle BEC$ exterior angle

---

[1] Non-deductive analysis step as in Prop. 31; because the reverse step, used in the synthesis, is a valid theorem, we can *conclude*, in the analysis, that B lies on that hyperbola. For if it does, the preceding steps of the analysis can be deduced.

$\Rightarrow \angle BEC = 2\angle BAE = 2\angle BAC$

on the other hand: $\triangle BED \cong \triangle BCD$ [I, 4]

$\Rightarrow \angle BEC = \angle ECB = \angle ACB$

3. Symperasma

We have shown that in $\triangle ABC$, $\angle ACB = 2\angle BAC$, and arc AC is divided by B in the ratio 2:1   [VI, 33].

Corollary (Not in Pappus' Text)

With Prop. 34a, one can trisect an acute angle $\angle AMC$.

Choose A on AM; draw circular arc with radius AM; it intersects MC in C (without loss of generality, C can be so chosen); draw chord AC. With Prop. 34a, construct B, dividing arc AC so that arc AB = 2 arc BC. Obviously, $\angle BMC = 1/3\angle AMC$.

7.4.2 Proof Protocol Prop. 34b

Task: On a given arc AC over chord AC, find B so that arc AB = 2 arc BC

1. Analysis
    1.1 Assumption: B has been found ($\angle ACB = 2\angle BAC$).
    1.2 *Apagoge*
    Extension of configuration: draw AB, BC; bisect $\angle ACB$, intersecting AB in D; draw perpendiculars DE, ZB.
    1.3 *Resolutio*
    B is *given*
    [We will need AD = DC, therefore we must have AE = EC[1]]
    E is *given* [midpoint of AC]
    AC: CB = AD: DB [VI, 3]
    AD: DB = AE: EZ [VI, 2/V, 16]
    $\Rightarrow$AC: CB = AE: EZ, i.e., AC: AE = CB: EZ [V, 16]
    since we must have AC = 2AE, CB must be 2EZ
    and $BC^2 = 4EZ^2$
    $BC^2 = BZ^2 + ZC^2$
    $\Rightarrow (BZ^2 + ZC^2): EZ^2 = 1: 4$
    This ratio is *given*.
    Because E and C are *given* as well,
    and BZ is to be a perpendicular onto AC,
    B lies on a uniquely determined hyperbola[2]:

---

[1] *Diorismos*, not given explicitly in Pappus' text.

[2] As in the case of Props. 31 and 34a, this last step of the analysis is non-deductive; its validity rests on the fact that the converse is a valid theorem.

on the hyperbola with focus C, directrix ED,
and eccentricity factor 2.[1]

2. Synthesis (reconstruction, not in Pappus' text)

2.1 Kataskeue/ekthesis

Bisect AC in E, draw perpendicular ED.

Construct hyperbola with focus C, directrix ED,
eccentricity factor 2, as described in *Coll.* VII, #237 Hu.

Let B be the point of intersection between arc AC and the hyperbola. Then B
divides arc AC in ratio 2:1.

2.2 Apodeixis

2 EZ = BC

[hyperbola: $4EZ^2 = BZ^2 + ZC^2$

but $BZ^2 + ZC^2 = BC^2$, (I, 47)]

$\triangle ABZ \sim \triangle ADE$; DB: AD = EZ: AE [VI, 2]

$\Rightarrow$ DB: AD = BC: AC [construction of E]

DC bisects $\angle ACB$ in $\triangle ABC$ [VI, 3]

$\Rightarrow \angle ACB = 2\angle ACD$

By construction, $\angle CAB = \angle ACD$

[I, 4 for triangles AED, CED]

This means that arc AB = 2arc BC [VI, 33].

---

[1] *Coll.* VII, #237 Hu, Jones (1986a, I, pp. 365–369, # 316–317) constructs such a hyperbola. Its
points B fulfill the conditions and proportions analyzed above. Therefore, we can conclude in the
analysis that B lies on this hyperbola.

II, 8 General Angle Division

8 Props. 35–38: General Angle Division and Applications

The transition to the general angle division implies a transition from solid to linear geometry. This is explicitly mentioned by Pappus. Some linear problems will arise from generalizing a plane or solid problem, and this is the case for the angle division.[1] The connection to problems of the second, and even the first kind remains transparent; in fact, Props. 37/38 is an analogue to IV, 10/11. Pappus makes no attempt to single out, via *diorismos*, which of the general cases would become solid or plane. The introductory sentence to Props. 35–38 even suggests that Pappus thought the character of a problem is sufficiently established when an analysis leads to conic sections: the problem is then taken to be solid in general. Again, this has consequences for the evaluation of analysis as a technique to determine the appropriate level of a problem/theorem in 42–44.[2] Apparently, it is limited in power and application, and regularly only used "after the fact," i.e., to subject existing arguments to critical evaluation.

context: angle division, regular polygon construction, generalization of lower level problems.
source: Nicomedes (?).[3]
means: I, II, III, IV, VI, XII, 2, *symptoma* of quadratrix and spiral.
method: synthesis for 35; reduction to 35 (analysis only) for 36–38.
format: problems.
historical significance/reception: /.
embedding in *Coll.* IV: motif "*symptoma*-mathematics of quadratrix and spiral": Props. 19–22 and 26–29, motif "angle section": Props. 31–34, motif "relation of arc to straight line": Prop. 26.
purpose: illustrate how linear problems arise from lower-level ones by generalization, in the context of mathematics of the third kind.

---

[1] Recall cube multiplication, Prop. 24: if Pappus thought it is analogous to general angle division, i.e., linear as a result of generalization, he was mistaken. He is, however, correct in his assessment that the general angle division is "linear."

[2] For singling out the plane and solid cases of general angle division, one would need an instrument comparable in power to Galois theory. As mentioned in the introduction to 31–34, Gauss was able to single out the plane cases. I know of no attempt for solid cases.

[3] *Procl in Eucl.* 272 Friedlein associates Nicomedes with a systematic study of the properties of the quadratrix. So Nicomedes is a possible source for Props. 35–41; cf. also Iambl. apud *Simpl. in Cat.* 192 Kalbfleisch, 65b Brandis. Within the present commentary, we cannot explore the hypothesis that Props. 35–41 are in fact taken from Nicomedes' book on the quadratrix. But that is at least a plausible possibility. Quite a number of connections between 25–27 and 35–41 can be detected, beyond the use of the *symptoma* of the quadratrix.

H. Sefrin-Weis, *Pappus of Alexandria: Book 4 of the Collection*,
Sources and Studies in the History of Mathematics and Physical Sciences,
DOI 10.1007/978-1-84996-005-2, © Springer-Verlag London Limited 2010

8.1   Prop. 35: Angle Division

The close connection between the motion-generated quadratrix and (inscribed) spiral manifests itself nicely in the close analogy of the angle divisions that the curves entail. Compare the following parallel proof protocols.

8.1.1   Proof Protocol Prop. 35

35a Quadratrix

Start with an arc LT, to be divided in a *given* ratio a:b
[implicit *diorismos,* not mentioned: to use the quadratrix, arc LT has to be at most a quadrant; otherwise, bisect, and compose after construction is complete]
1. Kataskeue
   Complete the quadrant BKLT, inscribe the quadratrix KAC.
   Draw the perpendicular AE onto BC.
   Divide AE in Z, so that AZ:ZE = a:b   [VI, 9].
   Draw the parallel ZD to BC.
   Draw the perpendicular DH onto BC.
   BD intersects the arc BKLT in M.
   M solves the problem.

2. Apodeixis
   Arc KT:arc LT = KB:AE, arc KT:arc MT = KB:DH
                                              [*symptoma* of the quadratrix]
   $\Rightarrow$ arc LT:arc MT = AE:DH = AE:ZE   [V, 22]
   $\Rightarrow$ arc ML:arc MT = $\angle$ABD:$\angle$DBC = AZ:ZE = a:b
                                              [VI, 33, V, 17, constr. of AZ:ZE]

35b Spiral

Start with an arc AC, center B, to be divided in H in a *given* ratio a:b

1. Kataskeue
   Complete the circle through A, C with center B.
   Inscribe in it the spiral with generator CB.
   It intersects AB in D.
   Divide BD in E so that DE:EB = a:b   [VI, 9].
   Draw the circle through E, center B.
   It intersects the spiral in Z.
   BZ intersects arc AC in H.
   H solves the problem.

2. Apodeixis
   Circumference:arc AC = BC:BD.
   Circumference:arc HC = BC:BZ

[*symptoma* of the spiral]
⇒ arc AC:arc HC = BD:BZ = BD:BE   [V, 16, V, 22]
⇒ arc AH:arc HC (= ∠HBA:∠CBH) = DE:BE = a:b
                                          [VI, 33, V, 17, construction of
                                          ED:BE].

8.2   Prop. 36: Equal Arcs on Different Circles

8.2.1   Proof Protocol Prop. 36

Start with two circles, centers E and Z, the one with center E is assumed to be the larger one. The task is to cut off arcs of equal length.
Reduction to Prop. 35 (analysis)
   1.1 Assumption: problem solved, arc AHB = arc CD (in length)
   1.2 *Apagoge*: extension of configuration;
   in the circle with center Z, construct arc CT,
   similar to arc AHB; then arc CT < arc CD.
   1.3 *Resolutio*
   Arc AHB:arc CT = $d1$:$d2$
   [equal parts of circumferences, cf. * in the proof of Prop. 26]
   ⇒ arc AHB:arc CT is *given*
   ⇒ arc CD: arc CT is *given* [arc CD = arc AHB].
   The problem has now been reduced to Prop. 35.

Sketch for a construction (not in Pappus): In the smaller circle, choose a sector CZD arbitrarily. With Prop. 35, divide it in the ratio of the diameters ($d1$:$d2$) of the given circles in T, and construct, in the larger circle, a sector AEB, similar to sector CZT (same angle). Then arc AB:arc CT = $d1$:$d2$ = arc CD:arc CT, therefore, arc AB = arc CD [V, 9].

8.3   Props. 37 and 38: Regular Polygon with Any Given Number of Sides

Prop. 37 constructs an isosceles triangle with angles at the base in a *given* ratio to the angle at the vertex (cf. IV, 10). In Prop. 38, the inference is drawn that we can inscribe in a circle a regular polygon with any prescribed number n of sides. For this task reduces to the construction of an isosceles triangle with a fixed vertex angle ($2\pi/n$), thus a *given* ratio of vertex angle to base angles (cf. IV, 11). As one can see, these two propositions are very closely related to the construction of a regular pentagon in book IV of the *Elements*, and also, more loosely, to the other constructions in IV. It is not at all unlikely that the question about general angle division originally arose in the context of attempts to inscribe regular polygons into a circle (cf. above, remarks on angle trisection in the introduction to Props. 31–34). If so, Pappus' choice of examples for mathematics of the second, and the third kind, as expanding beyond the generic limits of plane geometry was well-taken, and yields a well-rounded portrait. With the final pair of propositions, on general angle division, we return to a question that goes back to the beginning of a developmental line.

8.3.1   Proof Protocol Props. 37 and 38

1. Protasis
Task: construct an isosceles triangle with the angles at the base in a fixed ratio to the angle at the vertex.
2. Analysis
    2.1 Assumption: $\triangle ABC$ solves the problem.
    2.2 *Apagoge*: extension of the configuration.
    Extend AB, complete semicircle ACD, radius BC = BA; join CD.
    2.3 *Resolutio*

$\angle$CAB:$\angle$ABC is *given*	[by hypothesis, $\angle$B is vertex angle].
$\angle$ABC = 2$\angle$ADC	[III, 20].
$\Rightarrow$ arc CD:arc AC is *given*	[VI, 33].

The situation has been reduced to the construction of a semicircle ACD, to be divided in a *given* ratio in C. By appeal to Prop. 35, C is *given*, and the sought triangle is *given* in kind.

3. Synthesis
    3.1 Kataskeue/Ekthesis
    Draw EH, divided in Z in the *given* ratio; bisect ZH in T.
    Construct a semicircle over AD, center B.
    Divide the arc AD in C so that arc CD:arc AC = EZ:ZT.   [Prop. 35];
    $\triangle$ ABC solves the problem.
    3.2 Apodeixis

$\angle$DAC:$\angle$ADC = EZ:ZT	[by construction]
$\Rightarrow$ $\angle$DAC:$\angle$ABC = EZ:ZH	[III, 20]

Since any regular polygon in a circle can be divided into isosceles triangles where the angles at the base have a *given* ratio to the vertex angle, we can, with Prop. 37, inscribe a regular polygon with any prescribed number of sides into the circle.

II, 9 Quadratrix, Rectification Property

9 Props. 39–41: Further Results on *Symptoma*-Mathematics of the Quadratrix (Rectification Property[1])

This second set of problems of the third kind exemplifies mathematics when the problem is "linear" by nature. Props. 39–41, like Props. 35–38, belong to the second path of development for the mathematics of motion curves, the analytical track. The analysis focuses on reduction to the *symptoma* of the quadratrix, and to the rectification property in Prop. 26. A certain tendency for setting out the quadratrix in a separate auxiliary figure and arguing "parallel" can be detected, but with the slim observation basis we have, we cannot be certain that this is typical.

> context: ratio of arcs to straight lines; rectification of circle; commensurability for arcs.
> source: Nicomedes on quadratrix (?).
> means: VI, X (the latter for Prop. 41), Prop. 26.
> method: analysis-synthesis for 39 and 40, synthesis for 41.
> format: problems.
> historical significance/reception: /.
> embedding in *Coll.* IV: motif "quadratrix and properties": Props. 26–29; motif "*symptoma*-mathematics of motion curves": Props. 19–30; motif "incommensurable magnitudes": Props. 2/3.
> purpose: illustrate *symptoma*-mathematics of the third kind, where the problem is "by nature" linear (does not allow for plane/solid sub-cases).

9.1 Prop. 39: Converse of Circle Rectification

9.1.1 Proof Protocol Prop. 39

Task: find a circle, the circumference of which is equal to a *given* straight line c.
1. Analysis
    1.1 Assumption: circle a has been found,
    with circumference c laid out as a straight line.
    1.2 *Apagoge*: extension of the configuration
    Construct a circle b and rectify it with the quadratrix   [Props. 26/27].
    Result: straight line d, equal to the circumference of b.
    1.3 *Resolutio*
    radius (a): radius (b) = c:d
    [XII, 2, *Circ mens* I, V, 15, cf. * in proof protocol for Prop. 26]
    c:d *given*; radius (b) *given*   [by construction]

---

[1] My assessment differs slightly from Knorr's, who mentions these propositions in passing and connects them to the angle division property of the quadratrix (Knorr 1989, p. 214).

H. Sefrin-Weis, *Pappus of Alexandria: Book 4 of the Collection*,
Sources and Studies in the History of Mathematics and Physical Sciences,
DOI 10.1007/978-1-84996-005-2, © Springer-Verlag London Limited 2010

⇒ radius (a) *given*   [*Data* 2].
2. Synthesis (not in Pappus[1])
  2.1 Kataskeue
  Construct a circle b with radius r
  Rectify the circle with quadratrix, resulting in straight line d [Prop. 26]
  Construct r' with r:r' = d:c   [VI, 9].
  Draw the circle a with radius r'.
  It solves the problem.
  2.2 Apodeixis
  Circumference(b):circumference(a) = r:r'
  [XII, 2, *Circ. mens.* I, V, 15, cf. * in proof protocol Prop. 26]
  r = r'= d:c [by construction].
  d = circumference (b) ⇒ circumference (a) = c   [V, 9].

9.2   Prop. 40: Construct a Circular Arc over a Line Segment, in a Given Ratio

Prop. 40 obviously has a connection to the following longstanding question in
Greek geometry (cf. above, excursus on squaring the circle): are circle and straight
line comparable, i.e., can they be brought into a ratio? Note that the ratio for Prop.
40 underlies certain restrictions. In modern notation, it is obviously >1:1, because
the arc is always longer than the chord. Because of the set-up of the argument used
in Prop. 40, the ratio is also at most π:1 (no angles larger than 180° are considered in
Greek geometry).

9.2.1   Proof Protocol Prop. 40

1. Analysis
  1.1 Assumption: problem solved
  arc AB:AB equal to the given ratio
  1.2 Analysis proper
    1.2.1 Extension of configuration
    C midpoint of arc AB, X center of the circle, draw XC,
    R on AB;
    Auxiliary figure: quadrant ZHE of an arbitrary circle,
    quadratrix ZK, ∠EHL, equal to ∠CXA; points L, M,T, N.

---

[1] cf. Co p. 108/109 F

1.2.2 *Apagoge*

Arc AB:AB = arc AC:AR = arc LE:LM

[by construction[1]]

arc LE:TN = LH:HK

[*symptoma* of the quadratrix, Prop. 26]

Δ HLM with TN ∥ LM

TN:LM = TH:LH   [VI, 4]

arc LE:LM = TH:HK   [V, 16, V, 23]

1.3 *Resolutio*

arc AB:AB *given* ⇒ TH:HK *given*

HK *given* in length   [quadratrix]

⇒ TH *given* in length   [*Data* 2]

T on a uniquely determined circle, as well as on the quadratrix

⇒ T *given* ⇒ HL *given* ⇒ ∠EHL *given* ⇒ ∠CXA *given*

CX *given* in position [perpendicular bisector of AB]

A *given*, AX *given* in position   [*Data* 29]

⇒ X *given* ⇒ arc AB *given*

My reading of the final steps as presented here differs from Hultsch's (235, #3 Hu). He argues as follows: A is *given*; ∠AXB is *given*, therefore B is *given* (*Data* 90). Therefore, AB is given in position and length (*Data* 26); therefore, arc AB is *given* in length. In my opinion, this reading has two weak points. First of all, it tries to establish that AB is *given*, but AB is already postulated in the formulation of the problem. Furthermore, I would argue, it is not enough to show that arc AB is *given* in length. For that is already clear when TH is shown to be *given*. One has to show that arc AB is *given* in position as well in order to construct it.

2. Synthesis (reconstructed, only sketched in Pappus)

2.1 Kataskeue

Quadrant ZHE with quadratrix ZK.

Choose D on HZ so that DH:HK equals the given ratio   [VI, 9].

Draw the circle with radius DH, center H.

It intersects the quadratrix in T.

Draw the perpendicular TN, join TH (intersecting quadrant in L), and the perpendicular LM.

Construct the perpendicular bisector RX of AB (R on AB).

Transfer ∠HTN to A, onto AB (vertex A).

It intersects RX in X (i.e., X can be so chosen).

Draw the arc AB, through A, with center X.

By construction, it passes through B.

Arc AB solves the problem.

---

[1] One has to appeal to the proposition used already in Props. 26, 36, and 39: XII,2, *Circ. mens I*, V, 15: arcs in the same ratio as radii. Then consider similar triangles, half-chords, perpendicular on chord; the ratio of radii can be replaced with the ratio of half-chords. Cf. * in the proof protocol of Prop. 26.

2.2 Apodeixis
ΔARX ~ΔHTN ~ΔHLM
⇒ ∠AXR = ∠LHM
⇒ arc AC:AR = arc LE:LM   [VI, 33]
arc LE:LM = TH:HK   [quadratrix]
TH = DH
⇒ arc AB:AB = arc AC:AR = DH:HK (the *given* ratio)

9.3   Prop. 41: Incommensurable Angles

The incommensurability motif connects this proposition to Props. 2/3: construction
of two *irrational* lines. Properly speaking, incommensurability was defined for
straight lines, and it involves consideration of ratios. Since the *symptoma* of the
quadratrix, together with its rectification property, establishes an equivalence
between the ratio of two arcs and that of two straight lines, it can be used, in quite
an obvious way, to define "incommensurable" and "*irrational*" angles (or arcs).

Set out a quadrant and a quadratrix in it. To construct two incommensurable
angles, simply set out two incommensurable lines BH and BT on the side of the
quadrant, and corresponding perpendiculars ND, KE, as well as angles ∠EBZ,
∠DBZ, via the quadratrix. Then the ratio of the arcs (or angles) will be equivalent
to KE:ND, i.e., to BH:BT, by the *symptoma* of the quadratrix. And since BH, BT
are incommensurable, so are the angles.

II, 10 Analysis for an Archimedean Neusis

10 Props. 42–44: Analysis for an Archimedean *Neusis*/Example for Work on Solid Loci

10.1 General Observations on Props. 42–44

An Archimedean *neusis* from *SL* was mentioned in the meta-theoretical passage in connection with a violation of the homogeneity criterion. It was claimed that the *neusis* is solid, where a plane argument might have sufficed. In Props. 42–44, Pappus employs geometrical analysis in order to show that the Archimedean *neusis* is really solid, and also to present a solid argument that is, in his view, useful for many solid locus problems. That is, Props. 42–44 are once again designed from a methodological point of view, in a twofold way. First of all, they illustrate what was meant by the homogeneity criterion for methods of argumentation, and how one would proceed when demarcating plane from solid geometry via analysis. The *neusis* discussed in Props. 42–44 is related to the ones used in *SL* 5 –9, closest to *SL* 9. It is indeed solid, although Pappus' argument is not quite able to prove this, because geometrical analysis is limited in this regard; see discussion below. It seems as though Descartes was quite familiar with Pappus' attempts. His criteria for determining the appropriate level of a problem, though not as far-reaching as he himself hoped, connect to Pappus' attempts here and are vastly superior.[1]

Besides being an illustration, or implementation, of the method to test an argument for concurrence with the homogeneity criterion, Props. 42–44 are also meant to be exemplary, a model, or prototype for how one might deal with a whole class of solid problems. A certain tendency toward algorithmization can be detected. It is in line with similar observations on Props. 31–34 (cf. introduction to Props. 31–34, analysis in 31–34). Both groups together spell out in a rather concrete way, with a methodological emphasis, what Pappus stated about the solid kind of geometry, and its differentiation from plane geometry in the meta-theoretical passage.

context: homogeneity criterion, analysis used to determine level of an argument; solid locus problems/theorems.
source: Pappus, based partially on work by Aristaeus.
means: Data, II,V,VI, *Con.* I.
method: analysis.
format: problem.
reception/historical significance: no reception is attested, but Props. 42–44, together with 31–34 and a few examples from *Con.* V are our only examples for ancient mathematical

---

[1] Cf. Descartes (1637, pp. 383–402) (180–219 Smith/Latham). In the expanded Latin edition of 1659 (Schooten), Prop. 43 is discussed on pp. 174 ff., whereas pp. 34–35 report some Cartesian remarks on the division of problems of the second from those of the third degree.

H. Sefrin-Weis, *Pappus of Alexandria: Book 4 of the Collection*,
Sources and Studies in the History of Mathematics and Physical Sciences,
DOI 10.1007/978-1-84996-005-2, © Springer-Verlag London Limited 2010

arguments with solid loci; the arguments in the extant *Konika* have the synthesis only. Thus, 42–44 are important as historical source texts. Also, Props. 42–44 are testimonies for a use of Greek geometrical analysis which has so far been underestimated: it could be employed, systematically, to determine the inherent level of an already existing mathematical argument, i.e., it could be employed for methodological evaluation.

embedding in *Coll.* IV: connection to the remark on an Archimedean *neuses* in *SL*, in the meta-theoretical passage; motif "Archimedes": 13–18, 19–22, 30, 35; motif "analysis of solid loci": 31–34; motif "analysis as a method in connection with *symptoma*-mathematics": 28/29; motif "analysis": 4–12; 31–34; motif "*neusis*": 23–25, 31–33.

purpose: illustrate how analysis can be used to determine the level of a theorem or problem, when that level is not obvious from the means employed explicitly; illustrate an exemplary path of reasoning when working with solid loci.

literature: Baltzer apud Hu II, pp. 1231–1233, Heath (1921, II, pp. 386–388), Jones (1986a, pp. 573–577), Knorr (1978b, 1978a, 1986, 1989, p. 228 with #25 [1]), Tannery (1912, I, pp. 300–316), Zeuthen (1886, pp. 263–265).

In what follows, some general remarks will be given on the following topics:

1. Background of 42–44: Archimedean *neusis*, criticized for being solid.
2. Purpose of 42–44: show, via analysis, that the *neusis* is solid; also, the analyses are said to be useful for many other solid locus problems (i.e., they are examples for how to work in the second kind of geometry).
3. Analysis as a criterion for determining the level of a problem: its ingredients.
4. Limits and gaps for the analysis-criterion.
5. Typification as a feature of analysis of solid loci.
6. Aristaeus as a possible source for 42–44.
7. Possible alternatives, avoiding the *neusis*, for *SL* 18.

10.1.1   Criticism of Archimedes' Use of a Neusis in SL

Archimedes' *SL* contains, as 1–11, preliminary lemmata on the spiral as motion curve. They are separated off from the treatise proper, as "lambanomena," and used in *SL* 12ff. as quasi-archai, with the definitions placed between these lambanomena and the actual treatise. *SL* 12 ff. become *symptoma*-mathematics, meeting in themselves the most rigorous standards for geometrical argumentation, and avoiding any reference to motions. *SL* 5–9 use *neuses*, without giving an explicit construction.[2] These *neuses* are indeed solid, i.e., they require for their construction conic sections or solid loci, unless additional limiting conditions apply. *SL* 18 appeals to the *neuses* in *SL* 7 and *SL* 8 within an exhaustion proof.[3] The criticism voiced by mathematicians after Archimedes

---

[1] The last-mentioned reference concerns a solid *neusis* by Al –Jurjani, which, according to Knorr, is very close to Props. 42–44 and to the Archimedean *neuses* in *SL* 5–9. Al-Jurjani's complete solution is purely synthetic, whereas Pappus only provides an analysis, given that his purpose is not an actual solution of the *neusis* per se.

[2] Heath argues that since Archimedes uses only existence, not construction, his argument is not solid; I doubt that this line of reasoning would have impressed an ancient mathematician, since existence arguments by way of appeal to the continuity principle were not used in geometrical argumentation.

[3] *SL* 18: The circumference of a circle circumscribed around a spiral of first rotation is equal in length to the subtangent for a tangent in the endpoint of the first rotation.

is anachronistic in the following sense: it could only be formulated after Apollonius' work on conic sections and on plane versus solid analysis which made a systematic discrimination between plane and solid locus problems (including plane and solid *neuses*) possible. The criticism is therefore certainly post-Archimedean. Archimedes himself could not have used the standards it rests upon, and probably considered *neuses* themselves as a legitimate argumentative tool.[1] But the objection is, in principle, not without base, or beside the point: the *neuses* in *SL* 5–9 are solid, in retrospect, if you will. Another question is whether it is possible to prove *SL* 18 with plane means only. See below.

10.1.2 Pappus' Purpose in Props. 42–44, and the Content of 42–44

As Pappus declares, he intends to give an analysis for (one of) the Archimedean *neuses*, so that the reader will not be puzzled when going through Archimedes' book on spiral lines. In light of the meta-theoretical passage, this obviously means that Pappus intends to show that the *neuses* are solid, and to do so via analysis. And this is indeed what happens in Props. 42–44. Furthermore, Pappus claims that the arguments he is going to give are useful for many other solid problems, too. On this, see below, #5. Pappus' intentions in 42–44 have often been misunderstood. Hultsch and Eberhard were of the opinion that 42–44 either do not target Archimedes, or do not qualify as critical evaluations (cf. Hu ad locum). It is true that 42–44 do not specifically address *SL* 7, but the argument, though most closely related to *SL* 9, is applicable in an analogous way to *SL* 7 and 8. It is exemplary for an analysis of Archimedes' *neuses*. Perhaps Pappus chose, out of several possible analyses, the one that best serves for "many other solid problems". He chose one that is most closely connected to the *neusis* for the angle trisection (which is, in fact, one of the two problems to which all solid problems reduce). Knorr (1978) is correct in pointing out that the *neusis* in 42–44 is most closely connected to the *neusis* for angle trisection. However, he furthermore claims that the argument is therefore not directed at Archimedes' *neuses* in *SL*, but at the *neusis* in *Lib. ass.*VIII. This almost amounts to assuming that Pappus was not aware which book by Archimedes he was looking at. Against Hultsch, Eberhard, and Knorr, and with Zeuthen, Tannery, Heath, and others, I regard it as certain that Props. 42–44 target Archimedes, *SL* (which is not to say anything yet as to their validity). Furthermore, Knorr (1978, 1978b, 1986) reads Props. 42–44 as a misguided attempt on Pappus' part to provide a plane constructive solution for the *neusis*. The fact that the analyses in 42–44 do not provide such a solution, but would lead straightforwardly to a solid one, is taken by Knorr as a sign of utter confusion on Pappus' part. It must be stated that, at least in this respect, Knorr's reading has no basis in the text. Pappus simply does not announce, and does

---

[1] Cf. above, remarks on *neusis* in the introduction to Props. 23–25.

not attempt, a plane solution in Props. 42–44.[1] He announces, and gives, an analysis. The analysis leads to conics/solid loci, which is in complete accordance with what Pappus has said about the *neuses* in the meta-theoretical passage. The majority of scholars have acknowledged the solid, and intentionally solid character of Pappus' arguments in Props. 42–44, and the fact that they target *SL* 5–9.

10.1.3   Analysis in 42– 44 as a Criterion for Establishing the "Solid" Nature of the Neusis

Props. 42 and 43 give an *apagoge* only of an analysis. Under certain general conditions, a certain point must lie on a hyperbola (Prop. 42) and on a parabola (Prop. 43). No diorismos to specify plane, and filter out impossible cases is given, and no *resolutio*, either. Prop. 44 then proceeds to the analysis of a *neusis*. This analysis leads to the conditions investigated in Props. 42 and 43. The crucial point lies on a parabola and on a hyperbola. It is constructible via solid loci, in general. Note, once again: an analysis, and an analysis only is presented; not a solution, and that is also not the point. No synthesis, no construction, no diorismos, and no proof are forwarded. A *diorismos*, as reconstructed, e.g., in Tannery (1912) and in Baltzer apud Hu,[2] reveals that a complete analysis, with the intent of ultimately constructing the *neusis*, leads to several cases, some solid, some plane, some unsolvable. Pappus is content to have shown that the *neusis* is, by nature, and unless specifying conditions are brought to bear, solid, because geometrical analysis leads to solid loci.[3] Has he shown that, however, and if so, in what sense?

10.1.4   Limits and Gaps in the Pappus' Account

In fact, Pappus' criterion is not sufficient to prove that the *neusis* is solid. His analytical method does, in principle, not work infallibly in a general way. For Greek geometrical analysis works on specific configurations only, and one can never guarantee, a priori, that one has used all the information that might lead to a specifying condition, pushing the level of the problem down. And this means that analysis can successfully prove demarcation of the level of a problem "upward" only, in this case show that the problem, if solvable at all, is at most solid. It cannot prove that all information has been exhausted, and therefore it cannot demarcate "downward," i.e., show rigorously that no plane method would suffice. That, however, was the goal.

---

[1] Cf. a similar misunderstanding of the analyses in Props. 28/29 as attempted solutions.

[2] Cf. Tannery (1912, p. 307/308), Baltzer apud Hu (Hu appendix, pp. 1231–1233). A complete analysis, with the intent of ultimately constructing the *neusis*, leads to several cases, some solid, some plane, some unsolvable; Zeuthen (1886, pp. 263–265) gives a complete analysis-synthesis, with ancient means; cf. also Heath (1921, II, pp. 386–388).

[3] Compare the introductory phrase to Props. 35–38. Pappus seems to believe that an analysis leading to conics actually shows that a theorem is solid.

So what has Pappus achieved with his analysis? He has achieved a high degree of plausibility. He has clarified the situation at hand, provided grounds for an argument to either proceed to a constructive proof of the *neusis*, or to specify further. And this is essentially what geometrical analysis can do, even within mathematics (cf. above, excursus on analysis-synthesis in the introduction to Props. 4–12). It does not provide a proof, it provides grounds for argumentation.

For the meta-theoretical question here, Pappus (or whoever produced such analyses of existing *neusis* arguments) would have to count on it that if additional information is available in the configuration, someone else will detect it, and show that one can push the level of the problem one down in this specific case. If over a longer period of time (and the time between Apollonius and Pappus is quite long) no one has come up with an argument that shows that additional restrictive conditions are implied, the thesis gains plausibility. The method expects, it seems, an argumentative context of continued investigation and discussion. It does not give a final answer in the sense of a proof; such an answer would be reached only if we get to the plane level. Still, it is not useless. It may very well reflect what actually did go on in Hellenistic mathematics, in the field of the investigation of solid locus problems and in meta-theory via analysis (investigation in the framework of a well-defined discourse, with standard tools and topoi, but both open-ended and oriented toward concrete problems, with proof character assigned only to synthetic arguments, where the truth and generality of a conclusion can be asserted, but remains relative to the principles set down as starting points). Such a view on the nature of analysis for meta-theoretical questions is perfectly in line with an understanding of inner-mathematical analysis as a structured and systematic investigative tool that does not have proof character, but yields material insight into the constitutive ingredients of a question at hand (as a heuristic method, essentially). It is not necessary that analysis have proof character in order for it to be a successful and valid mathematical technique. Neither is that required on the meta-level. Geometrical analysis as a means to determine the level of a problem is not watertight. It is, however, practicable and provides argumentative grounds both for the claimed level of a proposition and for further investigation.

One way of determining "downward" in a concrete case would be to actively search for conditions under which the problem would become plane. Then one would be able to "capture" some of the plane cases. But again, as long as one cannot be sure one has exhausted all possible additional specifications, one would not be sure about a particular case, unless one knows it is plane. Plane cases of the above *neusis* exist, as Tannery has shown, and they are accessible to the means available to the ancients. According to Jones (1986, p. 530) attempts at separating out plane cases from higher general problems were made (though they could not be made in a completely exhaustive systematic way, and were only satisfactory if leading to a plane locus for a particular case). Above, in the discussion of Prop. 32, an example for a plane case differentiated out was mentioned. It is discussed in Zeuthen (1886), and in Descartes, as a case where a solid locus itself reduces to a plane locus (188–191 Smith/Latham, cf. above, comments on Prop. 32, plane case). It is not at all implausible that Apollonius devoted much of his energies and attention to a project of systematically exhausting all construction problems that are plane, and that would mean that one has "proved," for any other problem, that it is (at least) solid. This way, one might proceed, on the basis of an exhaustive classification of all possible concrete plane cases, toward a rigid demarcation between plane and solid problems. Perhaps this is what motivated Apollonius in his minor works. The question cannot be pursued here.

Another limitation of Pappus' method of determining the appropriate level of an argument is that it was apparently only applied to the question of plane versus solid arguments. There are no traces, and Pappus also makes no attempt in this direction, of an operationalization of analysis for the demarcation downward for problems of the third kind (cf. above, Props. 35–38). Apparently, there were only techniques available that moved generally in the area where Apollonius had also worked: plane and solid loci, and their investigation via analysis, with a view to finding the simplest construction means possible. While analysis was used for working on loci on surfaces (extension of the field of application for Apollonian techniques), this project did not get far enough to explore systematically the connections to the solid loci, or the plane loci.

Finally, the following gap in Pappus' account (though not in his method of analysis as a criterion for the level of an existing problem solution) has to be acknowledged. Perhaps it is due to the fact that the manuscript is damaged and incomplete at the end of *Coll.* IV. There is no discussion of the plane problem in Apollonius, where Apollonius uses a solid construction: normal to a parabola in *Con.* V, 58/62. This problem was mentioned alongside with Archimedes' *neusis* in the meta-theoretical passage.[1]

10.1.5   Solid Loci in Props. 42–44, in Comparison to Props. 31–34

Pappus claims that the analyses in Props. 42–44 are useful for many solid problems. This is not an implausible claim. According to Zeuthen (1886, p. 272) and Knorr (1986, pp. 300 ff.), investigation of *neuses* was an important part of ancient work with conics. Furthermore, the *neusis* is closely related to the *neusis* for angle trisection in *Lib. ass* VIII. Generally speaking, since all solid problems reduce to either angle trisection or two mean proportionals, reduction to standard configurations – of which the *neusis* in Props. 42–44 could very well have been one – related to one of these problems would have been an effective strategy in working on solid problems.[2] On the treatment of solid loci compare also the remarks on analysis in Props. 31–34 in

---

[1] Zeuthen (1886, pp. 284–288) and Tannery (1912, pp. 302–305) give a construction, and an argument why the problem would be classified by Pappus as plane. For an argument, with ancient means, showing that the locus in question becomes plane see the above-mentioned passage from Zeuthen, and Knorr (1986, pp. 319–321).

[2] Zeuthen (1886, pp. 272–278) gives a survey of problems for which a complete analysis-synthesis for 42–44 might have, or could have, been used. No ancient sources survive.

the introduction to Props. 31–34. Perhaps Pappus' "standard" analysis, claimed to be useful for a number of solid problems, is even representative of how the ancient mathematicians worked in this field.

Props. 31–34, and 42–44 are our only remaining sources on ancient analytical treatment of solid loci. Of course, Pappus may have given a very idiosyncratic picture, but it may have been one that could make sense of the actual practice. Thus, it would seem to be a reasonable task to look for common features in these documents from Pappus, as parts of his overall portrait of solid geometry. Such features may have been typical for ancient work on solid loci in general. A detailed discussion cannot be given here. In my opinion, it is not implausible that it would result in a picture that fits with Zeuthen's 1886 book on the ancient treatment of conics in many essential respects. The following general observation must suffice for the purposes of the present commentary. It looks as though reduction to standard catalog configurations was a typical strategy. This reduction works with the *symptomata* of conics (not the definitions). There is a certain preference for reduction to *Con.* I: basic *symptomata* of conics. One decisive point is singled out, and is shown to be constructible as the intersection of a conic section and either a circle or another conic section. This strategy would obviously contribute to a procedural standardization, and facilitate the reduction to standard examples or configurations. A problem can be brought to the point, as it were, and there is a good chance that the resulting reduced configuration was captured in one of a set of standard examples (specific parallels between Props. 31–34, and 42–44).[1]

The last step of the analysis is non-deductive, but valid because the reverse in synthesis would be an appeal to a theorem. The same characteristics were observed in the analysis of Props. 31 and 34.

In detail, the parallels are closest to 34a; the same phase of analysis is used in 34a and 42/43, and the reduction is to the same set of basic *symptomata*: *Con.* I, 20/21.

10.1.6 Aristaeus as a Possible Source

Pappus himself claims responsibility for 42–44 in the form in which he presents it. However, this does nor exclude that he selected and edited another source. For there are, as in the case of Prop. 34, traces of a pre-Apollonian treatment of conics/solid loci in Props. 42–44. Tannery (1912, p. 308), Knorr (1986, p. 323 ff.), and Jones (1986, pp. 572–584) have argued that Pappus used an argument by Aristaeus as source for his analyses in Props. 42–44. In what follows, I summarize Jones' arguments. Pappus had no access to any other major comprehensive pre-Apollonian work on solid loci besides Aristaeus. Props. 42–44 contain only the analyses, in contrast to synthetic constructions already available before Apollonius (this argument is perhaps not entirely compelling, given that Pappus would have given only the analysis, even if his source had the synthesis, also). The definitions used for the loci

---

[1] Cf. also Knorr (1989, pp. 94–100) on Menaechmus's cube duplications.

coincide with the pre-Apollonian *symptomata* for the curves. However, this layer of technical vocabulary is combined with later, Apollonian terminology: (e.g., parabola, hyperbola). Jones believes that Pappus himself is responsible for this fusion of Aristaean analyses with Apollonian techniques and labels. The situation is analogous to 34a versus 34b, and this observation makes it plausible that 34a is also by Pappus. The degree to which Pappus (or an anonymous post-Apollonian author) revised the material from Aristaeus cannot be determined. It is possible that we have, in Props. 42–44 (much as we do in Prop. 34b) an indirect testimony on Aristaeus' solid loci.

A similar picture as in Props. 31–33 and 34a/34b emerges in Props. 42–44. The mathematics of the second kind was analytically dominated. Apollonius' contribution was the last word, the completion of the theory in this field. But his contributions did not completely replace the earlier work in this area. The earlier works, such as Aristaeus's solid loci, remained valid and useful. In this respect, Apollonius' role was unlike Euclid's, whose *Elements* replaced all predecessors in plane geometry.

10.1.7   Alternatives for the Neusis

Though Props. 42–44 do not in fact establish this beyond doubt, Pappus and the Hellenistic mathematicians who evaluated Archimedes's argument were right: the *neuses*, as employed by Archimedes in *SL* 18, are solid. According to Pappus, the critics also gave a proof for *SL* 18 using only plane means. Unfortunately, that solution, or purported solution, does not survive. The Hellenistic mathematicians may have tried to implement further limiting conditions in the *neusis* problems *SL* 7 and *SL* 8, as applied in *SL*18, so that they become plane, or – and that is more likely – they may have tried to provide a plane argument to replace the *neuses* altogether within the proof of *SL* 18. In what follows, I will briefly discuss two suggestions by historians of mathematics as to what alternatives might have been put forth.

(a) Tannery and Heath: *SL* 18 with plane means, avoiding the *neusis*
   Tannery (1912) and Heath (1921) propose a reformulation of *SL* 18[1]: assume that the circumference is already rectified, and that it has been laid out as a segment perpendicular to the generator of the spiral at the endpoint of the first rotation. Then the line connecting its endpoint with the endpoint of the generator is a tangent to the spiral. As Knorr has pointed out,[2] this is no longer the same theorem as *SL* 18. Among other things, it no longer answers to Pappus' description of the theorem to be "saved": find a straight line equal to the circumference of a circle. It is hard to see what one would gain from replacing *SL* 18 with this alternative. For further objections cf. Knorr (1978).

[1] Tannery (1912, pp. 309–316), Heath (1921, II, pp. 556–561), cf. Zeuthen (1886, pp. 278–279, #2).
[2] Knorr (1978, p. 81).

(b) Knorr (1978) and Heath: change *neusis* argument in *SL* 7 and *SL* 8

Heath had already argued that Archimedes should not have been criticized, because he only assumes the existence of the *neusis*, not an actual construction. This amounts to saying that instead of a *neusis* construction, we should see him as operating with an implicit appeal to the principle of continuity. Knorr takes this suggestion a step further and argues that we could save Archimedes' argument, in Archimedes' own style, if we replace the implicit appeal to the continuity principle by an explicit convergence argument.[1] While it is true that such an argument would be in line with typically Archimedean procedures for heuristics, and that Archimedes appears to have been experimenting with mathematical formulations for limiting processes, I doubt that Knorr's suggestion would have satisfied the Hellenistic critics, or even Archimedes himself.

(i) There is no evidence that Archimedes' steps toward convergence arguments were picked up, developed, and integrated into geometry beyond heuristics in antiquity. We saw that this aspect of his work on motion curves was not taken up. In fact, Archimedes himself eliminated all traces of quasi-mechanical heuristics from the geometrical parts of his published works, giving orthodox exhaustion proofs, with no recourse to convergence processes, instead. It is very likely that Archimedes thought that the assumption that the *neuses* must exist is sufficiently supported by the principle of continuity. But everyone would have granted that the *neusis* must exist, just as a line the length of the circumference must exist. That does not render a construction argument superfluous. If it did for Archimedes, it would have made *SL* 7 and *SL* 8 themselves superfluous. Rather than trying to "save" them this way, he might have eliminated them altogether, and used a convergence/existence argument in *SL* 18 directly. Thus, even if such an argument was put forth by a Hellenistic mathematician, in order to "save" the *neuses*, it is doubtful that such an argument would have been deemed superior to a constructive solution (even if by conics). Finally, such an argument would not have been judged as plane, or equivalent to a plane argument (solvable by means of circle and straight line) – which is what the Hellenistic mathematicians were looking for.

It seems as though we can, at present, not say what the Hellenistic plane argument for *SL* 18, which Pappus claimed was put forth by Archimedes' critics, would have looked like. Pappus may have been wrong, it may have been invalid, or not really plane. In all likelihood it did not look like the suggestions put forth in secondary literature so far.

Let us now turn to Pappus' analysis of the *neusis*, Props. 42–44.

---

[1] Knorr (1978, pp. 93f.); cf. also Dijksterhuis (1987, pp. 139–140) on the background of *SL* 5–9, and Dijksterhuis (1987, pp. 268–274) for a reconstruction similar to Knorr's.

10.2   Props. 42–44: Analysis of an Archimedean Neusis

10.2.1   Proof Protocol Prop. 42

AB, CD, DE $\perp$ AB, CD:DE = a:b *given* (assume CD $\geq$ DE).
   Claim: E lies on a uniquely determined hyperbola.

1. *Apagoge*
   Extension of configuration:
   Draw CZ $\perp$ AB, rectangle ZDEH.
   Construct ZK, ZT so that CZ: ZK = a: b, ZK = ZT [VI, 9].
2. *Resolutio*
   T, K *given*; CZ: ZK *given*.
   $(CD^2 - CZ^2):(ED^2 - ZT^2)$ *given*.
   $EH^2: (KH \times HT)$ *given*.
   Appeal to *Con.* 21 [converse, non-deductive[1]], yields:
   E must lie on the hyperbola through T with diameter TK,
   latus rectum t (t:HK = EH^2:(TH $\times$ KH)), and ordinates parallel to AB.

10.2.2   Proof Protocol Prop. 43

AB *given* in length and position, DC $\perp$ AB, AC $\times$ CB = $t \times$ CD (t *given*). Claim:
Then D lies on a uniquely determined parabola.
1. *Apagoge*
   Extension of configuration:
   Bisect AB in E.
   Construct EZ, Z so that t $\times$ EZ = EB^2 [II, 14].
   Rectangle DHEC.
2. *Resolutio*
   Z *given*, EZ *given*; by construction: EB^2 = t $\times$ EZ *given*.
   $EC^2 = t \times ZH = DH^2$.
   Appeal to *Con.* I, 20 [converse, non-deductive[2]], yields:
   D must lie on the parabola with vertex Z, diameter EZ,
   parameter t, and ordinates parallel to AB.

10.2.3   Proof Protocol Prop. 44

The text of Prop. 44 is badly damaged. As in the translation, I follow the reconstruction by Hultsch, drawing on his corrected diagram in the appendix to Hu
(pp. 1231–1233). Baltzer's corrections are implemented there.

---

[1] Cf. Props. 31 and 34a; there, too, the last step of the analysis appealed to the converse of a theorem.
[2] Cf., again, Props. 31 and 34a for an analysis with a non-deductive last step.

Start with a circle ABC, chord BC, A *given*.

Task: draw AE, D on BC and AE, E on circle, ED *given* in length (*neusis*).

1. Analysis

  1.1. Assume AE has been so placed, ED has the *given* length.

  1.2. *Apagoge*: extension of the configuration

  Draw DZ ⊥ BC, DZ = AD (Z will, in general, not lie on the circle).

  1.3. *Resolutio*

    1.3.1 AD:DZ = 1: 1

    Z lies on a uniquely determined hyperbola   [Prop. 42].

    1.3.2.BD × DC = AD × DE   [III, 35]

    BD × DC = DZ × DE, and DE is *given*

    ⇒ Z lies on a uniquely determined parabola   [Prop. 43]

    ⇒ Z is *given*

The text breaks off here; presumably, one would now say that D is *given*, as foot of the perpendicular from Z onto BC, and thus AE is *given*. Pappus has in fact only presented the analysis, not a solution, to the *neusis*, to show that the problem is solid, because, in general, the construction of the auxiliary point Z will involve a parabola and a hyperbola. For a synthesis, leading to a *neusis* related to *SL* 5–9 cf. Knorr (1978b), Heath (1921, II, pp. 386–388), Tannery (1912, pp. 307–308), Baltzer apud Hultsch 1231–1233. We get several solutions/cases. The *resolutio* in Prop. 44 only shows the constructibility of one of the solutions. This was enough for Pappus' purposes. For the limitations of his argument, see the introduction.

Coll. IV ends rather abruptly, if with a concluding phrase (probably by the copyist). The text of *Coll.* IV shows signs of deterioration (in addition to manuscript damage) at the end of the book. It seems that we are missing at least the allegedly plane argument for *SL* 18, and perhaps also the plane argument for the Apollonian problem of finding the normal to the parabola, as well as the conclusion.

Appendices

Appendix: The Diagrams in the Present Edition, and Vat. gr. 218

As stated in the introduction to Part I, I have modeled the diagrams in the present edition, wherever possible, on the figures in Vat. gr. 218 (**A**). For diagrams that refer to circles and straight lines only, moderate adjustments were sufficient. For the ones referring to higher curves, more drastic revisions became necessary, and in a few cases I was compelled to deviate completely from **A**, and draw the diagrams afresh.

For the "plane" diagrams, I worked from a scan of the original figure in **A**, redrew it, and afterwards adjusted the location of individual points. **A**'s figures are extremely schematic. As a result, the relative position of points is most often rendered accurately, but congruent lines and angles do not appear as congruent, parallels not as parallel, and right angles not as right angles. This can cause severe difficulties when one attempts to use the diagrams as they are intended: as argumentative devices in the proofs. The degree of abstraction one has to continually perform when going through the arguments is simply too high. I therefore resorted to shifting individual points within the sketch diagrams, so as to produce configurations that cause less disturbances. In other words, I made congruent lines look congruent (with some amount of tolerance), parallels like parallels, etc. This procedure, of course, entailed shifting all the points on connecting lines as well. With these modifications, the plane diagrams are still modeled on the figures in **A**.

For the figures containing the Archimedean spiral (Props. 19–22, and 35b), I proceeded similarly, preserving the overall frame and adjusting the spiral line as well as angles created in the figure. For the figures referring to the conchoid, the quadratrix, conic sections, and curves in space, I was unable to work from **A**. All these curves are represented in **A** as circular arcs in the plane. I therefore drew new figures, using both **A** and Hultsch's edition as reference points. For Prop. 44, the "prototype" used was the diagram in the appendix to Hultsch's edition.

In what follows, I will briefly describe in what way my diagrams differ from the ones given in **A**. I will also reproduce **A**'s diagram for the limit case in Prop. 15.

Prop. 1: diagram taken over from **A**.

Prop. 2: diagram taken over from **A**, C, T, and E moved.

Prop. 3: diagram taken over from **A**, L moved.

Prop. 4: diagram taken over from **A**, E, H, and N moved, EK = EL.

Prop. 5: diagram taken over from **A**, M, E, L, and H moved.

Prop. 6: diagram taken over from **A**, A, E, H, and Z moved, T, B added.

Prop. 7a: diagram taken over from **A**, all points except A and Z moved.

Prop. 7b: diagram taken over from **A**, all points except D and C moved.

Prop. 8: the circles with centers A and B were made equal; this resulted in a repositioning of all other points; R added.

Prop. 9: diagram taken over from **A**, D moved.

Prop. 10: diagram taken over from **A**, B, and Z moved, center of encompassing circle renamed H (for N); H and O eliminated.

Prop. 11: diagram taken over from **A**, B, D, E, H, K, and Z moved.

Prop. 12: diagram taken over from **A**, E, L, Z, and T moved.
Arbelos: diagram taken over from **A**, semicircle over AC instead of BC.
Prop. 13: diagram taken over from **A**, N, C, A, and L moved, K added.
Prop. 14a: diagram taken over from **A**, A, M moved.
Prop. 14b: diagram taken over from **A**, D added.
Prop. 14c: diagram taken over from **A**.

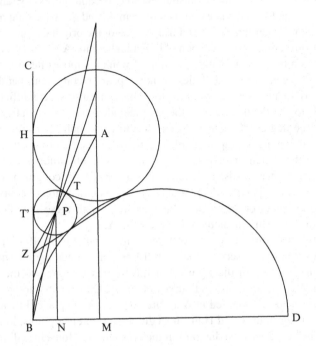

Prop. 15a: circles taken over from **A**, all other lines moved, C added.
Prop. 15b: the diagram in A is missing a number of necessary points; outer circle
 and semicircle over CB taken over from **A**, all other lines and points
 adjusted or corrected.
Prop. 15c: figure missing in **A**, reconstructed.
Prop. 15: limit case: The following is a reproduction of the (flawed) diagram in **A** for
 the case that instead of a semicircle over BC, we have a tangent in B; cf.
 appendix Hu p. 1227 for a correct diagram and a proof.
Prop. 16a: diagram taken over from **A**, TZ ∥ BC.
Prop. 16b: diagram taken over from **A**, TZ ∥ BC, A center of circle TZ, N foot of
 perpendicular from P, E and O with perpendicular ending in S added.
Prop. 16c: diagram taken over from **A**, circle with center P moved, semicircle with
 center A, diameter equal to BD, points renamed: T, Z, N for H, Y, Z
 respective, S added.
Prop. 16d: diagram taken over from **A**, circle with center A moved slightly, D for S.
Prop. 17: diagram taken over from **A**.
Prop. 18: diagram taken over from **A**, perpendiculars onto AC added.
Prop. 19/20: diagram taken over from **A**, spiral adjusted.

Prop. 21a: diagram taken over from **A**, arc CD = arc CA, spiral, arcs ET, ZH adjusted.

Prop. 21b: diagram taken over from **A**, spiral adjusted, H and T switched.

Prop. 22a: diagram taken over from **A**, spiral adjusted, BZ accordingly.

Prop. 22b: diagram taken over from **A**, spiral adjusted.

Conchoid: redrawn, in **A** the line is represented by a circular arc.

Prop. 23: redrawn, in **A** the conchoid is represented by a circular arc.

Prop. 24: redrawn, adjustments too numerous; in **A**, incomplete corrections were added by the second hand.

Prop. 25: redrawn, horizontal unequal lines instead of vertical equal ones, order A–C–D–B instead of A–B–C–D.

Quadratrix: redrawn, square ABCD, quadrant DAB, quadratrix adjusted.

Prop. 26a: redrawn, square CBAD, quadrant DCB, quadratrix adjusted.

Prop. 26b: redrawn, square CBAD, quadrants DCB and KCZ, quadratrix adjusted.

Prop. 26c: redrawn, square CBAD, quadrants DCB and KCZ, quadratrix adjusted.

Prop. 28: redrawn, quadrant CBA, three-dimensional figure.

Prop. 29: redrawn, sector CBA, spiral BA, three-dimensional figure.

Prop. 30: redrawn, spherical spiral for right-hand side diagram; quadrant CDA, arc AZ with center C for left-hand side.

Prop. 31a diagram taken over from **A**, Q renamed as Z.

Prop. 31b: diagram taken over from **A**, LH moved, circle with radius DK, hyperbola DH adjusted, m added.

Prop. 32a: diagram in **A** not used; DE = 2BA.

Prop. 32b: diagram taken over from **A**.

Prop. 32c: diagram taken over from **A**, BD $\perp$ BC, E and A adjusted accordingly.

Prop. 33: **A** has two identical figures, one each for analysis and synthesis, representing the hyperbola by two circular arcs with cusp in D; redrawn, k added, second E eliminated.

Prop. 34a: **A**'s diagram is an isosceles triangle ABC, with E and D distributed almost equally on AC, and BZ as well as the hyperbola BH missing; redrawn.

Prop. 34b: diagram taken over from **A**, the segment was made smaller than a semicircle, B, D, and Z moved.

Prop. 35a: **A**'s diagram had a full circle, Z is mislabelled; redrawn: quadrant TBK with quadratrix KC.

Prop. 35b: diagram from **A**, T eliminated, H and A moved, arc ZE adjusted, D added.

Prop. 36: circle with center E taken over from **A**, point A moved, K added, circle with center Z, arc CTD redrawn, T instead of B.

Prop. 37: diagram taken over from **A**, EH added.

Prop. 39: **A**'s diagram has two equal circles a, b and two vertical lines c, d; redrawn.

Prop. 40: redrawn; quadrant EHZ, quadratrix ZTK, arc ACB with center X, o and p added.

Prop. 41: redrawn, quadrant CBA with quadratrix AEDZ, point names N and K corrected.

Prop. 42: the figure in **A** is missing the hyperbola; redrawn.

Prop. 43: redrawn.

Prop. 44: drawn afresh; no diagram for Prop. 44 in **A**.

Bibliography of Works Cited

Primary Sources

Apollonius

Apollonii Pergaei Opera quae supersunt, Vol. I/II, ed. J. Heiberg, Leipzig 1891–1893

W. Lawson, *The Two Books of Apollonius Pergaeus Concerning Tangencies*, London 1771

G.G. Haumann, *Versuch einer Wiederherstellung der Bücher des Apollonius von Perga von den Berührungen*, Breslau 1817

Apollonii de tactionibus quae supersunt ac maxime lemmata Pappi in hos libros Graece, nunc primum edita e codicibus inscriptis, cum Vietae librorum Apollonii restitutione, ed. M. Camerer, Gotha 1795 (German translation 1831)

Apollonius of Perga on Cutting off a Ratio, trans. E.M. Macierowski, ed. R.H. Schmidt, Fairfield 1987

Apollonius, Conics, books V–VII. The Arabic translation of the lost Greek original in the version of the Banu Musa, edited, with translation and commentary, by G. Toomer, New York/Berlin/Heidelberg/Tokyo 1990

Archimedes

Archimedis Opera Omnia cum commentariis Eutocii, Vol. I–III, ed. J. Heiberg, Leipzig 1910–1915

Archimedis Opera Omnia cum commentariis Eutocii, Vol. IV, ed. Y. Dold-Samplonius et al., Stuttgart 1975

Archimedes. Werke, trans. A. Czwalina, Darmstadt 1983

Aristotle

Aristotelis Analytica priora et Posteriora, ed. W. D. Ross, Oxford 1964

Diocles

Diocles on Burning Mirrors. The Arabic translation of the lost Greek original, ed./trans. G. Toomer, Berlin/Heidelberg/New York 1976

Euclid

Euclidis Opera Omnia, Vol. I–V, eds. J. Heiberg/E. Stamatis, Leipzig 1969 ff.

Euclidis Opera Omnia, Vol. VI, eds. J. Heiberg/W. Menge, Leipzig 1896

Euclid. The Thirteen Books of Euclid's Elements, Vol. I–III, trans. T.L. Heath, Cambridge 1926[2] (Dover reprint 1956) [This work is cited as Heath 1926]

Die Data von Euklid, trans. C. Thaer, Berlin/Göttingen/Heidelberg 1962

Euclid's Data, transl. C. M. Taisbak, Copenhagen 2003

Heron

Heronis Opera, Vol. I, ed. W. Schmidt, Leipzig 1899, with Suppl., Leipzig 1899
Heronis Opera, Vol. II, fasc. 1, eds. L. Nix/W. Schmidt, Leipzig 1900
Heronis Opera, Vol. III, ed. H. Schöne, Leipzig 1903
Heronis Opera, Vol. IV–V, eds. W. Schmidt/J. Heiberg, Leipzig 1912–1914

Pappus

Pappi Alexandrini Mathematicae Collectiones, trans. F. Commandino, Pesaro 1588, revised edition Bologna 1660 [This work, in its later edition, is referred to as Co]
Pappi Alexandrini Collectionis quae supersunt, Vol. I–III, ed. F. Hultsch, Berlin 1876–1878 [This work is referred to as Hu]
Pappus d'Aléxandrie. La Collection Mathématique, Vol. I–II, translated with notes by P. Ver Eecke, Brügge/Paris 1933 [This work is cited as Ver Eecke 1933b]
A Critical Edition of the Text of the Collection of Pappus of Alexandria (Books II–V), ed. A. Treweek, Diss. London 1950 (unpublished) [This work is referred to as Tr in the apparatus to the Greek text and the notes to the translation]
J. Torelli Veronensis Geometrica, J. Torelli, Verona 1769 [contains *Collectio* IV, chapters 45–52 (# 30–35), on the quadratrix; this work is cited as To]
Die Sammlung des Pappus von Alexandrien, griechisch und deutsch, Halle 1871 [only Bd.1, containing *Collectio* VII and VIII, were published]
The Commentary of Pappus on Book X of Euclid's Elements. Arabic text and translation, ed. G.Junge/W. Thomson, Cambridge, MA 1930
R. H. Hewson, *The Geography of Pappus of Alexandria*: A translation of the Armenian fragments; in: *Isis* 62 (1971), pp. 187–207
Pappus of Alexandria. Book 7 of the Collection, Vol. I–II, edited with translation and commentary by A. Jones, Berlin/Heidelberg/New York 1986 [This work is cited as Jones 1986a]
Pappos, Synagoge IV. Ein spätantiker Querschnitt zur klassischen griechischen Geometrie, translation, with notes and commentary by Heike Sefrin-Weis, Diss. Mainz 1997, published in microfiche format 1998

Proclus

Procli Diadochi in Primum Euclidis Elementorum Librum, ed. G. Friedlein, Leipzig 1873

Simplicius

Simplicii in Aristotelis Physica Commentaria, ed. H. Diels, *CAG* IX–X, Berlin 1882–1895
Simplicii in Aristotelis Categorias Commentaria, ed. C. Kalbfleisch, Berlin 1907

Secondary Literature

Becker, Oskar (ed.). 1965. *Zur Geschichte der griechischen Mathematik*. Darmstadt
Becker, Oskar. 1973². *Mathematische Existenz. Untersuchungen zur Logik und Ontologie mathematischer Phänomene*. Tübingen

Becker, Oskar. 1975. *Grundlagen der Mathematik in geschichtlicher Entwicklung*. Frankfurt (reprint of the 1964 Freiburg edition)

Behboud, Ali M. 1994. Greek geometrical analysis. *Centaurus* XXXVII: 52–86

Berggren, John L. 1984. History of Greek mathematics. A survey of recent research. *HM* 11: 394–410

Bombelli, Raffael. 1929. *L'Algebra, parte maggiore dell' aritmetica divisa in tre libri*, ed. E. Bortolotti. Bologna

Bortolotti, Ettore. 1923. La trisezione dell' angolo e il caso irreducibile della equazione cubica nell' algebra di Rafael Bombelli. *Rendiconti delle sessioni della R. Academia delle scienze dell' Istituto di Bologna*, Anno Academico 1922/1923. Bologna

Bos, Henk J.M. 1981. On the representation of curves in Descartes's Géométrie. *AHES* 24: 295–338

Bos, Henk J.M. 1984. Arguments on motivations in the rise and decline of a mathematical theory: the "construction of equations" 1637–c.a. 1750. *AHES* 30: 331–380

Bos, Henk J.M. 2001. *Redefining geometrical exactness: Descartes's transformation of the early modern concept of construction*, New York

von Braunmühl, A. 1892. Historische Studie über die organische Erzeugung ebener Curven von den ältesten Zeiten bis zum Ende des achtzehnten Jahrhunderts. In *Katalog mathematischer und mathematisch-physikalischer Modelle, Apparate und Instrumente*, ed. Dyck, Walter, 54–88. Munich

Buchner, Ferdinand. 1824. *De arbelo Archimedeo*. Elbing

Cantor, Moritz. 1894–1900. *Vorlesungen über Geschichte der Mathematik,* Vol. I–II. Leipzig

Casey, John. 1882. *A sequel to the first six books of the elements of Euclid, containing an easy introduction to modern geometry with numerous examples*. Dublin

Chasles, Michel. 1839. *Geschichte der Geometrie*. Halle (reprint Sändig 1993)

Chasles, Michel. 1875. *Aperçu historique des méthodes en Géométrie*. Paris (reprint of the 1837 edition)

Cornford, Francis M. 1932. Mathematics and dialectic in the Republic V–VII. *Mind* 41: 37–52 and 173–190

Descartes, René. 1659. *Geometria*, translated, with commentary and additional essays from various authors by Schooten, Frans. Leiden

Descartes, René. 1954. *The Geometry*, edited with translation and notes by D.E. Smith/M.L. Latham. New York (Dover reprint of the 1925 edition; contains a facsimile of the original French text; quoted as Descartes 1637)

Dijksterhuis, Eduard J. 1987. *Archimedes*. Princeton

Dyck, Walter (ed.). 1892. *Katalog mathematischer and mathematisch-physikalischer modelle, apparate und instrumente*. München (reprint Hildesheim 1994)

Elfering, Kurt. 1975. *Die Mathematik des Aryabhata I*. München

Etienne and Roels, J. 1986. Deux aspects particuliers du problème des moyennes dans Pappus d'Aléxandrie. *Revue des questions scientifiques* 167: 179–198

de Fermat, Pierre. 1679. *Varia Opera mathematica*. Toulouse

de Fermat, Pierre. 1922. *Œuvres de Fermat*, eds. P. Tannéry and C. Henry, Vol. I–IV. Paris 1891–1912

de Fermat, Pierre. 1923. *Ad locos planos et solidos isagoge*, deutsch:*Einführung in die ebenen und körperlichen Örter von P. Fermat*, trans. H. Wieleitner. Leipzig

Fowler, David. 1992. An invitation to read book X of the *Elements*. *HM* 19: 233–264

Gauss, Friedrich. 1801. *Disquisitiones Arithmeticae*. Leipzig

Gauss, Friedrich. 1889. *Disquisitiones Arithmeticae deutsch*, transl. H. Maser. Berlin

Gulley, Norman. 1958. Greek geometrical analysis. *Phronesis* III: 1–14

Gulley, Norman. 1962. *Plato's theory of knowledge*. London

Hankel, Hermann. 1874. *Zur Geschichte der Mathematik in Altertum und Mittelalter*. Leipzig (reprint Hildesheim 1965)

Heath, Thomas L. 1921. *A history of Greek mathematics*, Vol. I–II. Oxford

Heath, Thomas L. 1926^2. *The thirteen books of Euclid's elements, Vol. I–III*, trans. T.L. Heath. Cambridge (Dover reprint 1956) [see under primary sources, Euclid]

Heath, Thomas L. 1970. *Mathematics in Aristotle*. Oxford (1949^1)

Heiberg, Johann L and Zeuthen, H.G. 1906/07. Eine neue Schrift des Archimedes. *Bibliotheca Mathematica 3. Folge 7. Band*. 321–363

Hilbert, David and Samuel Cohn-Vossen. 1932. *Anschauliche Geometrie*. Berlin

Hintikka, Jaakko. 1973. *Time and necessity*. Oxford

Hintikka, Jaakko and Unto Remes. 1974. *The method of analysis*. Dordrecht

Hintikka, Jaakko and Unto Remes. 1976. Ancient geometrical analysis and modern logic. In *Essays in memory of Imre Lakatos*, ed. R.S. Cohen et al. Dordrecht

Hofmann, Joseph E. 1990. *Ausgewählte Schriften*, Vol. I–II, ed. Ch. Scriba. Hildesheim

Hogendijk, Jan P. 1981. How trisections of the angle were transmitted from Greek to Islamic geometry. *HM* 8: 417–438

Hogendijk, Jan P. 1984. Greek and Arabic constructions of the regular heptagon. *AHES* 30: 197–330

Hogendijk, Jan P. 1986. Arabic traces of lost works of Apollonius. *AHES* 35: 185–253

Hogendijk, Jan P. 1989. Sharaf-al-Din-al-Tusi on the number of positive roots of cubic equations. *HM* 16: 69–85

Jones, Alexander R. (ed.). 1986a. *Pappus of Alexandria. Book 7 of the Collection, Vol. I–II, edited with translation and commentary by A. Jones*. Berlin/Heidelberg/New York [see under primary sources, Pappus]

Jones, Alexander R. 1986b. William of Moerbeke, the Papal Greek manuscripts, and the Collection of Pappus of Alexandria in Vat. Gr. 218. *Scriptorium* 40: 16–31

Junge, Gustav. 1936. Das Fragment der lateinischen Übersetzung des Pappus-Kommentars zum 10. Buch Euklids. *Quellen und Studien zur Geschichte der Mathematik, Astronomie und Physik, Abt. B* (Studien), III: 1–17

Junge, Gustav and Thomson, William (eds.). 1930. *The Commentary of Pappus on Book X of Euclid's Elements. Arabic text and translation*. Cambridge, MA [see under primary sources, Pappus]

Katz, Victor. 1993. *A history of mathematics*. New York

Knorr, Wilbur. 1975a. *The evolution of the Euclidean elements*. Dordrecht

Knorr, Wilbur. 1975b. Archimedes and the measurement of the circle. A new interpretation. *AHES* 15: 115–140

Knorr, Wilbur. 1978a. Archimedes and the spirals. The heuristic background. *HM* 5: 43–75

Knorr, Wilbur. 1978b. Archimedes' neusis constructions in spiral lines, *Centaurus* XXII: 77–98

Knorr, Wilbur. 1978c. Archimedes and the elements: proposal for a revised chronological ordering of the Archimedean corpus. *AHES* 19: 219–290

Knorr, Wilbur. 1978d. Archimedes and the pre-Euclidean proportion theory. *Archive Internationale d'Histoire des Sciences* 28: 183–244

Knorr, Wilbur. 1981. The hyperbola-construction in the Conics, Book II. Ancient variations on a theorem of Apollonius. *Centaurus* XXV: 253–291

Knorr,Wilbur. 1982. Observations on the early history of the Conics. *Centaurus* XXVI: 1–24

Knorr, Wilbur. 1983a. La croix des mathématiciens. The Euclidean theory of irrational lines. *Bulletin of the American Mathematical Society, New Series* 9: 41–69

Knorr, Wilbur. 1983b. On the transmission of geometry from Greek into Arabic. *HM* 10: 71–78

Knorr, Wilbur. 1985. Euclid's tenth book: An analytical survey. *Historia Scientiarum* 29: 17–35

Knorr, Wilbur. 1986. *The ancient tradition of geometric problems*. Boston

Knorr, Wilbur. 1989. *Textual studies in ancient and medieval Geometry*. Boston

Lakatos, Imre. 1978. *Mathematics, science, and epistemology*. Cambridge

Mahoney, Michael S. 1968. Another look at Greek geometrical analysis. *AHES* 5: 319–348

Mäenpää, Petri. 1997. From backward reduction to configurational analysis. In *Analysis and Synthesis in Mathematics*, ed. M. Otte and M. Panza. Dordrecht

Molland, Andrew G. 1976. Shifting the foundations: Descartes's transformation of ancient geometry. *HM* 3: 21–49

Mueller, Ian. 1981. *Philosophy of mathematics and deductive structure in Euclid's elements*. Cambridge, MA

Netz, Reviel. 1999. *The shaping of deduction in Greek mathematics*. Cambridge

Netz, Reviel. 2000a. *The transformation of mathematics in the early Mediterranean world: From problems to equations*. Cambridge

Netz, Reviel. 2000b. Why did Greek mathematicians publish their analyses? In *Memorial volume for W. Knorr*, eds. H. Mendel, J. Moravscik, and S. Patrick, 139–157. Stanford

Netz, Reviel. 2004. Eudemus of Rhodes, Hippocrates of Chios and the earliest form of Greek mathematical text. *Centaurus* XLVI: 243–286
Neugebauer, Otto. 1975. *A history of ancient mathematics, Vol. II.* Berlin/Heidelberg/New York
Newton, Isaac. 1752. *Arithmetica universalis*, London
Newton, Isaac. 1972. *The mathematical papers of Isaac Newton*, ed. D.T. Whiteside, Vol. II, Vol V. Cambridge
Rehder, Wolfgang. 1982. Die Analysis und Synthesis bei Pappus. *Philosophia Naturalis* XIX: 350–370
Robinson, Richard. 1936. Analysis in Greek geometry, *Mind* N.S. 46: 464–473
Rudio, Ferdinand. 1892. *Archimedes, Huygens, Lambert, Légendre: Vier Abhandlungen über die Kreismessung.* Leipzig
Rudio, Ferdinand. 1907. *Der Bericht des Simplicius über die Quadraturen des Antiphon und des Hippokrates.* Leipzig
Saito, Ken. 1986. Compounded ratio in Euclid and Apollonius. *Historia Scientiarum* XXXI: 25–59
Sezgin, F. 1974. *Geschichte des arabischen Schrifttums, Bd. 5.* Leiden
de Sluse, René F. 1659. *Mesolabum.* Liège
de Sluse, René F. 1668. *Mesolabum*, 2nd, expanded edition. Liège
Steele, Sir Arthur D. 1936. Über die Rolle von Zirkel und Lineal in der griechischen Mathematik. *Quellen und Studien zur Geschichte der Mathematik, Astronomie und Physik Abt. B (Studien)* III: 287–369
Suardi, Conte G. 1752. *Nuovi istrumenti per la descrizione di diverse curve antiche e moderne.* Brescia
Szabó, Arpad. 1974. Working backwards and proving by synthesis. In *The method of analysis*, eds. J. Hintikka and U. Remes, 118–130. Dordrecht
Szabó, Arpard. 1974/1975. Analysis und synthesis. Pappos II, p. 634ff. Hultsch. *Acta Classica Universitatis Scientiarum Debrecensis*, X/XI: 155–164
Taisbak, Christian M. 1982. *Coloured quadrangles.* Copenhagen
Taisbak, Christian M. (ed.). 2003. *Euclid's Data.* Copenhagen
Tannéry, Paul. 1912. *Mémoires scientifiques de P.Tannéry*, eds. Johann L Heiberg, and Hieronymus G. Zeuthen. Toulouse
Toomer, Gerald. 1990. *Apollonius, Conics, books V–VII. The Arabic translation of the lost Greek original in the version of the Banu Musa*, edited, with translation and commentary, by G. Toomer. New York/Berlin/Heidelberg/Tokyo [see under primary sources, Apollonius]
Treweek, Athanasius P. (ed.) 1950. *A critical edition of the collection of Pappus of Alexandria (Books II–V)* [see under primary sources, Pappus]
Treweek, Athanasius P. 1950. In search of the mss. of Pappus. *Classical Association of New South Wales Proceedings*: 35–38
Treweek, Athanasius P. 1957. Pappus of Alexandria. The manuscript tradition of the Collectio Mathematica. *Scriptorium* 11: 195–233
Tropfke, Johannes. 1923². *Geschichte der Elementarmathematik* IV. Leipzig
Unguru, Sabetai. 1974. Pappus in the thirteenth century in the Latin West. *AHES* 13: 307–324
Van Roomen, Adriaan. 1596. *Problema Apollonianum*, Würzburg
Ver Eecke, Paul. 1933a. La mécanique des Grecs d'après Pappus d'Alexandrie. *Scientia* LIV: 114–121
Ver Eecke, Paul. 1933b. *Pappus d'Alexandrie. La collection mathématique*, Vol. I–II. Brügge/Paris [see under primary sources, Pappus]
Vieta, Franciscus. 1970. *Opera mathematica*; reprint of Frans Schooten's 1646 edition, by J. E. Hofmann, Hildesheim [contains Algebra, sive logistica specifica; Isagoge in artem analyticam; Liber VIII Variorum de rebus mathematicis responsorum, supplementum geometriae]
Viète, François. 1600. *Apollonius Gallus.* Paris
Whiteside, Derek T. (ed.). 1972. *The mathematical papers of Isaac Newton.* Cambridge [see Newton 1972]
Witte, Carl 1813. *Conchoides Nicomedeae aequatio et indoles.* Göttingen
Zeuthen, Hieronymus G. 1886. *Die Lehre von den Kegelschnitten im Altertum.* Kopenhagen (reprint Hildesheim 1966)
Zeuthen, Hieronymus G. 1903. *Mathematik im 16. und 17. Jahrhundert.* Leipzig

Index